HANDBOOK OF

TEXTILE TESTING AND QUALITY CONTROL

Handbook of TEXTILE TESTING AND QUALITY CONTROL

By

ELLIOT B. GROVER,

Abel C. Lineberger Professor of Yarn Manufacturing

and

D. S. HAMBY,

Burlington Industries Professor of Textiles

Department of Fiber and Yarn Technology
School of Textiles North Carolina State College
Raleigh, North Carolina

INTERSCIENCE PUBLISHERS, INC.

a Division of John Wiley & Sons, Inc. New York • London • Sydney

First Published 1960

Library of Congress Catalog Card Number 60-11026

6 7 8 9 10

ISBN 0 470 32901 7

PRINTED IN THE UNITED STATES OF AMERICA

PREFACE

The *Handbook of Textile Testing and Quality Control* has been designed foor use by executives, technical personnel, and students. Each test has been discussed following the general pattern of why it is important, factors that influence the test results, and the technique of making the test.

The discussion of why the test is important is primarily for the executives to aid them in understanding why certain tests are preferred under a given set of conditions and why other tests may be necessary under different circumstances. The technical discussions of the tests are intended as an aid for the quality control group in locating and correcting irregularities in processing. The sections devoted to discussions of testing techniques are for the use of laboratory technicians and students.

The writing and assembling of such a handbook is a job of immense proportions. It is humanly and physically impossible to include all testing and quality control techniques in use throughout the textile industry. Under these circumstances, the authors have confined their efforts to cotton and synthetic fibers and to the cotton and modified cotton systems of processing.

The authors are indebted to many individuals and corporations for suggestions and material included in this volume. The authors are also indebted to Professor Ronald A. Fisher, Cambridge; to Frank Yates, Rothamsted; and to Messrs. Oliver and Boyd, Ltd., Edinburgh, for permission to reprint Tables XLI, XLIII, and XLIV from their book, *Statistical Tables for Biological, Agricultural and Medical Research.*

ELLIOT B. GROVER
D. S. HAMBY

CONTENTS

CHAPTER I

Introduction to Textile Testing and Quality Control

Textile testing is the application of engineering knowledge and science to the measurement of the properties and characteristics of, and the conditions affecting, textile fibers, yarns, and materials. It involves the use of techniques, tools, instruments, and machines in the laboratory for the evaluation of the properties of these different forms of textiles.

Quality control is concerned with the evaluation of test data and its application to the control of the textile process, raw materials, intermediate products, and final product. It is concerned not only with quality level and the cost of maintaining this level, but also with the presentation of tangible values to measure quality and changes in quality.

Neither textile testing nor quality control alone can serve fully the textile technician, the mill, or the industry, but together they form a very strong team with dynamic possibilities. Testing provides the background and data, and quality control applies the results.

If the textile industry is viewed objectively, it will be seen that the ultimate effort, depending on the product, is to provide the best quality possible at the lowest cost. Other factors are involved, and these cannot be underestimated. For example, styling must be attractive to the customer in household and apparel goods; deliveries, stability, and other business practices must be of high order if the mill is to hold its customers; labor as well as public relations must be good or else other efforts are damaged. But, in the long run, the ultimate aim is to improve or increase the ratio of quality to cost, for both mean customer approval and financial success. In the improvement of the ratio of quality to cost, the mill must view its own problems in manufacturing and customer relations before taking action. In some cases, it might be preferable to aim for improved quality with no increase in costs. In other cases where quality is stabilized at a satisfactory level, the objective should be to reduce the cost of meeting this quality level. In other cases, the product may be too good for the market, and both the quality level and cost can be reduced. It is through testing and quality control techniques that the changes desired can be tried, for without tangible values provided by a quality control program, any effort to alter quality or cost amounts to tampering with the unknown.

Textile testing has attained an important position in the textile industry: it is just as applicable to the analysis of a finished fabric as it is to the raw material; it is as useful for the measure of a housefold fabric as for any Army fabric; it is as necessary to the cotton spinner in controlling the quality of his product as it is to the rayon producer in controlling the quality of his product; and it is as valuable a tool in the hands of the textile manufacturer as it is in the hands of the research technician.

The many forms of textiles all differ in their characteristics, just as people differ in personality. One type of yarn may be strong, another heat resistant, a third elastic, a fourth able to withstand wear, and still another may possess a combination of these. A knowledge of these physical properties means in reality a knowledge of how external forces or conditions affect the yarn or fabric. The laboratory equipped for textile testing is the proving ground for the determination, measure, and comparison of such properties.

The interpretation of results is as important as the actual physical tests. It is not enough to measure a single property of a yarn; it is also necessary to know what other forces or conditions influence the result. For example, if a cotton yarn is tested for strength, it is necessary to know the extent of the influence of atmospheric moisture, since cotton yarn, being hygroscopic, is strengthened by increases in its moisture content. Or, the elastic limit of a rayon yarn may be of more importance to a weaver than its ultimate tensile strength. To understand the interrelationship of the many properties of textile materials, a knowledge of the many individual test methods and procedures is essential.

General Aspects of Testing and Quality Control

Textile testing is broad in scope. It can include, for instance, the means for determining the properties of a fiber, a yarn, or a fabric. It can include the analysis of the properties of a known or an unknown material. It can be used to measure the outside factors that influence test results.

A program of quality control should include a testing program involving the performance on a periodic basis of certain routine tests designed to measure the characteristics of the raw or processed material. The data obtained here can be analyzed statistically. The control desired might be aimed at meeting:

1. Standards established by an individual organization. For example, the staple length of raw cotton, the evenness of picker laps, the size of roving, the strength of yarns, the number of filaments in a rayon yarn.
2. Established scientific specifications. For example, the specifications set up by the American Society for Testing Materials, the U. S. Army Quartermaster Corps, or other established authorities.
3. Market requirements or standards. For example, meeting the requirements for width, ends, picks, and weight for certain staple fabrics listed in trade periodicals.
4. Consumer needs or demands. For example, to forecast the effectiveness of a material to meet consumer needs for wear or for dye fastness.
5. The needs of improving the ratio of quality to cost.

Finally, the proper quality control program envisions improved inspection methods: The frequency and number of tests required to measure or determine quality levels.

Control of Operating and Testing Conditions. In order for test results to be reliable, instruments must be in calibration. Failure to maintain proper calibration of balances, scales, strength test machines, or other instruments will introduce serious errors in the results obtained, thus nullifying all efforts of proper control. Such a condition can lead to costly errors in manufacturing. By the same token, failure to train technicians properly can result in similar errors. For example, it will be found that two operators using the same instrument to get some one measurement on the same sample could arrive at two sets of data differing in average, spread or distribution of individual values, or both. Thus, errors of operators due to lack of proper training in the correct technique, or due to carelessness or fatigue, can cause trouble in any quality control program.

Failure to maintain proper atmospheric conditions, or to make proper corrections to standard conditions from existing conditions, can introduce errors in results. For example, if cotton yarn is tested at a high level of, say, 80 per cent relative humidity, its weight and strength will be high in comparison to what it should be at standard conditions.

It should be understood that variations exist in any textile process. Similarly, raw material, material in process, and the finished product all contain variations. As long as these variations are normal or within some system of chance causes, conditions of control exist. However, if abnormal variations occur that are due to assignable cause, then the control program spots such a condition and corrects it. Errors that are due to measuring equipment and instruments (lack of calibration), to operators (wrong technique, fatigue, transcription mistakes), or to operating conditions (failure to check atmospheric conditions) increase the burden of the quality control program in spotting variations due to assignable causes in the manufacturing process. One of the primary functions of a testing and quality control program is the establishment of routine programs for proper maintenance of operating and testing conditions.

Laboratory Analysis of Textile Materials. This work is concerned with the physical and chemical measurement of the characteristics of textile materials in any form. It involves not only the control tests mentioned above, but also specialized procedures and tests. For example, this work involves the physical analysis of new fabrics, the measurement of special characteristics of fiber, yarn, or fabric, and the chemical tests for measurement of such properties as fiber composition, dye characteristics, and finish characteristics. In other words, the testing laboratory acts as a service department for the entire organization: for instance, the purchasing department may need an analysis of the competitive value of some supply item, or the designing department may need an analysis of the blend and structure of the yarn in a competitor's sample of fabric.

Machine Performance Tests. These tests are concerned with the measurement of the quality of machine performance, as, for example, the determination of the effect of variation in speed of the card licker-in on the amount of waste removed. The results of tests in this general category form a basis for engineering judgment for the improvement of standards of manufacture. Changes as a result of machine performance tests may have a direct bearing on operative work load. Optimum conditions found as a result of these tests should be the basis for additional time and motion study. Testing is primarily concerned with the evaluation of quality of the product resulting from changes in machine performance, and quality control techniques offer the means for evaluating the significance of the tests. Over all, the aim is better utilization of equipment. In a similar manner, if a decision is to be made as to the competitive value of two machines, each similar as to function but produced by different machine builders, the testing and quality control organization can evaluate relative performances. Thus, management has a very valuable tool at its disposal to supply information on vital questions to aid in making correct decisions.

Facilities and Tests

The program for successful testing and quality control must consider the following factors: space and arrangement of space, tests to be performed and the equipment for these tests, personnel to perform the tests and evaluate the results, and the methods and procedures for sampling, performing, evaluating, and applying the results. The latter item is in reality the core of the entire subject.

The space alloted to the textile testing laboratory in the plant should be separated from the manufacturing division, but accessible to and by it. It could be located between or near the administrative center and the production departments. The laboratory rooms should have controlled atmospheric conditions, i.e., a conditioning system capable of maintaining standard conditions of 70° F. temperature and 65 per cent relative humidity. Plenty of wall space for cabinets, work tables of the proper height, good lighting conditions, and an abundance of electrical outlets are other requirements.

Fig. 1.1. Typical testing laboratory.

The minimum size of a laboratory suitable for a small mill would be about 650 square feet. Areas up to 1500 square feet can be arranged for efficient operations and are sufficiently large to handle a wide variety of tests. If needs exceed areas of 1500 square feet, it would be wise to consider dividing the area into two or more units. Tables can be placed around a portion of the perimeter of the area. Strip wiring, consisting of special molding with outlets spaced every 36 inches and at a level of 36 to 40 inches from the floor should be run around the perimeter. Laboratory tables should be 30 inches high. Table tops of black bakelite-type composition will stand up admirably, and the black gives a good contrast with fibers and yarns. Lighting should be plentiful and evenly distributed with a minimum of 50-foot candles at any work area. Insofar as atmospheric conditioning is concerned, the problems are so varied that no definite recommendation can be made. For small areas, commercial type units are now available from the larger manufacturers. For large areas, a competent engineer should be consulted for recommendations. In either case,

Fig. 1.2. Another view of a testing laboratory.

in planning the laboratory, provision should be made for insulating the area from its surroundings. Walls and ceiling (and if feasible, the ceiling beneath the floor) should have a minimum of four inches of some good insulation material that is not susceptible to mold formation.

The selection of quantity and type of equipment must be governed by the type of products to be tested and the extent of control desired. A study of some of the needs is given on the following pages.

Personnel should include trained individuals to perform the tests and persons capable of properly interpreting and analyzing the results. Laboratories are excellent centers for the indoctrination and training of potential managerial and technical personnel. Many mills have come to use the laboratory as the proving ground and the reserve for young college graduates whom they plan to use in supervisory work. The variety of materials passing through the laboratory allows the personnel to become familiar with the products of the plant at various stages and learn something of their characteristics. The process of collecting samples from the production of the manufacturing departments presents an opportunity for the novice to make the acquaintance of supervisory and production help.

With the fulfillment of the above requirements, physical testing becomes the means for measuring the characteristics of materials, for controlling the processing of these materials, and for increasing our knowledge of the character of textiles.

Routine Tests Performed in the Industry

An analysis was made of the tests performed and facilities for making these tests for several mills in the Southern area. The results of this survey are given herewith to show in general the trends in quality control through the use of the testing laboratory.

It should be noted that this analysis is based on the following conditions. (*1*) Only those mills are included which reported having laboratories. (*2*) Only physical tests are listed, with few exceptions. (*3*) Both spinning and spinning-weaving mills are included. (*4*) All mills included in the survey manufactured fine and coarse goods made from yarns spun on the cotton system. (*5*) The fact that tests other than those listed might be performed by some of the units.

In the majority of mills, the testing and quality control group is in a central laboratory headed by a laboratory supervisor or chief chemist, depending on the type of plant and product. In most of the mills, the laboratory supervisor reports to the general superintendent, in many to the director of research and development, and in other mills to the general manager or person having equivalent authority.

In many cases the mill superintendent authorizes changes deemed necessary by the laboratory; however, in some mills, the laboratory supervisor or his superior authorizes changes in the plant.

In integrated companies, a central laboratory often serves the group. In two-thirds of the cases, the laboratories operate on one shift only; in the balance, either a skeleton crew or no crew is used on other shifts.

Nearly all mill laboratories service the fabric design departments by supplying analysis details. In three-fourths of the mills, laboratories reported doing research and development also; only about a fourth employ special laboratories for this. At the same time, the testing laboratories are very closely coordinated with other departments, such as the research, machine development, fabric development, standards, and similar departments. About two-thirds of the mills reported using quality control charts of one type or another.

A tabulation of the routine tests that are performed in mills processing mainly on the cotton system has been prepared and is given in Table 1.1. This listing has been compiled from the results of the mill survey referred to above and from informal investigations of trends and practices in the textile industry. For each particular test are given the instruments or methods used in performing the test. No attempt has been made to list the instruments in any order of preference, for in many cases the objectives desired by the different technicians vary to such an extent that the choice of instrument is more or less fixed.

The question that always arises when a mill plans to establish a laboratory is just what equipment is needed. The answer can only be given when all the facts are known: type of production, space allocation, and extent of quality control desired. For the average weaving mill spinning its own yarn on the cotton system and selling in the grey, the two lists of instruments following Table 1.1 can be considered as the nucleus for well-equipped operations.

TABLE 1.1

Routine Tests in Mill Processing

Test	Method or instrument
	Cotton Fiber Tests
Cotton color	Nickerson-Hunter cotton colorimeter
Fiber cross section	Hardy microtome and microscope; photomicrograph camera
Fiber fineness	Sheffield Micronaire; Arealometer (also maturity)
Fiber length	Fibrograph; Suter-Webb sorter
Fiber maturity	Actual count under microscope (caustic soda); differential dye technique; Causticaire method with Sheffield Micronaire (also fineness); Arealometer (also fineness)
Fiber strength	Pressley tester; Scott-Clemson tester; Stelometer; Scott IP—4 with special jaws; Instron tester
Nep potential	Nepotometer
Staple length and grade	Cotton classer
	Preparatory Tests
Comber waste removal	Suter percentage scale; actual weights
Evenness: card sliver; comber sliver, drawing sliver	Saco-Lowell sliver tester; Uster evenness tester; Pacific evenness tester; Brush uniformity analyzer
Evenness: roving	Uster evenness tester; Pacific evenness tester; Brush uniformity analyzer
Lap evenness	Saco-Lowell lap meter; template and scales; Uster Varimeter
Lap (or room) moisture	Aldrich regain indicator; oven tests on cotton
Moisture regain	(See fabric tests)
Nep count	Actual count at card web, using nep board
Oil application in opening or picking	Measure at point of application; chemical analysis of product
Size: card sliver, comber sliver, drawing sliver, laps	Actual weight of short lengths (1–5 yards) from grain scales
Size: roving	Weight on grain scale of length (12-yard) made on reel

Table 1.1 (*Continued*)

Test	Method or instrument
	Yarn Tests
Grade	Yarn board comparison with A.S.T.M. D-13 (U.S.D.A.) Standards; Seriplane visual comparison
Moisture regain (or content)	Emerson oven; Brabender semi-automatic oven; Hart moisture meter
Twist: ply yarn	Untwist on hand or power-driven twist tester (also, contraction due to twist)
Twist: single yarn	One-inch untwist on hand-operated twist tester; 10-inch untwist-twist on hand or power-driven twist tester
Yarn elongation	See yarn strength
Yarn evenness	Uster evenness tester; Pacific evenness tester; Brush uniformity analyzer; visual examination
Yarn number: beams	Calculation from ends and net weight
Yarn number: skein lengths	Skeins by hand reel or automatic reel; weight on grain scales or yarn number on direct reading quadrant
Yarn number: short lengths	Length by measure and weight on precision balances (chemical or torsion types); length on Universal ruler and number on direct reading torsion balance
Yarn strength: single strand	Inclined plane tester; pendulum-type tester; Moscrop tester (multiple ends); Uster single strand tester; Instron. (Elongation possible simultaneously)
Yarn strength: skein	Pendulum-type tester with 120-yard skein
Yarn tension (in processing)	Sipp-Eastwood Tensometer; Brush tension analyzer
	Fabric Tests
Abrasion	Wyzenbeck; Taber; Schiefer abrasion tester; Stoll Universal tester
Air permeability	Frazier; Gurley
Crease or wrinkle	Monsanto wrinkle recovery tester; TBL tester; Drape-Flex stiffness tester

(*Table continued*)

Table 1.1 (*Continued*)

Test	Method or instrument
	Fabric Tests (continued)
Crimp in yarn	Brighton crimp tester; inclined plane tester; Instron; hand measurement; special devices using chain-loading principle
Fabric construction	Actual count with pick glass or low-powered microscope; ravelling ends (use hand tally for count)
Fabric thickness	Randall and Stickney thickness gauge; Schiefer Compressometer (by Frazier)
Fabric weight	Weight of short or long lengths; weight on precision scales of small samples cut by template
Moisture regain	Emerson oven; Brabender semi-automatic oven; Hart moisture meter
Starch content	Removal by Diastafor or solvents
Strength: breaking	Grab or strip test on pendulum-type testers or Instron
Strength: burst	Mullen burst tester; ball burst attachment for Scott pendulum tester
Strength: tear	Trapezoid or tongue test on pendulum-type tester or Instron
	Other Tests
Fiber drag (drafting force)	Draftometer
Fiber fineness: continuous filament types	Vibroscope
Package density	Suter Manville tester
Yarn and fabric visual analysis	Projectina; microscope

Small Instruments and Tools

1. Scissors and shears, including pinking shears
2. Steel rules (12-inch), yard sticks, steel tapes, and measuring sticks to 72 inches
3. Pick-out needles and combs
4. Tweezers

5. Small velvet-covered boards
6. Aluminum templates. Sizes frequently used, inches: ($1\frac{1}{4} \times 6$, $1\frac{1}{2} \times 6$, 4×6, 6×6, 3×8, 4×4, 4×10)
7. Stands and holders for skeins
8. Stop watch
9. Spare pens and ink (for charts)
10. Cutting blocks and cutting dies
11. Hand tally (Veeder-Root)
12. Nep-counting boards
13. Magnifying glass on stand

Basic Equipment for Laboratory, Spinning-Weaving Mill

1. Psychrometers: sling (with spare matched thermometers) and recording types
2. Drying oven, such as Emerson or Brabender type
3. Fibrograph for fiber length analysis
4. Pressley or Scott-Clemson tester for fiber strength
5. Sheffield Micronaire for fiber fineness
6. Saco-Lowell lap meter, Uster Varimeter
7. One-yard template for measuring one-yard lengths of sliver
8. Roving reel
9. Yarn reels for 120-yard skeins: one hand and one power-driven, Fidelity Machine Company or Alfred Suter Company
10. Balances of different types: two calibrated in grains (Whitin) for weighing sliver, roving, and yarn; torsion-type yarn-numbering balance; Chain-O-Matic balance in grains; Toledo scales in pounds and tenths to about 10-pound capacity; Shadowgraph by Exact Weight Scale Company with 1500-grain capacity
11. Quadrant-type direct yarn numbering scales by Alfred Suter Company or direct reading Shadowgraph for yarn number from skein
12. Evenness tester: Uster or Brush; for coarse work or wool blends, Pacific tester and Saco-Lowell tester
13. Inclined plane single strand tester, Model IP-2, by Henry L. Scott Company

14. Yarn winding unit for preparing black boards for yarn appearance, Standard Mill Supply Company
15. Seriplane by Alfred Suter Company
16. Set of yarn appearance standards from A.S.T.M.
17. Twist tester with attachment for untwist-twist test from U. S. Testing Company, Henry L. Scott Company, or Alfred Suter Company
18. Four-inch pick counter with attached tally; 2-inch pick counter
19. Abrasion tester: Schiefer, or Stoll Universal
20. Thickness gauge: Randall and Stickney, or Schiefer Compressometer
21. Combination yarn and cloth tensile tester, pendulum type, by Henry L. Scott Company
22. Tensometer to 75-gram capacity
23. Binocular microscope mounted on illuminated field
24. Calculating machine

Supplementary List of Equipment for Advanced Quality Control Program

1. Suter-Webb sorter for cotton fiber length evaluation
2. Shirley Analyzer for non-lint evaluation
3. Nepotometer (Wright Machinery Company) for predicting nep characteristics of cotton; also U.S.D.A. nep test machine
4. Nickerson-Hunter cotton colorimeter (Macbeth) for cotton color
5. B & L microscope and photomicrograph camera
6. Suter percentage scale for comber noil percentage
7. Torsion-type precision balances (Roller Smith Company) in 3, 25, and 100 mg. capacities
8. Inclined plane tester, Model IP-4, for heavy cord single strand test
9. Additional pendulum tester of proper capacity, so that one tester for yarn and one for fabric
10. Ball burst attachment for above pendulum tester
11. Mullen burst tester with dual range
12. Uster single strand tester
13. Brush tension analyzer

14. Monsanto or TBL crease tester
15. Schiefer Compressometer
16. Air permeability, Frazier
17. Densometer, air-flow type, by W & L Gurley
18. Package density, Suter-Manville tester
19. Polarized light microscope
20. Projectina for viewing yarn or fabric
21. Static tester
22. Fiber blender (U.S.D.A. type)
23. Instron Tester (if research contemplated)

It should be noted that mills handling wool or other natural fibers would require other instruments not shown here and would not require all those listed.

CHAPTER II

Accuracy of Measurement

All too often the results of textile measurements are expressed in terms that indicate an accuracy which is not justifiable on the basis of proper sampling, reading of instruments, and calculations. Interpretation of these results and subsequent action is influenced by the expressed degree of reliability of results and may cause unwarranted changes to be instituted in manufacturing processes. This chapter will attempt to outline basic recognized principles of measurement and calculation.

Vernier Scale

A common auxiliary measuring device found on many instruments is the vernier scale. The vernier is an auxiliary sliding or movable scale so designed as to increase the accuracy of the reading of the instrument by proportioning the smallest division of the instrument scale.

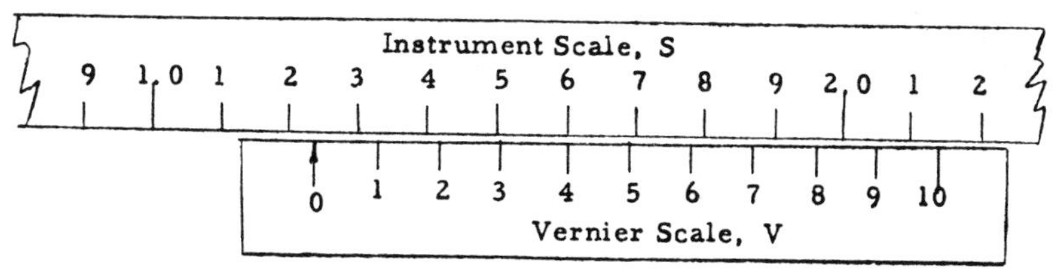

Fig. 2.1. Principle of the vernier scale.

The instrument scale S, shown in Fig. 2.1, has primary divisions at 1.0 and 2.0. It has secondary divisions between the primary divisions (1, 2, 3, . . ., 9) that divide the primary division into ten parts. The vernier scale V has a zero index, shown by the arrow, which indicates the reading of the instrument, but only within the accuracy of the stationary or instrument scale divisions. It will be noted that the zero index in the illustration points to a position approximately midway between 1.2 and 1.3. The vernier scale locates this position accurately in the following manner: Note that the division line "4" on the *V* scale coincides with the "6" division line on the *S* scale. This coincidence indicates that the *V* scale has been moved four-tenths of the distance between 1.2 and 1.3. Therefore, the accurate reading by use of the vernier is 1.24.

In a like manner, any position of the zero index between any of the secondary markings on the *S* scale can be determined by noting the coincidence of any two opposing lines and reading the vernier at this point to obtain the final digit.

The principle underlying the vernier depends on the condition that there are ten divisions of the *V* scale in a length equal to only nine divisions of the *S* scale. Therefore, each vernier division is nine-tenths of the length of each division of the instrument. So, by moving the *V* scale to the right one-tenth of a secondary division of the *S* scale, the "1" division on the scale will coincide with a division on the instrument scale. Similar increments of movement on the *V* scale to the right would result in its 2, 3, 4, . . . or 9 lines coinciding with some division of the *S* scale. As indicated in the example, the position of coincidence gives the final digit of the reading. Note that this digit should be read on the *V* scale, and not on the *S* scale.

Vernier scales also are designed with physical construction differing from the example cited. For instance, consider the micrometer calipers used for close measurements in the machine shop. But no matter what shape or form the vernier scale takes, the underlying principles are the same.

The fact that a machine or an instrument is set up with a vernier does not always mean that its use is justified: The instrument may not be built sufficiently accurate to justify the vernier, or might not warrant the use of the vernier. The technician should be cognizant of the limitations of the instruments he uses and should be able to calibrate such instruments. Incorrect calibration, poor instrument design, improper sampling, or other factors could easily vitiate and cancel out the accuracy supposedly obtainable and advertised for any given instrument.

Significant Figures

The degree of accuracy of a series of numbers in a calculation should be determined by the accuracy possible (i.e., the limitation of the accuracy of the measuring medium used in obtaining the data) and the common sense or practical accuracy desired. These conditions impose limitations on the number of significant figures that should be carried in any number or calculation.

Measurements are seldom, if ever, exact. Confusion exists in the minds of many people because they associate an exact number of units or a count (there are 100 pennies to a dollar) with the answer to a measurement; the number expressing the measurement is not exact, it is an approximation of a true value. For instance, if 100 people measure the distance from New York to Washington, there will be 100 different answers; no answer is likely to be exact, but all might be good approximations.

A better understanding of the application of the principles involved can be obtained by studying the relative and absolute errors in a series of numbers, such as those in Table 2.1.

TABLE 2.1
Relative and Absolute Error

Measurement	Relative error	Absolute error
612.	1 part in 612	1.
61.2	1 part in 612	0.1
6.12	1 part in 612	0.01
0.612	1 part in 612	0.001
0.0612	1 part in 612	0.0001
0.00612	1 part in 612	0.00001

These numbers are correct to three figures, that is, three significant digits. For example, the exact value of 61.2 would lie between 61.15 and 61.25. All the numbers listed above have the same *relative* error: namely, 1 part in 612. However, it will be noted that the *absolute* error in each case is different and depends on the position of the last significant digit. Some question may arise as to the significant figures in the numer 61,200. If this were written 612×10^2, the number of significant figures (3) would be obvious.

It is apparent from the above analysis that significant digits in each number are made up of all digits up to and including the first doubtful one. For example, in the number 61.2, the .2 is the doubtful figure but is considered as being significant.

Textile men have violated only too often the principles involved in the use of significant numbers. Consider a draft constant of 651. If calculated from actual gearing and if superfluous digits were not dropped, this constant might have been figured as 651.33. The last two digits (.33) are not desirable, give a false impression of accuracy, and do not add to the accuracy of calculations when the constant is used. For example, if using this constant, the size of the gear to give a draft of 21 is calculated to be 31.016.

$$\text{Teeth in draft gear} = \frac{651.33}{21} = 31.016$$

But a 31-tooth gear must be used. Now with 651 as the constant, a 31-tooth gear would still be used. The relative accuracy of the constant at 651 with 3 significant figures is 1 part in 651. Furthermore, with a constant of 651 and with the 31-tooth gear, the draft obtainable varies as follows:

$$\text{Lower limit of draft} = \frac{651}{31.5} = 20.6$$

$$\text{Upper limit of draft} = \frac{651}{30.5} = 21.4$$

The variation is about 8 parts in 200, or 26 parts in 651. This shows the undesirability and misconception of the degree of accuracy obtained by using 651.33. In other words, 1 part in 651 is more accurate by far than the accuracy of the actual draft, which is limited to an accuracy of 26 parts in 650.

If yarn sizes are obtained in the laboratory by use of a quadrant scale, measurements are read to the nearest quarter of a number. Thus, for $50\frac{1}{2}/1$ yarn, we know that the exact number lies somewhere between $50\frac{3}{8}$ and $50\frac{5}{8}$, giving an accuracy of $2\frac{1}{2}$ parts in 500, or 1 part in 200, or $\frac{1}{4}$ part in 50. It is interesting to compare this accuracy with that used in the illustration above of 1 part in 651. This comparison proves that the use of decimal places in draft constants is absolutely unnecessary.

Addition. In adding a series of numbers, if any doubtful digits exist in a column, then the accuracy of the total obtained from that column is doubtful. In such a case, the total should include only the sum of those columns which contain significant figures. This is also true for subtraction. In the example below, note that the 2 is increased to 3 since the following digit is 6.

```
  3.1416
  6.12??
840.0???
--------
849.2616
849.3      (Correct answer)
```

When digits are dropped, increase the last digit allowed to remain by 1 if the following digit is larger than 5; if this following digit is less than 5, no change is made. When the digit to be retained is odd, and if it is followed by 5 alone or by 5 followed by zeros, then the odd number is raised by 1, which makes it an even number. On the other hand, if the digit to be retained is even and is followed by 5, or 5 followed by zeros, then the even number remains unchanged.

In the following numbers, the digits in italics are to be rejected because in the previous calculation they are not significant.

26.3*5*	becomes	26.4
26.3*50*	”	26.4
26.4*50*	”	26.4
26.3*51*	”	26.4
26.4*51*	”	26.5
26.3*49*	”	26.3
26.3*499*	”	26.3

In rounding off numbers in arithmetic computations, it is wise to do this after computations are complete in order to avoid error.

Multiplication. When two numbers are multiplied together, the answer will be theoretically no more accurate than the least accurate of the two numbers. In actual practice, the product is allowed to appear more accurate, i.e., carries one more significant figure than is contained in the multiplicand or the multiplier, whichever is the least accurate. Consider the following examples involving measurements only.

(*1*)
$$\begin{array}{r|l} 7 & \\ \times 6 & \\ \hline 42 & \end{array}$$

(*2*)
$$\begin{array}{r|l} 7. & 2 \\ \times 6 & \\ \hline 43. & 2 \end{array}$$

(*3*)
$$\begin{array}{r|l} 7.1 & \\ \times 6.1 & \\ \hline 42.6 & \\ 7 & 1 \\ \hline 43.3 & 1 \end{array}$$

(*4*)
$$\begin{array}{r|l} 7.1 & 225 \\ \times 6.1 & \\ \hline 42.7 & 350 \\ 7 & 1225 \\ \hline 43.4 & 4725 \end{array}$$

Everything to the left of the vertical lines is significant, and by the definition of significant figures includes one doubtful figure. In example (*1*), the final answer of 42 is justified and correct. In example (*2*), strictly by definition, it would be proper to use 43 as the answer, in which case the 3 is a doubtful number. The reason for this is that the .2 and 6 are both considered as doubtful numbers. However, a practice sometimes used is to give the answer as 43.2, but when such examples arise, note that the .2 is the second doubtful number. If the 6 had been written as 6.0, then 43.2 is justified and correct, with the .2 as the only doubtful number by definition; the 43.2 is much preferred to 43.20, as the use of the two decimals gives a false impression of accuracy, and the use of the last zero is not justified.

In example (*3*), the correct answer is 43.3; the use of 43.31 means that two doubtful numbers have been included and would signify that the answer is accurate to 1 part in 4331 parts; obviously, this accuracy is not warranted by the two original numbers, 7.1 and 6.1, both of which indicate an accuracy of less than 1 part in 100.

In example (*4*), the correct answer is 43.4. Use of any more digits is not justified. This example brings up another consideration: The values dealt with are measurements; therefore, why has one measurement been carried to four decimal places, indicating an accuracy of 1 part in 71000, whereas the second measurement carries an accuracy of 1 part in 61? The object of this example is to indicate that in cases involving measurements, it is unnecessary to make very accurate measurements on one correlated factor, if it is either impossible or impractical to make measurements on the second correlated factor to the same degree of accuracy.

The method of multiplication used in example (*3*) is illustrated more fully below. The advantage of this method lies in the fact that numbers that are not significant can be dropped easily.

```
    3645
    2973
   ------
    7290|
    3280|5
     255|15
      10|935
  ---------------
   10836|585
```

Final answer = 10837×10^3

Division. The work of long division can be simplified by dropping one digit of the divisor at each step instead of adding a doubtful zero.

```
23,426 /  64,916 \ 2
          46,852
          ------
  2343 /  18,064 \ 7
          16,401
          ------
   234 /   1,663 \ 7
           1,638
           -----
    23 /      25 \ 1
              23
              --
     2 /       2 \ 1
```

Answer = 2.7711

It should be emphasized that a count (such as number of samples in a test, of a monetary value as in dollars and cents) is an infinitely accurate number and must be considered as such in any calculation. Therefore, always distinguish between count (there are 100 cents in a dollar) and measurement (the strength of a skein of yarn is 101 pounds); the count is exact and precise, and any desired number of significant figures can be used. The accuracy of the answer for any calculation involving count and measurement will depend upon the accuracy of measurement alone.

Averages. When a series of readings has been totaled and an average obtained, this average contains less error than any one reading. It is accepted practice, therefore, to carry the average to one more significant figure than the individual readings.

TABLE 2.2

Example of Taking an Average

	Sample	Skein strength, lbs.
	1	92
	2	98
	3	96
	4	96
	5	95
Total	5	477
Average		95.4 lbs.

Accuracy, Precision, and Systematic Error

The average of a column of figures does not describe fully the character of a set of measurements. A measure of their dispersion about the average must be obtained. This measure is either the range or the standard deviation.

The comparison of a set of data with some given standard will involve the classification of the data into one of three groups: accurate, inaccurate or precise. When data are inaccurate, the average and the dispersion both may be outside the standard limits; and when data are precise, the average may be outside standard limits, but the dispersion will be within the established limits. Precision connotes reproducibility, and accuracy means hitting close to the value desired.

To demonstrate these three concepts, consider the sets of data in Table 2.3, each set representing the laboratory sizings of six skeins from three different lots of yarn spun as 35/1.

TABLE 2.3
Inaccurate, Accurate, and Precise Data

	Inaccurate	Accurate	Precise
	32.8	34.5	33.2
	34.2	35.2	32.8
	36.8	35.8	32.2
	31.2	36.0	33.5
	32.0	34.5	34.0
	33.2	34.8	33.0
Total	200.2	210.8	198.7
Average	33.37	35.13	33.12
Range	5.6	1.5	1.8
Specified average	35.0±0.5	35.0±0.5	35.0±0.5
Specified maximum range	2.0	2.0	2.0

The first set of figures is inaccurate, inasmuch as the average and the variation of the readings are outside the established limits.

The second set of values is accurate, since the average and the variation are within the given limits.

The third group of data is precise: the average is outside the standard limits, but the variation is acceptable. The values of

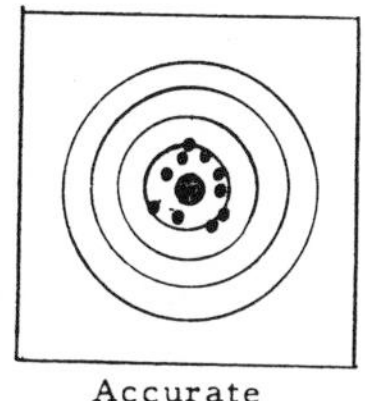

Accurate
(On the bull's eye)

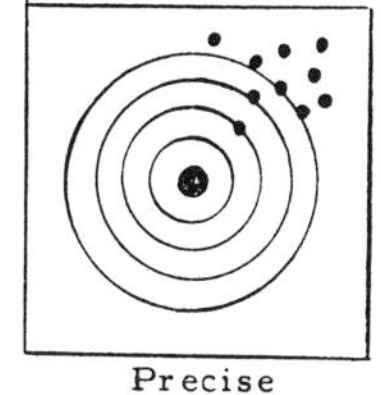

Precise
(Subject to systematic error)

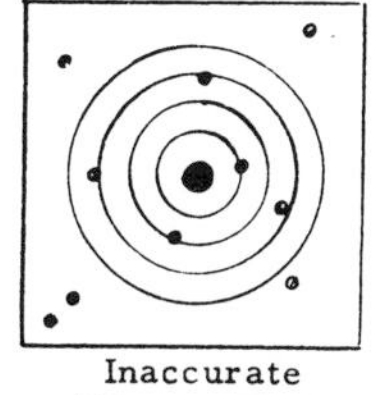

Inaccurate
(Hit or miss)

Fig. 2.2. Diagrammatic representation of accurate, precise, and inaccurate data.

the precise column are said to be subject to a systematic error, which in this case averages 1.8 units (or from 34.00 to 32.2). In other words, the error is consistent throughout the series. The yarn spun on the frame from which this lot was taken could be made acceptable by an appropriate draft gear change.

Inaccurate results are of no value, accurate results are acceptable and most desirable, and precise results can be of value if the extent or cause of the systematic error is known. These thoughts are shown diagrammatically in Fig. 2.2.

CHAPTER III

Presentation of Data

The attractive and effective presentation of data is essential if the data are to be of utmost value. The three basic methods of presenting data are to enumerate them in the text, to tabulate in the form of tables, and to plot graphically as charts.

The presentation of data in the text of a report is usually not recommended for several reasons, some of which are: the data are not easily comprehended; it is difficult if not impossible to locate trends in the data; and it is difficult to emphasize the individual figures or sets of figures.

Practically all data can be presented in the form of either tables or charts, and one or both of these methods is recommended. The tables are generally used when a large amount of data is to be presented and when there is little, if any, necessity of using charts or graphs. When tables are used in this way, it is helpful to extract the most important data, or any that are to be emphasized, and present them in a separate table. This practice is very helpful to the reader of the report or to the executive whose chief interests are the general basic facts and more important points presented in the report. With this arrangement, the summary table can be studied, and, if the reader wishes, he may also refer to the general tables for more detailed information.

Charts or graphs are of value when it is desired to show some relationship or comparison. When such a situation exists, it is practically impossible to show such relationships in a table. Even if it were possible to show relationship or trends by the use of a table, it would necessitate a considerable amount of study on the part of the reader to decipher. However, with the data plotted in chart form, it is necessary to take only one quick glance to comprehend the situation. This is extremely important from the viewpoint of the individual who does not wish to become involved in details and does not have the time to study long columns of numbers.

The mere presentation of data in chart or tabular form does not mean that no explanation of the data will be needed. In fact, it is usually necessary to devote one or more paragraphs in a report to the explanation of any chart or table.

Tables

Summary and General Tables. An example of a summary table is shown in Table 3.1. The general tables from which the data in Table 3.1 were extracted are Table 3.2 and Table 3.3. Note the ease with which the two yarns may be compared after the data is placed in summary form as in Table 3.1.

TABLE 3.1

Summary of Test Results for 24/1 and 30/1 Carded Cotton Yarns [a]

Tests	24/1	30/1
Skein strength (pounds)	90.0	69.7
Yarn number (English count)	23.9	30.4
Break factor	2147.0	2118.1
Single end strength (grams)	279.4	229.2
Twist (turns per inch)	15.7	18.8

[a] All yarn numbers and strengths corrected to a moisture regain of 7.5 per cent. Average of five bobbins, two tests per bobbin.

Construction of Tables. In the construction of a table, there are a few suggestions which, if followed, will aid in the presentation of a more attractive and effective table.

Title. Each table should have a title, preferably placed above the table. The title should state in a clear and concise manner what data are included in the table. Long titles should be avoided; however, clarity should not be sacrificed in order to shorten the title.

Numbering of Tables. When more than one table is to be used in a report, they should be numbered consecutively with the number appearing to the left of or above the title. The numbering simplifies

TABLE 3.2

Test Results for 24/1 Carded Cotton Yarn [a]

Bobbin number	Skein strength (pounds)	Yarn number (English)	Break factor (ct × st)	Single end strength (grams)	Twist (turns per inch)
1	94	23.8	2237	283	15.5
1	92	23.4	2153	278	15.8
2	87	24.2	2105	275	15.2
2	88	24.4	2147	266	15.9
3	91	23.4	2129	282	16.2
3	90	23.6	2124	290	15.7
4	89	24.3	2163	283	15.4
4	90	24.1	2169	288	15.8
5	88	23.6	2077	279	15.6
5	91	23.8	2166	270	15.9
Average	90.0	23.9	2147.0	279.4	15.7

[a] All yarn numbers and strengths corrected to a moisture regain of 7.5 per cent.

TABLE 3.3

Test Results for 30/1 Carded Cotton Yarn [a]

Bobbin number	Skein strength (pounds)	Yarn number (English)	Break factor (ct × st)	Single end strength (grams)	Twist (turns per inch)
1	68	31.8	2162	226	18.2
1	69	31.4	2167	229	19.0
2	72	28.9	2080	233	18.6
2	70	29.3	2051	230	18.9
3	69	30.2	2084	232	19.2
3	70	30.6	2142	231	19.0
4	73	30.8	2248	230	18.6
4	72	31.2	2246	226	18.8
5	67	29.6	1983	229	18.5
5	67	30.1	2017	226	18.9
Average	69.7	30.4	2118.1	229.2	18.8

[a] All yarn numbers and strengths corrected to a moisture regain of 7.5 per cent.

any references which may be made to the tables since it is necessary to use as the reference only the table number rather than the title.

Footnotes. Frequently it is necessary to make some explanation of one or more figures in a table, and to do this it is customary to use footnotes at the bottom of the table. When this is found necessary, the figures in the table and the footnote must be marked so as to identify the figures with the corresponding footnote. Markings commonly used are: *, **, †, ‡, etc., or by superscript such as [a], [b], [c] etc. The corresponding footnote is marked in the same manner as the figure and the same marking should not be used more than once in the same table unless one footnote is used to explain or discuss several figures. In the text of a report, it is acceptable to use numbers ([1], [2], etc.) to refer to footnotes, but it is not good practice to use numbers as a method of identification for footnotes to tables except in the stub (or items column) of a table.

Zeros. A zero should never appear in a table unless it is a computed value. If for some reason a reading or measurement were not made and there is no figure to enter into the table, the space should be marked by dots (. . .), dashes (– –), or some other convenient method, and an explanation offered in the form of a footnote.

Units. The units for the data in the table should be given. If the units are the same for all of the data, then it may only be necessary to include such units in the title. However, if this is not satisfactory, then it may be necessary to place the units at the head of each column or include them in the column by each figure or entry.

When percentages are used in the table, the term "per cent" alone should not be used. The basis used for calculating the per cent should also be included, such as, "per cent second-quality fabric" or "per cent increase in machine efficiency," etc.

Charts

The most common type of charts used to present industrial data are curves or line diagrams, which include frequency distribution and cumulative frequency distribution, and bar charts, which include component bar charts.

Construction of Curves or Line Diagrams. An example of a simple curve is shown by Fig. 3.1. This is one of the simplest types of line diagrams; however, there are a few rules or suggestions which should be followed so that the diagram will be most effective.

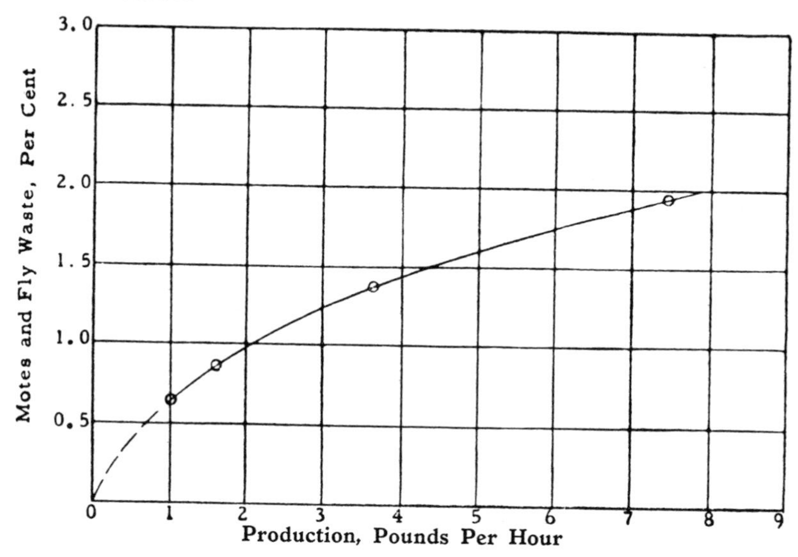

Fig. 3.1. Relationship of production to motes and fly waste at cotton carding with a constant licker-in speed. Cotton: Middling $1\frac{3}{32}$ ins.; licker-in speed: 800 r.p.m.; variable doffer speeds.

Selection of Values for the Vertical and Horizontal Scales. The improper selection of scale values can cause the reader to get a completely erroneous concept of the material being presented. If the scales are very short or close, the reader is likely to interpret the chart to mean that greater fluctuation exists than actually does, and over-expanded scales are likely to convey the impression that the relationship is unimportant or very small. The proper selection of scales may be very difficult at times; however, Croxton and Cowden (1) offer the suggestion that the scales be so chosen so that the movements of the curve which is to be emphasized is at approximately a 45-degree angle. (See Fig. 3.2.) This is certainly not to be followed blindly; it may be necessary

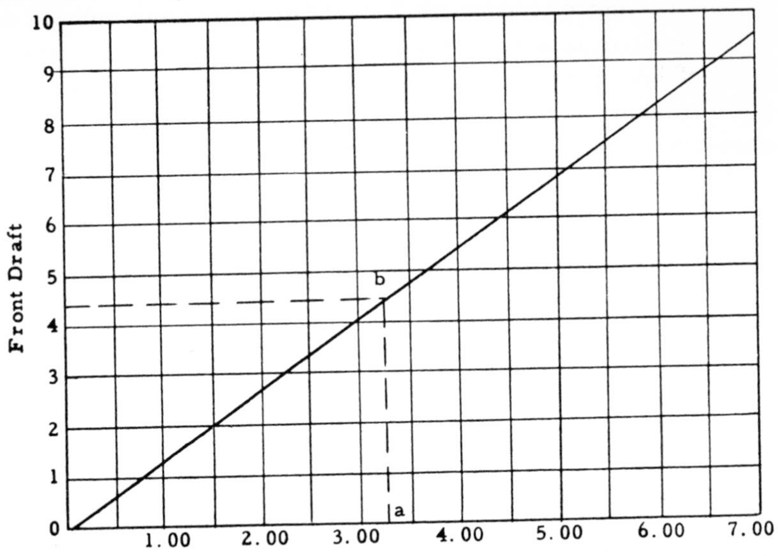

Fig. 3.2. Chart for determination of proper front draft on the Saco-Lowell J–3 roving frame using 11.5 grains per yard in folding zone. (Data courtesy of T. A. Hendricks, Graduate Fellow, School of Textiles, N. C. State College, 1951.) Example: If a 3.25 hank roving is to be delivered, the proper front draft would be 4.5 for minimum variation.

at times to use other scales or slopes to emphasize, minimize, or present a truer picture of the relationship.

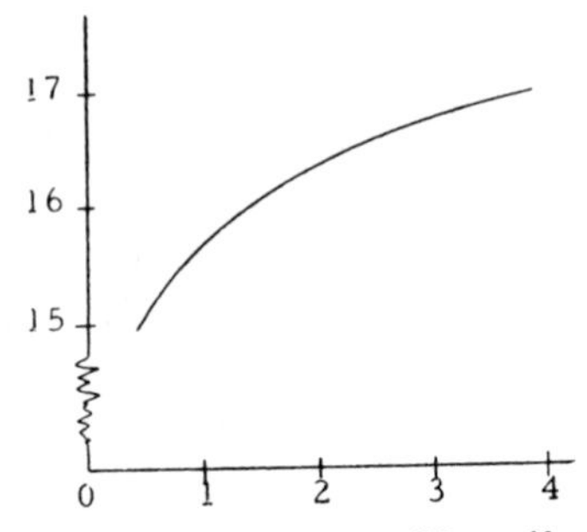

Fig. 3.3. Sample chart. (Note *Y* axis.)

The use of scales which begin with zero is usually recommended, especially on the Y axis. The use of scales which do not begin

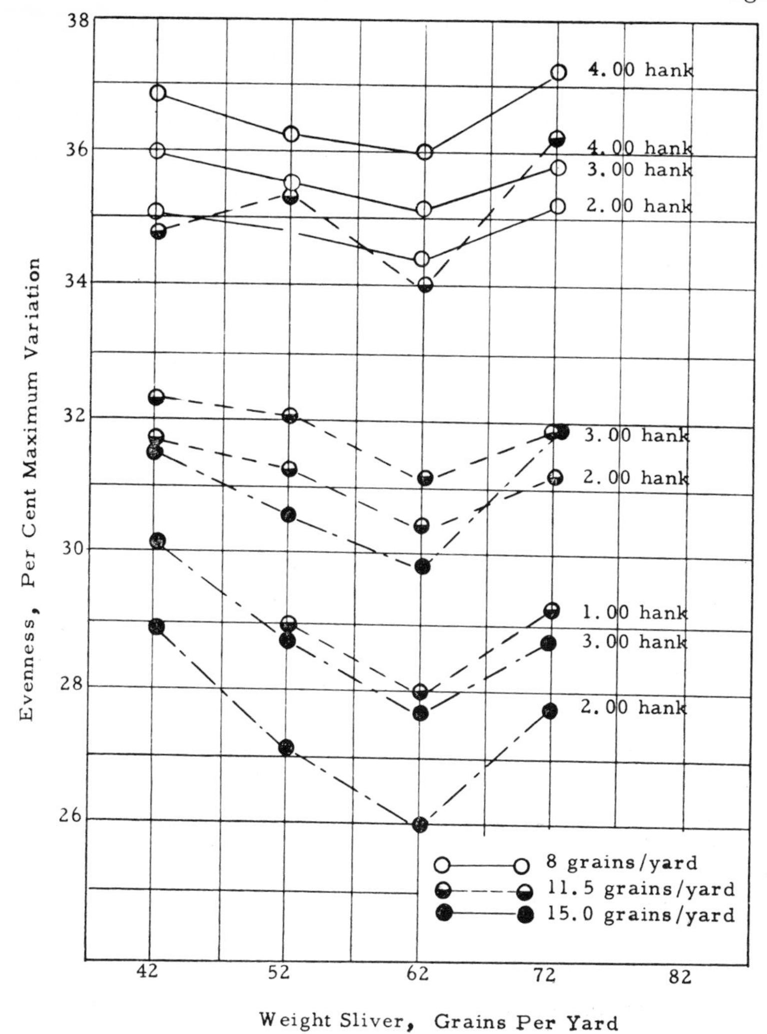

Fig. 3.4. Comparison of roving evenness and weight in folding zone with varying silver weight and hank roving.

with zero is likely to create an incorrect impression on the reader. A note of explanation or some other method of directing the attention of the reader to the fact that a scale does not begin with zero, should be used. One accepted method is to break the *Y* axis, such as is shown in Fig. 3.3.

The above suggestion is recommended for uses where the reader will inspect the chart for trends or relationships. However, for charts such as Fig. 3.4 which are to be used as a source of data, such as twist *vs.* contraction curves, it is not necessary to start the scales with zero since the curve is to be used primarily as a source of information and the shape or trend of the line is secondary.

Plotting Points and Drawing Curve. An examination of Fig. 3.1 will reveal that points indicated by a small circle (o) are drawn on the chart and a curve is passed through these points. This indicates that actual test data were entered on the chart; such points should not be drawn unless they represent actual test data. When no data are to be plotted and the curve is to be a representative or typical curve of some relationship, the points should then be omitted.

In connecting points that have been plotted, special care should be taken not to make any unwarranted assumptions as to the shape of the curve. That is, a straight line should not be drawn unless the data fit a straight line, and by the same token, a curve should not be drawn unless the data actually fit some particular type curve.

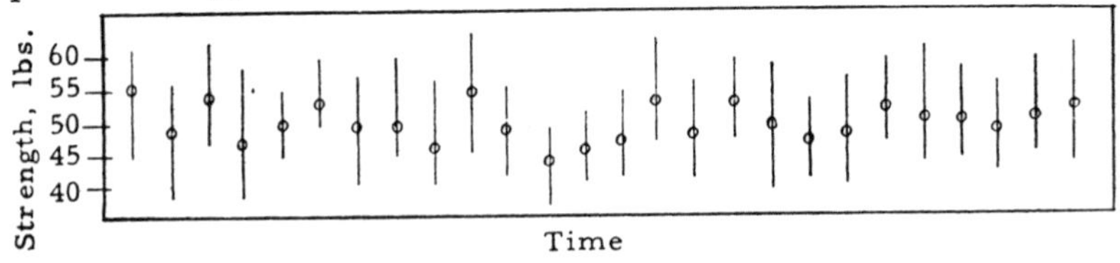

Fig. 3.5. Sample range chart.

When drawing more than two curves on the same chart, it will make a more attractive and legible chart if lighter lines are used. Also, when several curves appear on the same chart, it is necessary to distinguish each curve from the others. This can be done by the use of broken lines or by lettering the different curves.

Title. The title of a chart should meet the same requirements

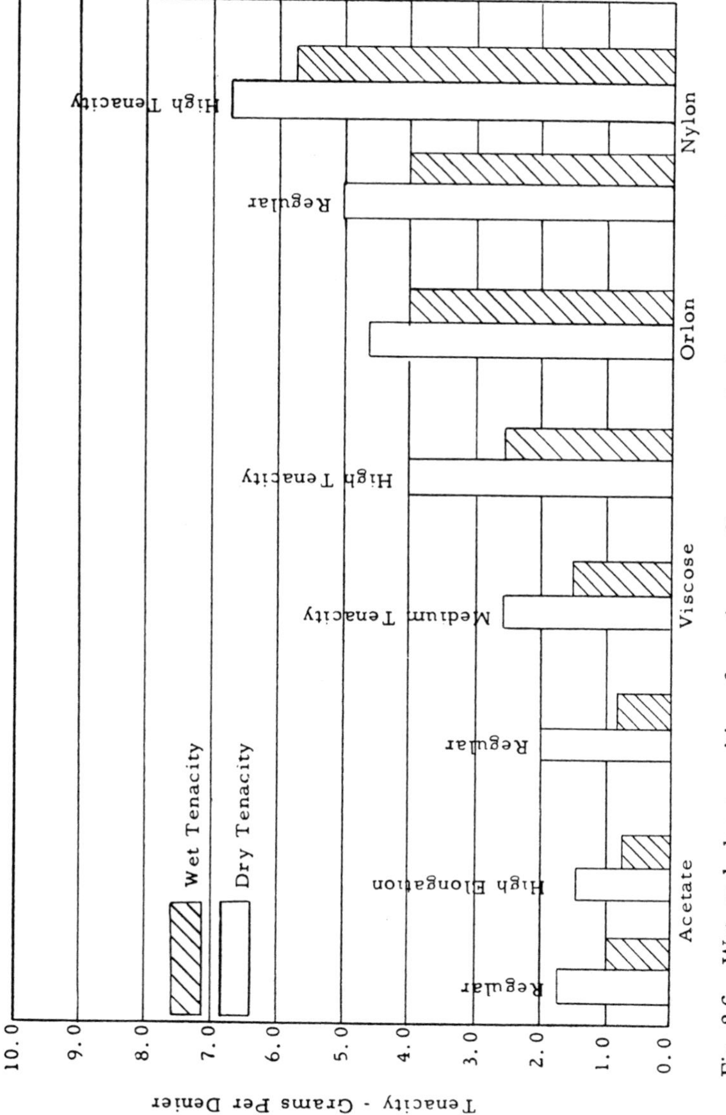

Fig. 3.6. Wet and dry tenacities of continuous filament synthetic yarns produced in the United States. Yarn tested for dry strength at standard conditions of 65% R.H., 70°F.

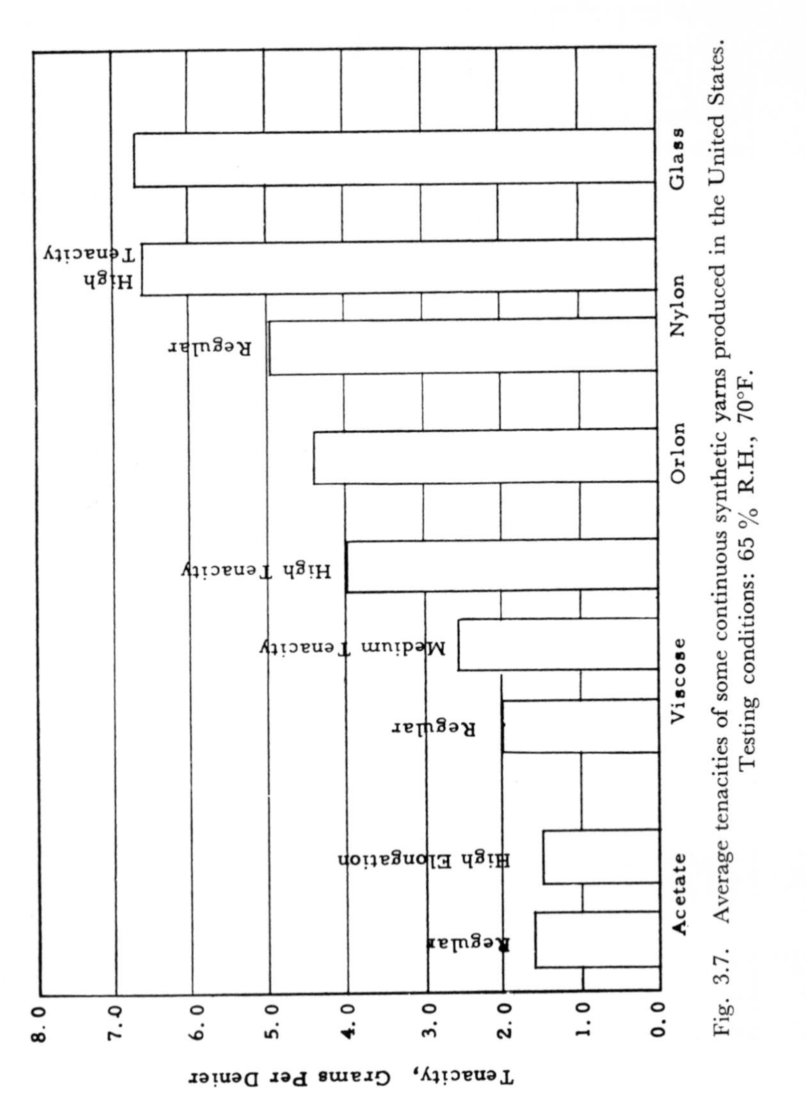

Fig. 3.7. Average tenacities of some continuous synthetic yarns produced in the United States. Testing conditions: 65 % R.H., 70°F.

as a title for a table with respect to clarity. The position of the title may be either above or below the curves, depending on the availability of space.

Construction of Range Charts. The type of chart shown in Fig. 3.5 may be used to good advantage either as a summary chart for reports or as a working chart for daily results. The heavy point represents the average of a set of data, and the length of the line through the point is controlled by the variation in the data.

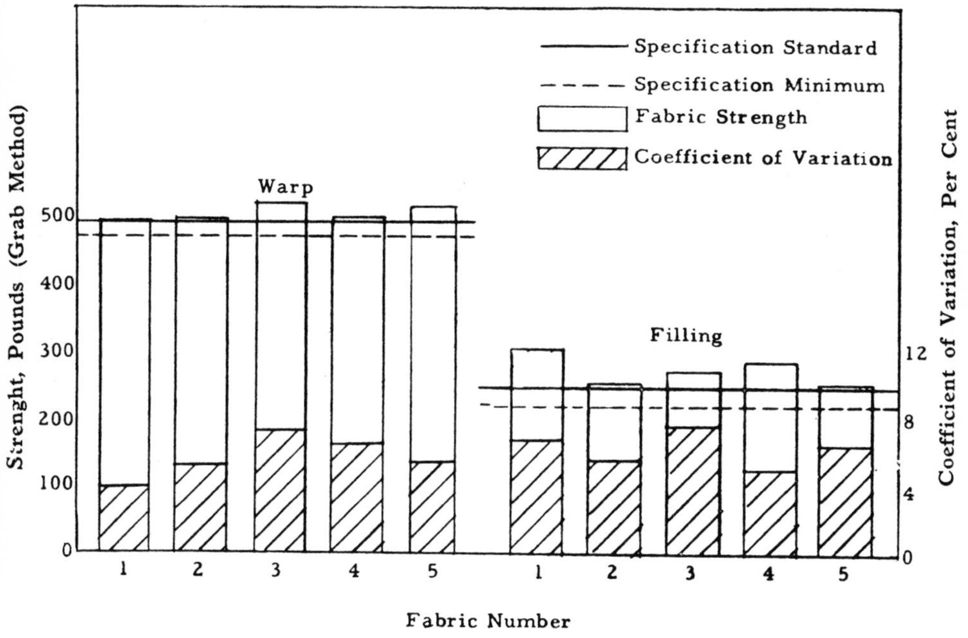

Fig. 3.8. Strength and coefficient of variation for strength of a cotton duck fabric tested at standard conditions.

For example, with the average value representing the mean of four measurements, the ends of the vertical line indicate the lowest and highest values of the four measurements. This chart is very useful when plotting $\bar{X}$ and R charts since both charts may be plotted on the same graph.

Construction of Bar Diagrams. For certain types of data it is advantageous to use bar charts and component bar charts to present a more vivid picture of the relationship between the factors involved. Examples of these are Figs. 3.6 and 3.7.

The bar charts should have titles, scales, source notes, footnotes, etc., the same as any other type of chart. However, there are a few suggestions which should be followed when constructing bar charts that do not apply to line diagrams or curves.

Size of Bars. The individual bars should not be out of proportion with respect to size. That is, they should not be extremely short and wide or very long and narrow. The distance between individual bars should not be less than one-half the width of the bar or greater than the full width of the bar. Guide Lines, shown in Fig. 3.8, are helpful if there are several bars and the chart is rather large.

References

1. Croxton and Cowden, *Applied General Statistics*, Prentice-Hall, New York, 1946.

CHAPTER IV

Analysis of Data

For any quality control system to be effective, the data collected by the testing laboratory must be analyzed and studied. The analysis of the data may include such simple statistical techniques as the average, standard deviation, range and coefficient of variation, or it may involve some of the advanced methods such as analysis of variance, correlation, and reliability of means. The simpler methods may be used by anyone with a fair knowledge of algebra; however, it is not recommended that anyone use the advanced techniques unless they have had some instruction in the use of such methods or have studied books which explain the techniques in detail.

As stated above, the average, or the mean as it is sometimes called, is a statistical technique, and it is the most commonly used statistical element. In the textile industry, as well as in all other industries, it is very common practice to calculate the average for a set of test data and accept that average as being the true representative of the product. Unfortunately, such an average will very seldom be the same as the product average and should be treated as only an estimate of the true average. Assume the average of a set of test data is a good estimate of the product average; actually, very little is yet known about the product if no further study of the test results is made. The average gives little or no indication as to the amount or type of variation in the quality of the material, nor does it indicate whether this variation is due to the process or is inherent in the product. To find the answer to such questions it is only necessary to calculate the standard deviation. With this one simple calculation, it is then possible to extract a considerable amount of additional information.

For example, suppose that the yarn number for a 30/1 yarn

had an actual average of 29.7/1 over a period of several weeks or months, and the standard deviation for the same period was 0.6. (The calculations for the standard deviation will be explained in the ensuing discussion.) Knowing only the average of 29.7, very little is actually known about the size of the yarn produced during that period. The overseer or superintendent may examine the test results and find that the yarn number varied from 27.0/1 to 31.9/1 and feel that the size does not have an excessive amount of variation. With the above set of conditions, that is, with an average of 29.7 and a standard deviation of 0.6, it is a simple matter to prove that the variation in yarn number was excessive if it ranged from 27.0/1 to 31.9/1. The techniques involved will be discussed in the following chapter, but it can be shown that approximately 68 per cent of all the yarn produced should have a size between 29.7 ± 0.6 (29.1/1 and 30.3/1) and that approximately 95 per cent of the yarn produced should size between $29.7 \pm 2(0.6)$ (28.5/1 and 30.9/1), and that practically 100 per cent should be between $29.7 \pm 3\ (0.6)$ (27.9/1 and 31.5/1). This means that with an average of 29.7/1 and a standard deviation of 0.6, the yarn should range in size from 27.9/1 to 31.5/1 in approximately the proportions calculated above and that this is the expected variation in size. Any yarn that does not have a size between 27.9/1 and 31.5/1 has too much variation, and an investigation should be made for any of the factors which will likely affect yarn number.

It is easier to understand now why the average is so limited in the amount of information that it gives about a set of data. At the beginning of the above discussion, the only information known about the yarn number was that the test results averaged 29.7. However, with the standard deviation known, it is possible to accurately determine the exact range over which the yarn number should vary, what part of the production will be within certain sizes, and when excessive variation is present. In the latter case, steps should be taken to better control the yarn size.

This is only one example of how the proper analysis and study of data will increase the value of test results and aid in producing a better quality product. In the following chapters some of the more common statistical techniques will be explained, and practical applications to the textile industry will be given.

Frequency Distribution

The definition of a grouped frequency distribution as given in the *A.S.T.M. Manual on Quality Control of Materials* is "a set of observations in an arrangement which shows the frequency of

TABLE 4.1

Skein Strength Tests in Pounds

73	72	75	74	72	75	73	73	76	78	74	73
74	75	73	76	75	73	75	76	76	75	74	76
72	75	75	74	76	75	76	75	75	73	76	74
76	71	76	73	74	74	74	72	72	76	77	74
75	76	74	74	77	71	72	76	75	75	77	75
77	73	77	75	73	75	74	75	74	72	74	74
74	77	75	75	74	76	73	73	73	75	74	73
73	74	73	71	78	73	75	75	76	74	75	76
76	73	75	74	76	74	74	71	75	75	73	74
70	72	74	73	74	72	78	75	77	75	73	76
75	75	77	74	77	74	75	74	74	74	76	76
73	74	75	76	74	77	76	73	72	77	75	79
74	78	73	74	76	74	75	75	76	72	75	73

$$\Sigma = 11622 \qquad \bar{X} = \frac{11622}{156} = 74.5 \qquad n = 156$$

occurrence of the values of the variable in ordered class." This means that if data or test results are plotted in the form of frequency of occurrence, a certain type curve will be formed: For example, assume the data in Table 4.1 to be skein strength results. If the data were plotted as shown in Fig. 4.1, it would show the frequency of occurrence of skeins with any given strength. From Fig. 4.1, it is seen that of 156 skein tests, one skein had a strength of 70 pounds, four had a strength of 71 pounds, eleven had a strength of 72 pounds, etc. The frequency is then plotted against the skein strength, and the frequency distribution is formed as shown in Fig. 4.2.

The relationship between the average, standard deviation, and frequency distribution is the basis for most of the elementary and advanced techniques that have been developed in the field of statistics. Some of these relationships and techniques will be discussed in detail.

	Number	Individual Frequency
70	1	1
71	1111	4
72	11111 11111 1	11
73	11111 11111 11111 11111 11111	25
74	11111 11111 11111 11111 11111 11111 11111 11	37
75	11111 11111 11111 11111 11111 11111 11111 11	37
76	11111 11111 11111 11111 11111	25
77	11111 11111 1	11
78	1111	4
79	1	1
	0 5 10 15 20 25 30 35 40	

Fig. 4.1. Frequency distribution from Table 4.1.

The data in Table 4.1 are hypothetical and for that reason the frequency distribution is perfectly symmetrical. In actual practices, a perfectly symmetrical curve is the exception rather

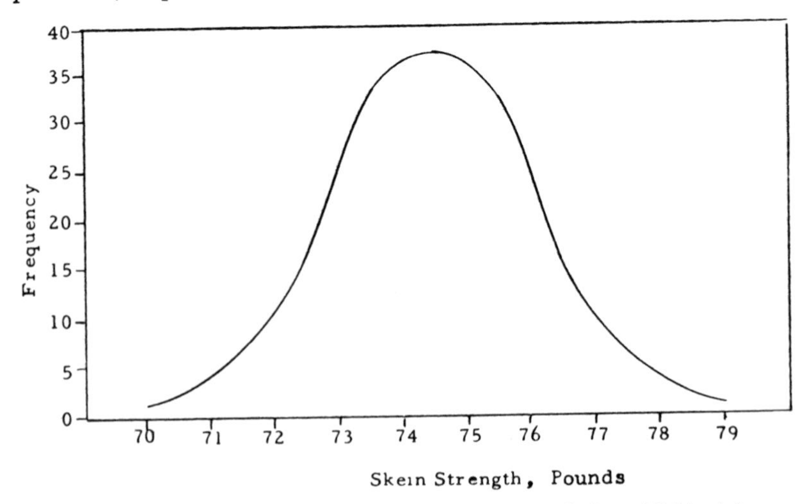

Fig. 4.2. Frequency distribution of skein strength from Table 4.1.

than the rule, and it is usually necessary to have several hundred tests to form a normal bell-shaped curve, even if the data under consideration were of the normal type.

Skewness. The term skewness describes the lopsidedness of a frequency distribution. The skewness, k, of a set of numbers may be calculated:

$$k = \frac{\sum_{i=1}^{n} (Xi - \bar{X})^3}{n\sigma^3}$$

where n = number of tests, σ = standard deviation.

For a symmetrical distribution, the value k is zero. If the long tail of the curve extends to the left, k is generally negative, and if the long tail extends to the right, k is generally positive. This

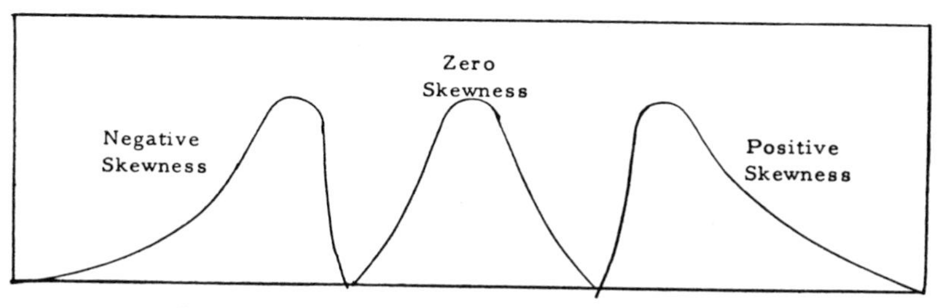

Fig. 4.3. Skewness of frequency distributions.

is shown in Fig. 4.3. The value k is not calculated as frequently as other statistical values, such as average and standard deviation. Generally, it is sufficient for the quality control engineer to know only the direction in which the skewness occurs. However, if two distributions are skewed in the same direction, it may then become necessary to calculate the skewness so as to know the relative magnitude of the skewness of the two distributions.

There are several measures which are available and can be used for analyzing data, such as: the median, the middlemost value in the distribution; the mode, the value which occurs most frequently; and others. However, the most commonly used measures are the average, standard deviation, coefficient of variation, and skewness.

Standard Deviation

The standard deviation, hereafter referred to as σ (read as "sigma"), of a set of numbers is defined as the root-mean-square deviation of the numbers from their average. In other words, the standard deviation is a measure of the amount of variation from the average of a set of data. For example, if both of two yarns have an average skein strength of 100 pounds, but with a standard deviation for the first yarn at 8 pounds and a standard deviation for the second yarn at 20 pounds, it would mean that the second yarn would have considerably more variation in skein strength than the first yarn. The first yarn is the better of the two. This is only one simple example of the value and use of sigma; more uses and examples will be discussed. The term "root-mean-square," which at first sight seems an unwieldy expression, in reality explains the deviation very completely. Because deviations from any average may be negative as well as positive, squaring these deviations removes all negative values. The mean puts the deviation on an average basis. The root brings the expression back to a linear basis from the power basis resulting from the squaring operation. Thus, sigma, the root-mean-square value, is an average measure of deviation unaffected by positive or negative values.

The units for sigma are the same as for the numbers used in calculating sigma. If data pertaining to pounds strength are being investigated, the units for sigma are pounds; if the data are for turns twist, the units for sigma are also turns, and so forth.

There are several different methods of calculating the standard deviation, depending upon the amount of data available and the intended use of the measurement. The amount of data is a factor which should be considered when choosing a method to calculate the standard deviation. If a large amount of data is available, it is generally referred to as a large sample size, and if a small amount is available, it is considered as a small sample size. Generally, if the sample size is larger than forty or fifty, the methods for large sample size are used. If less than forty or fifty observations are made, it is considered as a small sample size, and the methods

TABLE 4.2

Per Cent Elongation of 150 Denier, 40 Filament, Bright Viscose Rayon

(1) % elong., X	*(2)* X^2	*(3)* $(X - \bar{X})$	*(4)* $(X - \bar{X})^2$	*(5)* $(X - 16)$	*(6)* $(X - 16)^2$	*(7)* Range of gps. of 4	*(8)* Sub-group ave., $\bar{X}$
16.7	278.89	0.25	.0625	0.7	0.49		
17.8	316.84	1.35	1.8225	1.8	3.24		
16.6	275.56	0.15	.0225	0.6	0.36		
16.2	262.44	−0.25	.0625	0.2	0.04	1.6	16.8
15.9	252.81	−0.55	.3025	−0.1	0.01		
16.7	278.89	0.25	.0625	0.7	0.49		
17.3	299.29	0.85	.7225	1.3	1.69		
16.2	262.44	−0.25	.0625	0.2	0.04	1.4	16.5
17.3	299.29	0.85	.7225	1.3	1.69		
16.1	259.21	−0.35	.1225	0.1	0.01		
17.3	299.29	0.85	.7225	1.3	1.69		
16.5	272.25	0.05	.0025	0.5	0.25	1.2	16.8
16.6	275.56	0.15	.0225	0.6	0.36		
16.6	275.56	0.15	.0225	0.6	0.36		
16.8	282.24	0.35	.1225	0.8	0.64		
16.8	282.24	0.35	.1225	0.8	0.64	0.2	16.7
16.7	278.89	0.25	.0625	0.7	0.49		
15.8	249.64	−0.65	.4225	−0.2	0.04		
16.7	278.89	0.25	.0625	0.7	0.49		
17.3	299.29	0.85	.7225	1.3	1.69	1.5	16.6
17.0	289.00	0.55	.3025	1.0	1.00		
16.7	278.89	0.25	.0625	0.7	0.49		
17.1	292.41	0.65	.4225	1.1	1.21		
17.0	289.00	0.55	.3025	1.0	1.00	0.4	17.0
16.0	256.00	−0.45	.2025	0.0	0.00		
15.7	246.49	−0.75	.5625	−0.3	0.09		
16.2	262.44	−0.25	.0625	0.2	0.04		
16.0	256.00	−0.45	.2025	0.0	0.00	0.5	16.0
17.3	299.29	0.85	.7225	1.3	1.69		
16.3	265.69	−0.15	.0225	0.3	0.09		
16.4	268.96	−0.05	.0025	0.4	0.16		
16.1	259.21	−0.35	.1225	0.1	0.01	1.2	16.5
16.8	282.24	0.35	.1225	0.8	0.64		
16.1	259.21	−0.35	.1225	0.1	0.01		
16.4	268.96	−0.05	.0025	0.4	0.16		
17.1	292.41	0.65	.4225	1.1	1.21	1.0	16.6
16.3	265.69	−0.15	0.225	0.3	0.09		
16.1	259.21	−0.35	.1225	0.1	0.01		
15.8	249.64	−0.65	.4225	−0.2	0.04		
15.8	249.64	−0.65	.4225	−0.2	0.04	0.5	16.0
13.2	174.24	−3.25	10.5625	−2.8	7.84		
17.4	302.76	0.95	.9025	1.4	1.96		
16.2	262.44	−0.25	.0625	0.2	0.04		
15.8	249.64	−0.65	.4225	−0.2	0.04	4.2	15.6
16.4	268.96	−0.05	.0025	0.4	0.16		
16.6	275.56	0.15	.0225	0.6	0.36		
16.1	259.21	−0.35	.1225	0.1	0.01		
15.9	252.81	−0.55	.3025	−0.1	0.01	0.7	16.2
$\Sigma =$ 789.7	13015.51	.10	23.3000	21.70	33.11	14.4	197.3
$\bar{X} =$ 16.45	271.16	.0022	0.49	.45	.69	1.2	16.4
$n =$ 48							

outlined for small sample sizes are used. The exact dividing line between a large and small sample size is difficult to define. When it is doubtful which method should be used, it is always safe to use the calculations for the small sample size. The only difference between the two methods is whether n or $n - 1$ is used in certain of the formulas to be illustrated. When the value n reaches such a magnitude that $\sqrt{n}$ is for all practical purposes equivalent to $\sqrt{n-1}$, any of the methods may be used.

Large Sample Size

Method 1 (Sum of Squares of Individual Differences $\Sigma[X-\bar{X}]^2$) (Preferred method.) Use the data from Table 4.2, first, third, and fourth columns.

$$\sigma = \sqrt{\frac{\Sigma(X - \bar{X})^2}{n}}$$

where $(X - \bar{X})$ represents the value resulting from subtracting the average, $\bar{X}$, from an individual measurement, X and $\Sigma(X-\bar{X})^2$ represents the sum of the differences squared.

To calculate sigma by this method, follow the steps listed below. (In this example the n is the total number of measurements. As explained earlier in the chapter, for small sample sizes, it is necessary to use $n - 1$. When this is done, the formula becomes

$$\sigma = \sqrt{\frac{\Sigma(X - \bar{X})^2}{n - 1}}$$

It is always technically correct to use $n - 1$; however, with a large sample size, the results of using n or $n - 1$ are so nearly alike that for simplicity n is used.)

1. In column (*1*) list the individual test results, X, for per cent elongation. These are the original data.
2. Add the 48 values ($n = 48$) to get ΣX, which is 789.7.
3. Calculate the average, $\bar{X}$. $\bar{X} = (\Sigma X)/n = 789.7/48 = 16.45$.
4. In column (*3*), list the $(X - \bar{X})$ values.
5. Square the individual values in column (*3*), entering these values $(X - \bar{X})^2$ in column (*4*).

6. Add the $(X-\bar{X})^2$ values in column (*4*) to get $\Sigma(X-\bar{X})^2$ of 23.30.
7. Calculate the standard deviation by substituting in the formula as follows:

$$\sigma = \sqrt{\frac{\Sigma(X-\bar{X})^2}{n}}$$

$$= \sqrt{\frac{23.30}{48}}$$

$$= \sqrt{0.49}$$

$$= 0.70$$

Method 2 (Squares of Individual Readings, ΣX^2 and $[\Sigma X]^2$). Use the data from Table 4.2, first two columns.

$$\sigma = \sqrt{\frac{\Sigma X^2}{n} - \left(\frac{\Sigma X}{n}\right)^2}$$

where X represents each individual measurement; ΣX represents the sum of the individual measurements; ΣX^2 represents the sum of the squares of individual measurements; $(\Sigma X)^2$ represents the (sum of the individual measurements) squared; and n is the number of individual measurements.

To calculate sigma by this method, follow the steps listed below:

1. In column (*1*) list the individual test results, X, for per cent elongation. These are the original data.
2. Add the 48 values ($n = 48$) to get ΣX, which is 789.7.
3. In column (*2*), list the calculated values of X^2. (This can be simplified by using a squares table or calculator.) Add to get ΣX^2 of 13015.51.
4. Calculate the standard deviation by substituting in the formula as follows:

$$\sigma = \sqrt{\frac{\Sigma X^2}{n} - \left(\frac{\Sigma X}{n}\right)^2}$$

$$= \sqrt{\frac{13015.51}{48} - \left(\frac{789.7}{48}\right)^2}$$

$$= \sqrt{0.49}$$

$$= 0.70$$

Method 3 (Sum of Squares of Differences using Selected Number). Use the data from Table 4.2, first, fifth, and sixth columns.

$$\sigma = \sqrt{\frac{\Sigma(X - A)^2 - \frac{[\Sigma(X - A)]^2}{n}}{n}}$$

where A is some arbitrarily chosen number.

This method is similar to the one previously explained under the first method. However, it is simpler to use for certain types of data. The primary difference between the two methods is the number used to subtract from the individual measurements, X. In the first method, the value subtracted from the individual measurements was the average as shown by the expression $X - \bar{X}$. In this third method the average is not used, but some arbitrarily chosen number is used in place of the average. This number is selected so that the subtractions are simplified and can be done in much less time. In the example, the number 16 was chosen, and consequently subtraction is much simpler than when 16.45 was used. By choosing the proper number, subtractions can be done mentally without the aid of a calculator.

To calculate sigma by this method, follow the steps listed below:

1. In column (*1*) list the individual test results, X, for per cent elongation. These are the original data.

2. Choose a convenient number in the range of the original data. For this example, 16 was chosen.

3. In column (*5*) list the $(X - 16)$ values.

4. Add the $(X - 16)$ values to get $\Sigma(X - 16)$, which is in this case 21.70.

5. In column (*6*) list the $(X - 16)^2$ values.

6. Add the $(X - 16)^2$ values in column (*6*) to get $\Sigma(X - 16)^2$, which in this case is 33.11.

7. Calculate the standard deviation by substituting in the formula as follows:

$$\sigma = \sqrt{\frac{\Sigma(X - A)^2 - \frac{[\Sigma(X - A)]^2}{n}}{n}}$$

$$= \sqrt{\frac{33.11 - \frac{(21.70)^2}{48}}{48}}$$

$$= \sqrt{0.49}$$

$$= 0.70$$

It is necessary to use $[\Sigma(X - A)]^2/n$ in the equation above to correct for the fact that a number other than the average was used in the original calculations. Except for this correction, the equation would be the same as shown for Method 1.

It should be noted that all three methods illustrated for calculating the standard deviation of a set of numbers give the same answer, the only difference between the methods being the mathematical approach to the problem.

Method 4 (Range). Use the data from Table 4.2, first and seventh columns. In addition to the three methods previously described for calculating standard deviation, there is one available for estimating sigma for large sets of data. It is known as the range method and is used primarily in quality control work with industrial data where very large volumes of data are available. It is important to remember that this method should be used only when the distribution is normal and a sufficient amount of data is available.

To calculate the estimated standard deviation, which is designated by the symbol σ' (read as "sigma prime"), follow the steps listed below:

1. In column 1 list the individual test results for per cent elongation. These are the original data, and it is important when estimating sigma by this method that the data be listed in the order in which they were collected.

Assume for this example that the first four readings represent the elongation measurements from one bobbin, the second four readings represent the elongation measurements for the second bobbin, and so forth.

2. In column (7) list the range for each subgroup of four readings. The range is the difference between the highest value and the lowest value in the subgroup. For example, the range for the first subgroup is 17.8 − 16.2 or 1.6; for the second subgroup, it is 17.3 − 15.9 or 1.4, and so on.
3. Add the ranges in column (7) and calculate the average range, $\bar{R}$.

$$\bar{R} = \frac{\Sigma R}{n} = \frac{14.4}{12} = 1.2$$

4. Calculate the estimated deviation by substituting in the following formula:

$$\sigma' = \frac{\bar{R}}{d_2} = \frac{1.2}{2.059} = 0.58$$

It should be noted that the estimated sigma of 0.58 is not the same as the sigma calculated by the other methods of 0.70. This is due to the fact that the sample size is not sufficiently

TABLE 4.3

Constants for Calculating Standard Deviation by the Range Method

Subgroup size n	Factor d_2
2	1.128
3	1.693
4	2.059
5	2.326
6	2.534
7	2.704
8	2.8472
9	2.9700
10	3.0775

large for an accurate estimate of sigma by the range method.

In this preceding equation, $\bar{R}$ is a symbol for the average range. The range for each subgroup is designated by the symbol R, so that when the ranges are averaged, the symbol becomes $\bar{R}$. The symbol d_2 represents a series of constant factors, shown in Table

4.3, which are used when estimating the standard deviation by this method. The factor to be used is determined by the size of the subgroup. In this example, the subgroup size was 4; therefore, d_2 for an $n = 4$ was used; if the subgroup size had been 5, then d_2 for $n = 5$ would have been used, and so on.

The subgrouping of data will be discussed on the following pages; however, it should be remembered when using the range method that the efficiency of σ' is reduced considerably as the subgroup size increases. For this reason, the subgroup should range in size from 2 to 8 for the greatest efficiency (1).

Small Sample Size

When calculating the standard deviation for small sample sizes, the term and effect of the "degrees of freedom" enter into the picture. Degrees of freedom can be explained as the number of values which are free to vary from the mean, and are equal to $n - 1$. For example, with a subgroup of 4, given any three values of X and an average $\bar{X}$, the fourth X has no freedom to vary. Thus, with known X_1, X_2, X_3, and $\bar{X}$, all of which have the freedom to vary within different subgroups, the value of X_4 in each case is fixed and has no freedom to vary. The number of degrees of freedom in this case is 3.

There are two methods of calculating the standard deviation of small samples.

Method 5 (Squares of Individual Readings ΣX^2 and $[\Sigma X]^2$). Use the data from Table 4.4, first and second columns.

$$\sigma = \sqrt{\frac{\Sigma X^2}{n - 1} - \frac{(\Sigma X)^2}{n(n - 1)}}$$

The first four steps in the calculation are the same as for the large sample size, Method 2, in that the values of ΣX^2 and $(\Sigma X)^2$ are obtained initially.

The calculation for sigma varies from the former method in that the effect of degree of freedom is introduced by use of the factor $n - 1$.

The application of the formula for the data given is as follows, having calculated ΣX and ΣX^2 as suggested.

$$\sigma = \sqrt{\frac{66.92}{9} - \frac{(25.85)^2}{(10)(9)}}$$
$$= 0.10$$

TABLE 4.4

Turns Per Inch of Twist in 150 Denier, 40 Filament, Viscose Rayon

(1) Turns twist, X	(2) X^2	(3) $(X - 2.50)$	(4) $(X - 2.50)^2$
2.60	6.76	0.10	0.01
2.75	7.56	0.25	0.06
2.65	7.02	0.15	0.02
2.60	6.76	0.10	0.01
2.70	7.29	0.20	0.04
2.45	6.00	−0.05	0.00
2.55	6.50	0.05	0.00
2.40	5.76	−0.10	0.01
2.50	6.25	0.00	0.00
2.65	7.02	0.15	0.02
$\Sigma = 25.85$	66.92	0.85	0.17
$\bar{X} = 2.58$			
$n = 10$			

Method 6 (Sum of Squares of Differences using Selected Number. Use the data from Table 4.4, first, third, and fourth columns:

$$\sigma = \sqrt{\frac{\Sigma(X - A)^2 - \frac{[\Sigma(X - A)]^2}{n}}{n - 1}}$$

The similarity of the formula for standard deviation for small sizes to the corresponding formula for large sample sizes should be noted, for the only difference is the replacement of n under the square root sign by $n - 1$; again, this substitution compensates for sample size.

In the example given, the selected value of A is 2.50, which is much easier to handle than $\bar{X} = 2.58$. The calculation of $\Sigma(X - A)$ and $\Sigma(X - A)^2$ follows the same six steps as outlined in Method 3 for large sample sizes. The final calculation of standard deviation varies from that of Method 3 only by the replacement of $n - 1$ for n; this calculation is:

$$\sigma = \sqrt{\frac{0.17 - \frac{(0.85)^2}{10}}{9}}$$

$$= \sqrt{\frac{0.10}{9}}$$

$$= 0.10$$

As can be seen, the answers obtained by the use of the two methods are the same and either method may be used with confidence.

It should be noted that the range method of estimating the standard deviation is not included in the recommended methods for small sample sizes. As stated previously, this method is used only for large sample sizes (100 or over) and should never be used for small sample sizes.

Standard Deviation for Averages

All of the standard deviations calculated up to this point have been for individuals, in that the value obtained by these methods measures the variation of the individual readings. Quite frequently, and especially in the use of statistical quality control charts, it is desired to measure the variation of the averages of subgroups. Such a measurement gives a value defined as the standard deviation for averages. It is designated by the symbol $\sigma_{\bar{X}}$ (read as sigma sub-*X*-bar). The statistical formula for this is:

$$\sigma_{\bar{X}} = \frac{\sigma}{\sqrt{n}}$$

(σ' can be used in place of σ.)

In this relationship, the $\sigma_{\bar{X}}$ means the standard deviation for averages of subgroups of size n. For example, from Table 4.2 the standard deviation using the range method was found to be 0.58, which is the sigma for individuals. If it were necessary to know the $\sigma_{\bar{X}}$ for the averages of the subgroups of four which are shown in column (*8*) of Table 4.2, the calculation would be as follows:

$$\sigma_{\bar{X}} = \frac{\sigma'}{\sqrt{n}}$$

$$= \frac{0.58}{\sqrt{4}}$$

$$\sigma_{\bar{X}_{n=4}} = 0.29$$

The symbol $\sigma_{\bar{X}_{n=4}}$ indicates that 0.29 is the standard deviation for averages of subgroups of four measurements.

As shown in the above calculation, the standard deviation for averages, which is also known as standard error, is smaller than the standard deviation for individuals.

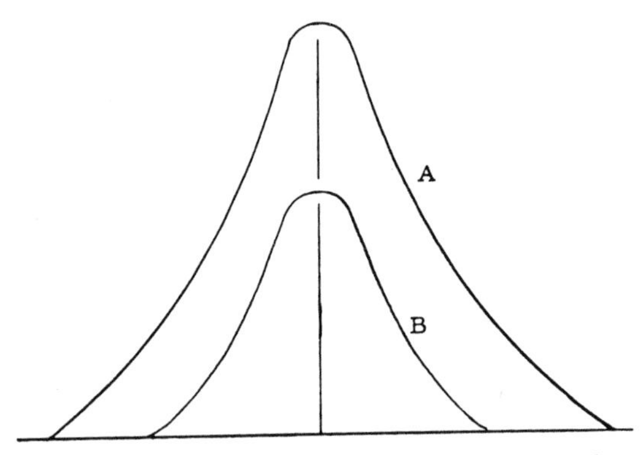

Fig. 4.4. Frequency distributions for *A*, individuals, and *B*, averages.

If five yarn skein strength tests were made from each of 100 bobbins and the 500 test results were plotted, a frequency distribution similar to the one labeled *A* in Fig. 4.4 would probably be formed, and it would be the distribution for the individual test results. Therefore, it would be known as the "distribution for individuals." However, if only tne averages for each bobbin, i.e., the subgroup averages, were plotted, a distribution similar to *B* in the same figure would result. Actually, the same yarn is represented by both curves, the only difference being that one curve represents the individual measurements and the other curve represents averages of bobbins.

The curve representing the averages will always include less area than the curve for individuals, which means that the variation between subgroup averages will always be less than the variation within individuals. The reason for this is that when a group of numbers is averaged, the entire group is then represented by one value, the average, which means that the variation within the subgroups for most practical purposes is eliminated and the only variation being shown is the variation between subgroups.

As can be seen in the formula

$$\sigma_{\bar{X}} = \frac{\sigma'}{\sqrt{n}}$$

with a larger subgroup the value n increases, and as a result, the standard error is smaller.

Application and Interpretation of Standard Deviation

As stated previously, the standard deviation is a measure of the dispersion, or variation, of a set of data about the average. For this reason, a group of data with a small sigma would have a smaller amount of variation than a group of data with a large sigma. In other words, a small sigma value means the product has greater uniformity than a product with a large sigma value.

Knowing the standard deviation and average for a population, it is possible to predict what per cent of the values will fall within certain limits as follows:

$\bar{X} \pm 1\sigma$ will include 68.27 per cent of the data
$\bar{X} \pm 1.645\sigma$ will include 90.0 per cent of the data
$\bar{X} \pm 1.960\sigma$ will include 95.0 per cent of the data
$\bar{X} \pm 2\sigma$ will include 95.45 per cent of the data
$\bar{X} \pm 3\sigma$ will include 99.73 per cent of the data

For example, as is shown in Fig. 4.5, if an average of 100 and a sigma of 5 are found for a set of skein strength data, then 68.27 per cent of the skein strengths in the group should be between 100 ± 5, or 95 to 105. Also, 95.45 per cent of the skeins should have a strength between $100 \pm 2(5)$, or 90 and 110; and 99.73 per cent of the skeins or all except three out of every thousand

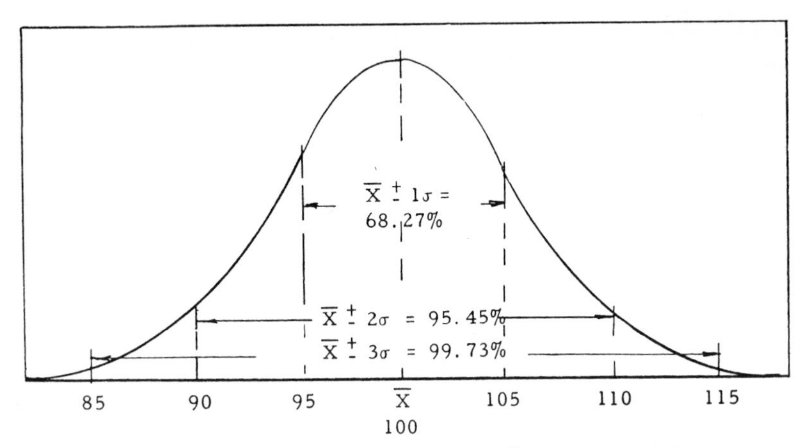

Fig. 4.5. Percentage of items included in $\bar{X} \pm 1$, 2, and 3σ limits.

should have a strength between $100 \pm 3(5)$, which is 85 to 115.

The statistical quality control charts which will be discussed in detail later are practically always based on the $\bar{X} \pm 3\sigma$ limits.

Coefficient of Variation. One of the most common applications for the standard deviation is in the calculation of the coefficient of variation, which is designated by the symbol V. The coefficient of variation is a measure of the relative dispersion and is useful in comparing the dispersions of two or more processes or materials or in comparing the same types of materials produced at different times.

In effect, the coefficient of variation is the percentage variation based on the average, as can be seen from the formula:

$$V = \frac{\sigma}{\bar{X}} \times 100$$

For example, if one yarn has an average skein strength, $\bar{X}$, of 89 pounds and a standard deviation, σ, of 3.6 pounds, and another yarn has an $\bar{X}$ of 105 pounds and a σ of 4.1 pounds, the yarn with the greater relative variation can be determined by comparing the respective coefficients of variation.

$$V = \frac{\sigma}{\bar{X}} \times 100$$

$$\text{First yarn, } V_1 = \frac{3.6}{89} \times 100 = 4.04 \text{ per cent}$$

$$\text{Second yarn, } V_2 = \frac{4.1}{105} \times 100 = 3.90 \text{ per cent}$$

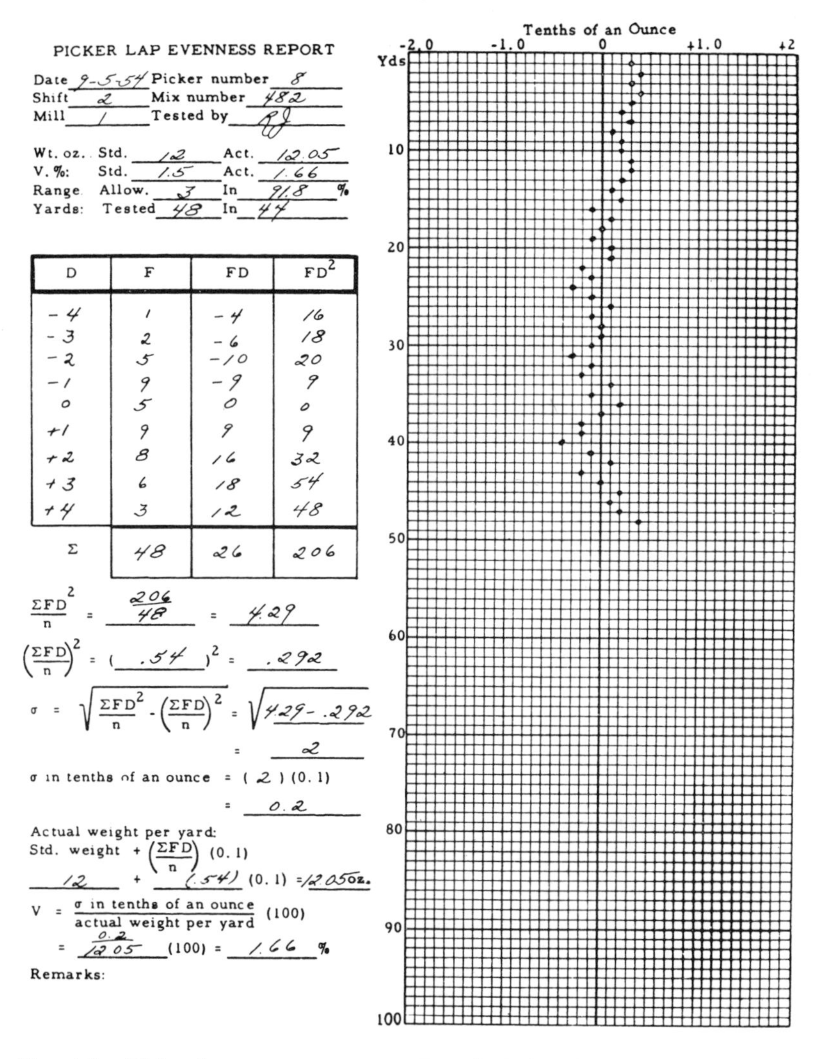

PICKER LAP EVENNESS REPORT

Date 9-5-54 Picker number 8
Shift 2 Mix number 482
Mill 1 Tested by

Wt. oz.: Std. 12 Act. 12.05
V. %: Std. 1.5 Act. 1.66
Range: Allow. 3 In 91.8 %
Yards: Tested 48 In 44

D	F	FD	FD^2
−4	1	−4	16
−3	2	−6	18
−2	5	−10	20
−1	9	−9	9
0	5	0	0
+1	9	9	9
+2	8	16	32
+3	6	18	54
+4	3	12	48
Σ	48	26	206

$\frac{\Sigma FD^2}{n} = \frac{206}{48} = 4.29$

$\left(\frac{\Sigma FD}{n}\right)^2 = (.54)^2 = .292$

$\sigma = \sqrt{\frac{\Sigma FD^2}{n} - \left(\frac{\Sigma FD}{n}\right)^2} = \sqrt{4.29 - .292}$

$= 2$

σ in tenths of an ounce = (2)(0.1)

= 0.2

Actual weight per yard:
Std. weight + $\left(\frac{\Sigma FD}{n}\right)$ (0.1)
12 + (.54) (0.1) = 12.05 oz.

$V = \frac{\sigma \text{ in tenths of an ounce}}{\text{actual weight per yard}} (100)$

$= \frac{0.2}{12.05} (100) = 1.66$ %

Remarks:

Fig. 4.6. Picker lap evenness report. D = deviation from zero, plus to the right, minus to the left; F = frequency of the deviations.

In comparing the two sigma values, it appears that the second yarn is more variable than the first yarn; however, the relative dispersion is greater for the first yarn than for the second.

The coefficient of variation is also very useful in calculating the proper sample size, and another very common use of the coefficient of variation in the textile industry is in connection with evenness testing. One example of this is in picker lap evenness. The form shown in Fig. 4.6 is used in recording and analyzing the measurements of lap evenness tests. The two important results in this analysis are the actual weight per yard and the coefficient of variation. Limits may be established for the coefficient of variation and in this way excellent control may be maintained over lap evenness.

Problem Section

TABLE 4.5

Weight Per Yard of Card Sliver

49.0	49.5	49.0	50.5	50.0	50.5
48.0	50.5	49.0	49.0	50.5	50.0
51.0	49.5	49.5	51.0	51.0	50.5
51.5	50.5	51.0	49.0	50.5	49.0
48.5	49.5	47.5	50.0	49.5	52.0
49.0	50.0	51.5	51.5	50.0	50.0
52.0	50.5	48.5	50.0	50.0	51.0
51.5	50.0	49.0	52.5	49.5	50.0
48.5	49.5	50.0	50.5	50.0	
51.0	50.5	49.5	51.0	50.5	
48.0	50.0	48.5	50.0	49.5	
49.5	49.5	50.0	49.5	50.0	

Practice Problem 1. a. Draw the frequency distribution curve for Table 4.5. b. Calculate the standard deviation by Methods 1 and 4. (Use $n = 4$ for subgroups.) c. Find what per cent of the data falls within the limits of $\pm 1\sigma$, $\pm 2\sigma$, and $\pm 3\sigma$. d. Calculate the coefficient of variation. e. Is the distribution skewed? If so, in what direction?

Reference

1. Dixon, W. J. and Massey, F. H., *Introduction to Statistical Analysis*, McGraw-Hill, New York, 1951.

CHAPTER V

Statistical Quality Control Charts

The so-called "statistical quality control chart" is probably the most commonly used statistical technique in the textile industry. During the comparatively short time that the charts have been in use in the field of textiles, they have been of great value and are now practically indispensable to the quality control engineer. There are several important advantages in favor of the use of control charts during the manufacture of textiles. These advantages shall be discussed in detail in one of the following sections.

Basis of Control Charts. Most control charts are constructed on the basis of "three sigma limits." This means that three standard deviations of the distribution involved are added to and subtracted from the average of the data to serve as control limits.

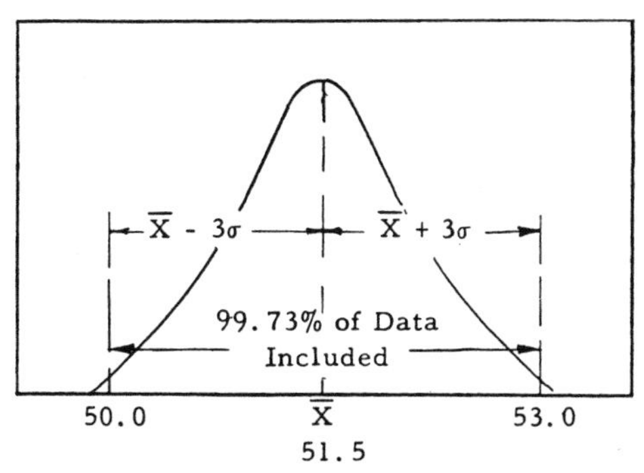

Fig. 5.1. Frequency distribution of finished sliver weight (grains per yard).

For example, suppose that a set of data representing the weight per yard of finisher drawing sliver had an average of 51.5 grains and a standard deviation of 0.5 grains. The upper limit of control becomes $\bar{X} + 3\sigma$, and the lower limit becomes $\bar{X} - 3\sigma$. In this example, the upper limit then is $51.5 + 3(0.5)$, or 53.0; the lower limit is 51.5 — 3(0.5), or 50.0. As previously discussed, the frequency distribution and the per cent of the observations included would be as shown in Figure 5.1.

By taking the $\bar{X}$ line and the three sigma lines and placing them in the horizontal position, they become a statistical quality control chart as shown in Fig. 5.2.

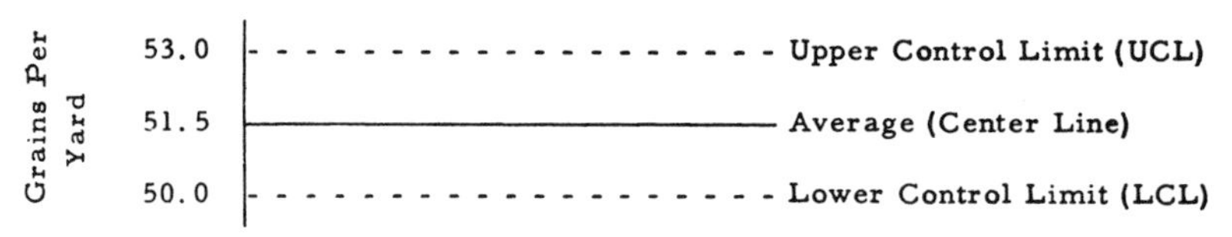

Fig. 5.2. Basic control chart.

If the data upon which the chart in Fig. 5.2 is based were representative, and if a state of statistical control existed when the data were collected, practically all finisher drawing produced in the future would have a weight between 50.0 and 53.0 grains, provided no change occurs in the process.

The control chart, therefore, eliminates guesswork in the evaluation of test data. It is impossible simply to examine test records and establish limits within which the date should fall without first analyzing the data from a statistical point of view. In any manufacturing process, the question always arises as to how much variation should be allowed. In textiles, common questions such as "how much picker lap variation should be allowed" or "how much variation in yarn strength should be expected for a certain type yarn," are arising frequently. The answer to this type of question is difficult if not impossible to give unless some analysis is made of available information. It has been common practice, not only in the textile industry but in all industries, for some member of the technical staff to casually examine test results and hazard a guess as to what the operating limits for a process should be. This is a very dangerous practice, for limits established in such a manner will most likely penalize the process or will give the process such a wide latitude of operation that it can go completely out of control before any corrective action is taken. The weight of the drawing sliver is a good example; it is extremely unlikely that the limits of 50 to 53 would have been set for that process unless a statistical analysis had been made of the data. If the limits had been arbitrarily set at 51 to 52, it would have meant that more was being demanded of the drawing process than it was capable of doing, and a considerable amount of gear changing would be done unnecessarily trying to maintain unreasonable limits. On the other hand, if the limits had been set at 49.5 to 53.5, that would have meant that the process was being permitted to introduce an excess of variation in the product without the knowledge of a need for correction. In other words, with the limits established on a scientific basis, the exact amount of expected variation in a product is fixed without guesswork and the limits can be used with confidence.

Control Limits. The control limits, which are commonly referred to as "action limits," are not necessarily specification limits, although any specification limits which may exist should be very close to the control limits. The control limits are the limits within which practically all of the test results from the process should fall. If the results do not fall between the limits (except three in 1000), then it is necessary to determine the reason for the excess variation. In other words, it is time for an investigation, or action, thus the name *action limits.*

The quality control chart has been compared to a paved highway. With the movement down the middle of the highway (about the center line), everything is peaceful (under control), but once the movement is off the road and onto the shoulders (out of control), then the road is rough, and the path should be altered and every effort made to get back onto the highway (get back into a state of control).

Specification Limits. Specification limits are the extremes that are set up in trade or engineering practice to determine the maximum variation acceptable for the particular item. For example, in a piece of fabric, Worth Street Rules specify that the width, weight, ends, picks, and so forth, must fall within definite limits for certain fabrics and that failure to do so shall result in penalty.

The chief difference between specification limits and control limits is that specification limits are dictated by the required performance of the product and the particular situation, whereas the control limits are an actual measurement of the performance of the product.

Where specification limits exist, it is always desirable to have them fall outside the control limits, that is, to be wider than the

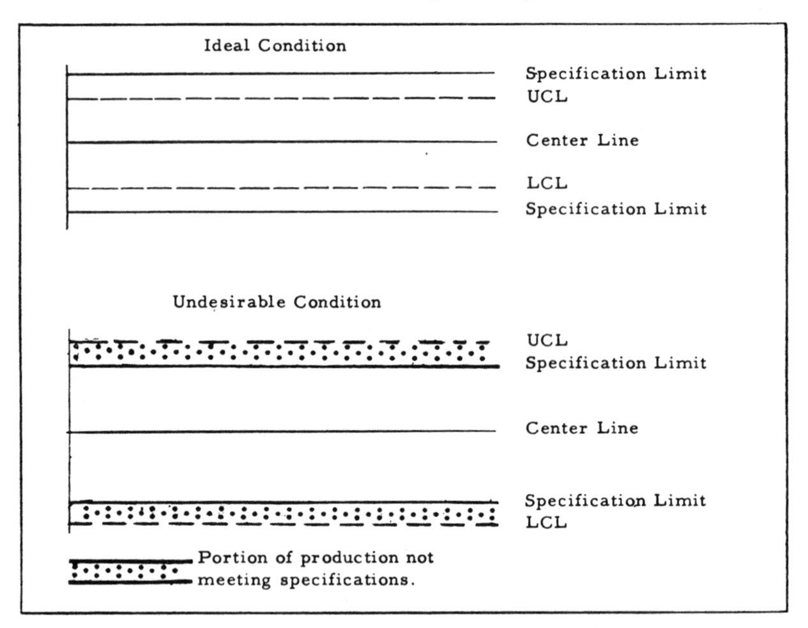

Fig. 5.3. Comparison of control limits and specification limits.

control limits. With this condition, the product will always be assured of meeting the requirements of the specification. If the specification limits are within or between the control limits, then only a certain per cent of the product will meet the specification. This is shown in Fig. 5.3.

Variation in Quality. There are two general types of variation measured by a quality control chart.

The variation which occurs between the control limits is known as *variation due to chance.* This variation is usually accepted as inherent in the process and may or may not be investigated. If it is known that all of the variation which can be removed from the product has been eliminated, then the variation between the control limits is accepted.

However, if it is felt that additional variation can be eliminated by some basic change in the process, then of course, the change is made. As the result of such changes and elimination of variation, the control limits should be brought closer together, which will mean a more uniform product.

The other type of variation which is measured by the control chart is *variation due to assignable causes.* This is variation occurring when a process (which was previously in a state of statistical control) has a test result which exceeds the limits. This situation indicates that there is something fundamentally wrong with the material tested and that the cause of trouble should be located and corrected.

The control chart method of identifying and separating the two types of variation is both quick and accurate. It eliminates expense and time which might otherwise be wasted in trying to distinguish between variation which can be eliminated and that which cannot be eliminated.

State of Statistical Control. The term *state of statistical control* is defined by the American Standards Association as:

> When the assignable causes have been eliminated from the production process to the extent that practically all the points plotted on the control chart remain within the control limits, the process is said to be in a state of *statistical control* with respect to the measure of quality under consideration. It may thus be concluded that greater uniformity can then be attained only through a basic change in the process itself.

Theoretically, only three points in every thousand are supposed to be outside the limits when a state of control exists; however, in actual practice a product or process is considered as under control even though a few more than three in every thousand exceed the limits. It is usually accepted that when a sequence of 25 to 30 samples fall within the limits, the process can be assumed to be in control without further delay.

Advantages of Using Quality Control Charts. *Aids in controlling quality.* The mere plotting and keeping of a chart will not control quality, nor will it aid to any great extent in the controlling of quality. However, if the chart is properly maintained and attention given to the careful analysis and interpretation of it, then it can be of great assistance in maintaining quality.

Variation in product can be reduced. With the use of control charts, all or practically all variation due to assignable causes can be eliminated, which will result in a controlled process and thus a more uniform end product.

Amount of variation in product can be predicted. With a state of statistical control existing, it is possible to predict with a high degree of accuracy exactly what per cent of the production will have a certain amount of variation or what per cent of the product will fall within certain limits.

Gives a reliable basis for action. If the control limits are exceeded, it is an immediate warning to all concerned that something has happened which is affecting the quality of the material. There is no guesswork as to when something is happening and when an investigation should be made. On the other hand, if the control limits are not exceeded, then that is definite assurance that nothing is wrong and no expensive investigations need to be made. Furthermore, with a proper background in the manufacturing process as well as in control chart technique, it is possible to predict trouble in advance by the analysis of trends on the chart.

Reduces testing and inspection costs. With controlled manufacture, as shown by the control chart, it is usually possible to show a substantial reduction in the cost of testing and inspecting of materials. This is due to the greater confidence which can be placed in the results of tests made on materials produced by a well-controlled process. It is also possible to calculate the exact number of tests which will satisfy certain requirements when the samples are to be taken from a controlled process.

Aids in locating trouble. With the proper subgrouping and identification of data, it is possible to locate causes of variation in quality much quicker and more accurately. This is true since the data will most likely be associated with a particular lot passing through a certain section of machines at a particular time.

Gives permanent pictorial record of quality. By simply keeping a control chart, it gives to the personnel involved, from section foreman to president, a permanent picture of past and existing quality, as well as an accurate prediction of expected future quality. It is impossible to get this picture from sheets and books of tabulated data.

Machine acceptance. One application for the control chart technique that is not directly related to production is the establishing or proving of machine specifications. If a machine manufacturer guarantees a machine to operate within certain tolerances, a statistical control chart can be used to ascertain if the machine is meeting the specification. If the control limits are equal to or less than the specifications, the machine is within the guarantee; however, if the control limits are larger than the specifications, it means that the machine is not performing as guaranteed.

Types of Control Charts. There are four types of control charts that are generally used in statistical control work. The type chart probably used more often than any of the others is the $\bar{X}$ chart; that is, the chart for averages. This type chart is based on the averages of subgroups of data which form the normal type of distribution. It will show the variation from subgroup to subgroup but will not show the variation within individual subgroups.

Since the $\bar{X}$ chart will not show the variation within individual subgroups, it is usually customary to plot the chart for ranges, the R chart, simultaneously with the one for averages. The R chart is usually plotted on the same page as the $\bar{X}$ chart, with the range of the subgroup plotted directly below the avarage of the same subgroup to facilitate interpretation of the charts. The above charts can be used very easily for yarn strength, yarn number, twist, roving size, sliver size, fabric strength, and so forth.

Where the chief interest is concerned primarily with inspection data, such as per cent seconds, the type of chart used is the $\bar{p}$ chart. The symbol p means fraction defective; therefore, the $\bar{p}$ chart would be for measuring the variation in fraction or per cent defective.

If the interest is centered around the actual number of defects rather than the per cent defective, then it is necessary to use the $\bar{c}$ chart. The symbol, c, means the number of defects per yard or cut of fabric, or the like.

Construction of Control Charts

Sampling. The sampling of any process must result in samples which are representative of the process. Any known variables such as type of material, methods of processing and types of equipment used, must be considered when setting up a sampling program. Each different cotton or fiber mix and each lot of sliver, roving, and yarn should be sampled individually and the samples not mixed. Other variables, such as humidity, shift and area, should also be considered.

Once the group from which the samples are to be taken has been established, it is important that the samples be selected in a random manner. If the samples are not chosen at random, the results are likely to be biased, which means that the data are of little if any value.

It is always better to take small samples frequently than to take large samples occasionally. For example, it is much more satisfactory to take twenty-five samples of five each than to take five samples of twenty-five each. The large infrequent sample sizes are not likely to show variations in quality since the average of the large sample size will tend to cover up and average the variation in the product. If small samples are taken frequently and the samples are marked as to type of material, machine from which sample was taken, shift, time, together with other pertinent identification, then it is possible to know the quality being produced at any time. With this arrangement, the quality control chart will be of the utmost value in aiding in the control of quality.

Rational Subgrouping of Data. For any control chart program to be successful, the data upon which the chart is based and all subsequent data must be divided into so-called "rational subgroups." The term "rational subgroups" means the grouping of the data or the forming of the subgroups so that the variation within the subgroup can be attributed only to chance or non-assignable causes. In other words, variation due to assignable causes should not be hidden within the subgroups.

The proper subgrouping of the data cannot be determined statistically or mathematically; it can only be done by engineering judgment and technical knowledge of the process being measured.

Control Chart for Averages. The control chart for averages, the $\bar{X}$ chart, is one of the most commonly used charts. It is used for data which will fit the normal distribution, and the data plotted on the chart are the averages of subgroups.

To construct an $\bar{X}$ chart, it is first necessary to determine the process average and standard deviation or to obtain an accurate estimate of these measures.

The data in Table 5.1 represent the skein strength for a 70/1 cotton yarn. If the data are considered as representative of the yarn being produced, then it is possible to calculate the control limits for the quality control chart. The three sigma limits would be calculated as shown below.

To calculate for chart control limits, follow the steps listed below:

1. Obtain $\bar{\bar{X}}$ (read as X double bar), which is the average of subgroup averages. For the example, this is 34.06.
2. Obtain $\bar{R}$ for subgroups. For the example this is 3.12.
3. Calculate the standard deviation, σ', for individuals, using the formula previously developed. (Note: values for d_2 are shown in Table 4.3 on page 50.)

$$\sigma' = \frac{\bar{R}}{d_2} = \frac{3.12}{2.059} = 1.515$$

4. Calculate the standard deviation, $\sigma_{\bar{X}_{n=4}}$, for averages of subgroups of four.

$$\sigma_{\bar{X}_{n=4}} = \frac{\sigma'}{\sqrt{n}} = \frac{1.515}{\sqrt{4}} = 0.76$$

5. Determine the upper and lower control limits, using the above sigma value.

$$\bar{\bar{X}} \pm 3\sigma_{\bar{X}} = 34.06 \pm 3(0.76)$$
$$\text{UCL} = 34.06 + 3(0.76) = 36.34$$
$$\text{LCL} = 34.06 - 3(0.76) = 31.78$$

TABLE 5.1

Skein Strength Test for 70/1 Cotton Yarn

Cone no.	Skein strengths, X 1	2	3	4	Average $\bar{X}$	Range R
1	34.5	33.0	33.5	34.0	33.75	1.5
2	30.5	36.0	33.5	33.0	33.25	5.5
3	33.0	33.0	35.5	37.0	34.62	4.0
4	39.5	37.0	33.5	32.5	35.62	7.0
5	32.0	34.5	35.5	32.0	33.50	3.5
6	33.0	35.0	34.5	34.0	34.12	2.0
7	31.0	32.5	32.5	31.5	31.88	1.5
8	30.0	30.0	32.5	34.0	31.62	4.0
9	31.5	33.0	36.0	32.5	33.25	4.5
10	32.5	35.0	34.5	32.5	33.62	2.5
11	38.0	37.0	35.0	35.0	36.25	3.0
12	36.5	34.5	34.0	35.5	35.12	2.5
13	35.5	36.0	33.5	33.0	34.50	3.0
14	33.0	31.5	34.5	33.5	33.12	3.0
15	34.0	33.5	34.5	33.0	33.75	1.5
16	32.5	33.5	35.5	31.5	33.25	4.0
17	37.0	35.0	36.0	37.0	36.25	2.0
18	35.5	36.0	34.0	34.5	35.00	2.0
19	35.0	34.5	34.5	33.0	34.25	2.0
20	30.5	31.5	34.0	35.5	32.88	5.0
21	34.0	32.0	33.5	32.0	32.88	2.0
22	35.5	33.0	35.5	32.0	34.00	3.5
23	34.0	34.5	35.0	34.0	34.38	1.0
24	33.5	36.0	35.0	35.5	35.00	2.5
25	33.0	35.0	37.0	38.0	35.75	5.0
					$\Sigma =$ 851.61	78
					$\bar{\bar{X}} =$ 34.06	
					$\bar{R} =$	3.12

There is a second method of calculating the three sigma limits which is very useful in saving time in those cases where repetition is involved. This method combines the values of d_2 and n to form a new constant, A_2. Analysis of the previous steps, *3*, *4*, and *5*, reveal that the three sigma limits are:

$$3\sigma_{\bar{X}_{n=4}} = \frac{3\sigma'}{\sqrt{n}} = \left(\frac{3}{\sqrt{n}}\right)\left(\frac{\bar{R}}{d_2}\right)$$

The three constant values, $3/\sqrt{n}d_2$, are calculated for different n values to give A_2.

$$A_2 = \frac{3}{d_2\sqrt{n}}$$

Values of A_2 are shown in Table 5.2. Using A_2, the control limits can now be set by

$$\bar{\bar{X}} \pm A_2\bar{R}$$

If step *5* in the previous sequence is repeated using $A_2\bar{R}$ in place of 3σ, the results are identical:

$$\bar{\bar{X}} \pm A_2\bar{R} = 34.06 \pm (0.729)(3.12)$$
$$\text{UCL} = 34.06 + (0.729)(3.12) = 36.33$$
$$\text{LCL} = 34.06 - (0.729)(3.12) = 31.79$$

With the center line, $\bar{\bar{X}}$, and the control limits calculated, the next step is to construct the chart (see Fig. 5.4A) and to continue plotting subsequent test results. It is very important that the data

TABLE 5.2

Table for Factor A_2

Sample size, n	Factor A_2	Sample size, n	Factor A_2
2	1.880	6	0.483
3	1.023	7	0.419
4	0.729	8	0.373
5	0.577	9	0.337
		10	0.308

be plotted in the order in which they are collected with respect to time. If several days are allowed to elapse and the data are then plotted at random with no regard for the order in which the samples were collected, it would be impossible to locate any trends in the process. Since this is one of the advantages of using this

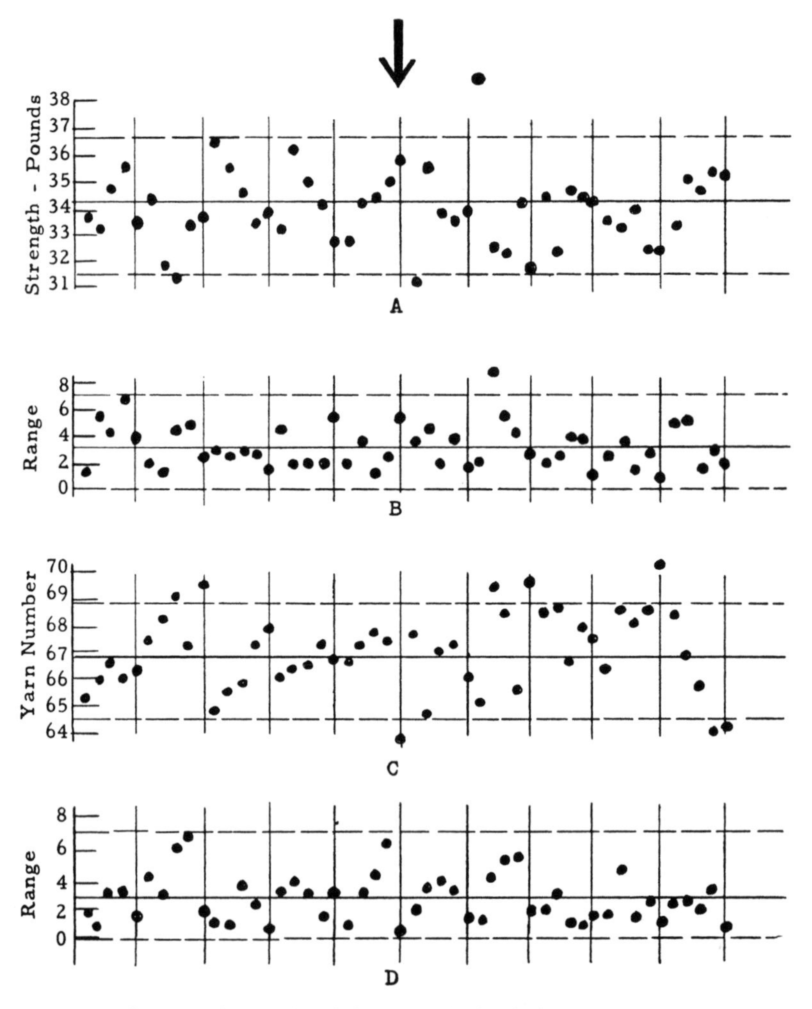

Fig. 5.4. Control charts. A: skein strength for 70/1 yarn, averages of subgroups of four; B: range for skein strength of 70/1 yarn, subgroups of four; C: yarn number for 70/1 yarn, averages of subgroups of four; D: range for yarn number of 70/1 yarn, subgroups of four. All control limits based on data up to point marked with an arrow.

method, it is necessary that this be kept in mind. One simple way to be assured always that the data are plotted in order is to plot the results as soon as the tests are completed. Another point which should be kept in mind is that the limits are based on a definite subgroup size and that all entries made on the chart should be for averages of the same size subgroup. For example, if the chart is based on a subgroup size of $n = 4$, then no other subgroup size should be used unless the limits are changed.

Since one of the factors which is likely to cause variation in the skein strength of yarn is the yarn number, it is usually customary to plot a chart for yarn number in connection with the skein strength chart. The data for the yarn number are shown in Table 5.3; the chart is Fig. 5.4c. The calculations involved are as follows:

$$\sigma' = \frac{\bar{R}}{d_2} = \frac{3.02}{2.059} = 1.467$$

$$\sigma_{\bar{X}_{n=4}} = \frac{1.467}{\sqrt{4}} = \frac{1.467}{2} = 0.734$$

$$\text{UCL} = 66.71 + 3(0.734) = 68.91$$

$$\text{LCL} = 66.71 - 3(0.734) = 64.61$$

Chart Arrangement. When drawing the charts on graph paper, some consideration should be given to the distance between the upper and lower limit lines. For neatness and uniformity, the scale should be so selected that the distance between the limit lines is not less than 1 inch or greater than $1\frac{1}{2}$ inches. This does not alter the usefulness or effectiveness of the charts, but it does give a much neater appearance to the work.

Control Chart for Ranges. The control for averages gives an excellent indication of the quality of the material from the process being studied. It shows the variation from subgroup to subgroup from day to day, etc.; it shows when something is basically wrong with the process, but it does not give any indication of the variation within the subgroup represented by the single average. For example, the averages of two subgroups could be identical and yet the difference between the highest and lowest values in each subgroup could be from a very small number for one subgroup to a very large number for the other subgroup. This naturally would not be detected by the $\bar{X}$ chart alone, and for this reason it is usually desirable to plot the range chart concurrently with

TABLE 5.3

Yarn Number Test for 70/1 Cotton Yarn

Cone number	Yarn number				Average	Range
	1	2	3	4		
1	65.0	65.8	66.0	64.0	65.2	2.0
2	66.6	65.2	66.2	66.0	66.0	1.4
3	68.8	67.0	66.0	65.5	66.8	3.3
4	64.2	65.0	67.2	67.7	66.0	3.5
5	67.0	65.3	66.0	67.2	66.4	1.9
6	66.2	65.5	67.8	70.0	67.4	4.5
7	68.0	70.0	67.0	68.0	68.2	3.0
8	71.0	72.0	67.4	66.0	69.1	6.0
9	66.5	64.0	66.8	71.0	67.1	7.0
10	69.5	70.5	69.0	68.5	69.4	2.0
11	64.6	65.0	65.5	64.2	64.8	1.3
12	65.8	65.0	66.0	65.0	65.4	1.0
13	63.8	67.0	65.0	67.8	65.9	4.0
14	68.0	65.5	67.0	67.8	67.1	2.5
15	68.0	68.0	68.0	67.6	67.9	0.4
16	67.0	65.5	64.0	67.5	66.0	3.5
17	64.0	66.8	66.2	68.0	66.2	4.0
18	66.0	65.6	66.2	68.8	66.6	3.2
19	66.4	66.8	67.0	68.2	67.1	1.8
20	68.7	67.4	66.0	65.0	66.8	3.7
21	66.2	66.8	67.4	66.8	66.8	1.2
22	66.0	67.5	66.5	69.0	67.2	3.0
23	66.5	66.0	70.2	68.2	67.7	4.2
24	70.0	67.0	69.0	63.8	67.4	6.2
25	63.0	63.8	63.0	63.4	63.3	0.8
				$\Sigma =$	1667.8	75.4
				$\bar{\bar{X}} =$	66.71	
				$\bar{R} =$		3.02

the $\bar{X}$ chart. The limits for the range chart of the skein strength data in Table 5.1 would be calculated as follows, using factors D_3 and D_4 (Table 5.4) which have been prepared for use with range.

$$\text{UCL} = (\bar{R})(D_4) = (3.12)(2.282) = 7.11$$
$$\text{LCL} = (\bar{R})(D_3) = (3.12)(0) = 0$$

The limits for the range chart of the yarn numbers shown in Table 5.3 would be calculated in exactly the same manner.

$$\text{UCL} = (3.02)(2.282) = 6.9$$
$$\text{LCL} = (3.02)(0) = 0$$

TABLE 5.4

Table for Factors D_3 and D_4

Sample size, n	D_3	D_4
2	0	3.268
3	0	2.574
4	0	2.282
5	0	2.114
6	0	2.004
7	0.076	1.924
8	0.136	1.864
9	0.184	1.816
10	0.223	1.777

Interpretation of Control Charts. Once the charts have been constructed and the process of entering the daily test results begins, attention must then be given to the interpretation of the charts. The most obvious deduction to be drawn from the charts is that when points fall outside the limits, the process is no longer in a state of statistical control. This should be followed immediately by investigation to determine the cause of the fluctuation in quality.

It is frequently possible to locate trends on the charts and thus make necessary changes before the process goes out of control

TABLE 5.5

Results of Yarn Number Tests of a 14/1 Cotton Yarn

Bobbin number	Yarn number				$\bar{X}$	R
1	13.48	13.83	14.66	14.37	14.08	
2	14.81	14.04	13.89	13.53	14.06	
3	15.13	14.18	13.83	13.66	14.20	
4	14.03	13.74	14.18	13.51	13.86	
5	13.11	14.37	13.79	13.30	13.64	
6	13.74	13.89	13.89	14.03	13.99	
7	14.84	13.79	14.01	13.32	13.99	
8	14.06	13.23	14.43	14.49	14.05	
9	13.68	14.27	14.39	15.22	14.39	
10	13.85	13.48	14.56	14.45	14.08	
11	14.33	14.56	13.74	13.81	14.11	
12	14.31	14.22	13.59	13.61	13.93	
13	13.76	14.03	13.97	13.35	13.78	
14	13.55	14.45	14.20	13.89	14.02	
15	13.23	13.68	14.51	13.85	13.81	
16	14.29	15.58	15.04	14.53	14.86	
17	15.36	14.01	14.43	13.83	14.41	
18	14.97	14.73	14.66	14.71	14.77	
19	14.25	13.81	14.03	13.97	14.02	
20	14.29	13.59	13.66	14.04	13.90	
21	13.66	14.16	14.08	13.61	13.88	
22	13.28	13.62	13.74	13.79	13.61	
23	13.62	13.30	13.66	13.57	13.54	
24	14.04	13.72	14.04	13.62	13.86	
25	14.81	14.49	14.49	14.95	14.68	
26	14.90	14.47	14.29	14.43	14.52	
27	14.62	14.53	13.85	14.75	14.44	
28	14.68	14.39	15.08	13.74	14.47	
29	15.27	14.73	14.08	16.34	15.10	
30	14.49	14.71	13.77	14.88	14.46	
31	13.40	14.58	14.12	14.52	14.16	
32	14.71	14.56	14.18	13.79	14.31	
33	13.91	14.08	13.50	16.10	14.40	
34	14.56	14.18	14.73	14.29	14.44	
35	14.18	14.53	14.33	14.68	14.43	
36	14.41	14.35	16.16	15.17	15.02	
37	13.68	13.66	14.18	14.31	13.96	
38	14.56	14.10	14.06	13.68	14.10	
39	13.77	15.02	14.41	14.58	14.44	
40	14.14	14.68	14.75	14.08	14.41	
41	14.45	14.29	12.99	13.33	13.76	
42	14.73	13.91	13.81	14.29	14.18	
43	14.22	13.64	13.85	14.29	14.00	

TABLE 5.6

Breaking Strength in Grams of Cotton Yarn

Bobbin number	Strength, grams				$\bar{X}$	R
1	160	140	136	149	146.2	
2	131	146	144	142	140.8	
3	143	146	149	151	147.2	
4	155	146	140	139	145.0	
5	148	157	147	161	153.2	
6	142	143	145	139	142.2	
7	144	146	147	134	142.8	
8	137	145	132	144	139.5	
9	163	150	153	145	152.8	
10	143	152	130	144	142.2	
11	149	139	147	135	142.5	
12	143	140	148	152	145.8	
13	147	140	143	156	146.5	
14	146	142	132	139	139.8	
15	149	146	137	143	143.8	
16	152	138	145	151	146.5	
17	145	143	156	161	151.2	
18	138	144	143	145	142.5	
19	146	147	135	142	142.5	
20	136	145	138	152	142.8	
21	145	139	130	134	137.0	
22	144	146	151	147	147.0	
23	148	155	140	145	147.0	
24	136	149	145	143	143.2	
25	139	148	149	152	147.0	
26	153	151	150	153	151.8	
27	146	154	146	153	149.8	
28	141	154	148	154	149.2	
29	140	149	156	129	143.5	
30	142	154	161	143	150.0	
31	161	133	149	147	147.5	
32	144	140	144	158	146.5	
33	151	135	148	126	140.0	
34	135	159	140	147	145.2	
35	151	150	143	142	146.5	
36	157	153	142	151	150.8	
37	146	142	154	144	146.5	
38	151	158	161	151	155.2	
39	147	154	140	154	148.8	
40	154	137	154	143	147.0	
41	144	162	158	155	154.8	
42	131	147	151	150	144.8	
43	148	154	156	140	149.5	

or correct the causes of variations in quality before they become extreme. Generally speaking, a minimum of three values that show a definite slope one way or the other can be interpreted as being a trend.

It is also possible to have significant changes in quality with only a few points falling outside the limits. When the data cease to be symmetrically distributed around the center line and begin to cluster close to the control limits, it indicates the level of quality has shifted, and steps should be taken to correct the situation.

By examining Fig. 5.4A, B, C, and D, some trends and explanations for the points being out of the limits may be found.

Problem Section

Practice Problem 1. a. Calculate the $\bar{X}$ and R control limits

TABLE 5.7

Tenacity of 150-Denier Viscose Rayon at Standard Conditions

Bobbin number	Tenacity, grams/denier					$\bar{X}$	R
1	2.1	2.2	2.2	2.2	2.2		
2	2.2	2.2	2.2	2.1	2.2		
3	2.3	2.3	2.3	2.2	2.2		
4	2.2	2.2	2.2	2.2	2.2		
5	2.1	2.2	2.2	2.1	2.2		
6	2.2	2.1	2.2	2.2	2.2		
7	2.3	2.3	2.3	2.3	2.3		
8	2.2	2.2	2.1	2.2	2.1		
9	2.2	2.2	2.3	2.2	2.2		
10	2.2	2.3	2.2	2.3	2.3		
11	2.2	2.3	2.2	2.1	2.3		
12	2.2	2.2	2.1	2.2	2.2		
13	2.1	2.1	2.1	2.1	2.1		
14	2.2	2.2	2.2	2.2	2.1		
15	2.2	2.2	2.2	2.2	2.1		
16	2.3	2.3	2.3	2.3	2.3		
17	2.2	2.1	2.2	2.2	2.2		
18	2.1	2.1	2.1	2.1	2.1		
19	2.2	2.2	2.2	2.2	2.2		
20	2.2	2.2	2.2	2.2	2.2		

for the data in Table 5.5 and construct the control charts. b. Does the process exhibit a state of statistical control?

Practice Problem 2. a. Calculate the $\bar{\bar{X}}$ and R control limits for the data in Table 5.6 and construct the control charts. b. Determine if the process exhibits a state of statistical control.

Practice Problem 3. a. Calculate the $\bar{\bar{X}}$ and R control limits for the data in Table 5.7 and construct the control charts. b. Determine if the process exhibits a state of statistical control.

CHAPTER VI

Analysis by Defects

The types of charts and data studied to this point have been for the normal type of distribution, covering such items of control as lap and sliver size, yarn number and strength, fabric strength, and the like. In all these, samples are selected at random and tests made on the samples to measure the particular qualities being studied. By selection of the proper number of samples, and by the statistical analysis of the data, a knowledge of the lot represented is obtained. Many of these tests are destructive: for example, strength tests destroy the samples for commercial use. It is for this reason among others that samples, rather than the entire lot, are tested. However, in reviewing the methods of analysis to here, it must be realized that the judgment concerning a lot is based on an analysis of representative samples. The method used so far is known as analysis by variables.

Another method of analysis is available to the quality control engineer, and this involves the attributes of the individual items comprising the lot. It is an analysis of the number of defective pieces or the number of defects per piece in a lot or run of material. In some types of processes it is known as the "go, no-go" type of analysis; in textiles it is the number of defects per yard or cut, the per cent second-quality fabric, the per cent rejected laps, the per cent second-quality cones, and the number of seconds produced. In short, it is the analysis of data that can be classified in one of these groups: (*1*) Per cent or fraction defective, (*2*) The number of defective units or pieces; (*3*) The number of defects per unit or piece.

Fraction or Per Cent Defective ($\bar{p}$ charts)

If during some working period a total of one-hundred cuts of fabric are inspected, of which ten are of inferior or second quality, then there are ten per cent defective cuts, or one-tenth of the cuts inspected are defective. Similarly, ninety per cent are good or effective. This illustrates in the simplest way what is meant by the per cent defective, a value signified in statistical analysis by $\bar{p}$ (read as p bar).

Just as control charts were designed for $\bar{X}$ and R, so can control charts be designed for the fraction defective, thus enabling the textile mill to apply quality control methods to inspection data where material is checked for bad or defective work. In dealing with fraction or per cent defective, the standard deviation is calculated by the following formula.

$$\sigma = \sqrt{\frac{\bar{p}(1 - \bar{p})}{n}}$$

where $\bar{p}$ is fraction defective, and n is the average number of units inspected at each period.

The upper and lower control limits follow the same statistical limits as set for $\bar{X}$ charts, namely, the three sigma limits. Thus, the limits for $\bar{p}$ control charts are

$$\bar{p} \pm 3\sigma$$

$$\text{or,} \quad \bar{p} \pm 3\sqrt{\frac{\bar{p}(1 - \bar{p})}{n}}$$

The data in Table 6.1 represent routine inspection of a standard construction fabric. The procedure in setting up the control chart (Fig. 6.1), using the data, is as follows:

1. For the period selected, obtain the total units inspected; in this case, it is 17648.
2. Determine the total number of defective units, which is 897 in the example.
3. Calculate fraction (or per cent) defective. This is the center line on the chart.

$$\bar{p} = \frac{897}{17648} = .051 \text{ (or } 5.1\ \%)$$

4. Determine the average value of n, which is 504 in the example. (See discussion on allowable variation for n.)

5. Using three sigma values, the control limits are calculated as follows:

$$\bar{p} \pm 3\sqrt{\frac{\bar{p}(1 - \bar{p})}{n}}$$

$$0.051 \pm 3\sqrt{\frac{0.051(1 - 0.051)}{504}}$$

$$0.051 \pm 3(0.0098)$$
$$0.051 \pm 0.029$$

Control limits:

$$\text{UCL} = 0.080 \text{ or } 8.0\%$$
$$\text{CL} = 0.051 \text{ or } 5.1\%$$
$$\text{LCL} = 0.022 \text{ or } 2.2\%$$

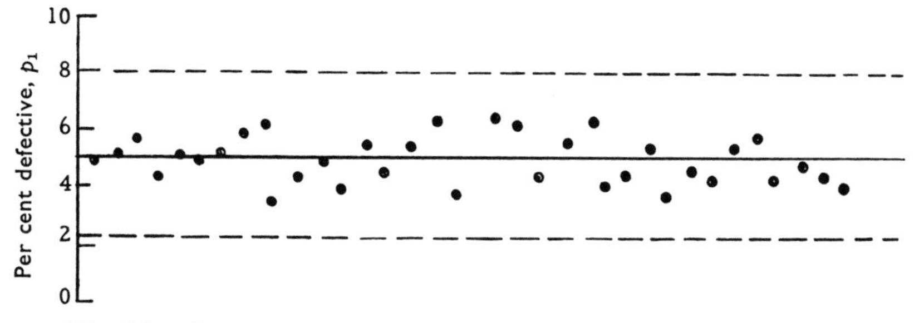

Fig. 6.1. Control chart for fabric inspection data shown in Table 6.1.

Value of Sample Size, *n*. It should be noted that the n used in the above calculations was the average number of cuts per day inspected. The average number was used because the actual number of cuts inspected daily varied in a normal way. This can be done for simplicity without decreasing the effectiveness of the charts, provided, however, that the largest sample size in the group does not exceed the smallest sample size by 20 per cent of the smallest sample size, or $n \text{ max} < 1.2\, n \text{ min}$. This saves a considerable amount of work when the sample sizes vary a small amount from day to day.

When the sample sizes do vary by more than the allowable amount, then it becomes necessary to adjust the control limits, but only the control limits, and not the center line. The center line should remain constant since it represents the average *per cent* defective, which does not change with the sample size unless it is reduced to an extremely small number. For example, if $\bar{p}$

TABLE 6.1

Inspection Results for Finished Fabric, 35-Day Run

Cuts [a]	No. of secs.	% Secs.	Cuts [a]	No. of secs.	% Secs.
500	25	5.0	500	29	5.8
510	27	5.3	490	30	6.1
513	30	5.9	497	20	4.0
493	22	4.5	505	22	4.4
497	26	5.2	500	28	5.6
503	25	5.0	505	19	3.8
505	27	5.3	500	24	4.8
498	30	6.0	495	22	4.4
509	32	6.3	515	29	5.6
512	19	3.7	510	30	5.9
505	22	4.4	508	21	4.1
498	25	5.0	506	25	4.9
500	20	4.0	508	23	4.5
500	29	5.8	500	20	4.0
515	24	4.7			
513	30	5.8	$\Sigma = 17648$	897	
509	33	6.5			
511	20	3.9	$\bar{p} = \frac{897}{17648} = 0.051$ or 5.1%		
510	35	6.9			
508	32	6.3	$n = \frac{17648}{35} = 504$		
500	22	4.4			

[a] Cuts of fabric inspected per day.

is found to be 10 per cent, and the sample size is reduced from 1000 to 500, $\bar{p}$ should remain constant; however, if the sample size is reduced to 10, then the $\bar{p}$ will change. This is true because a small sample size appears better than the lot more often than it appears worse than the lot, and for this reason a small sample size tends to give a smaller value for $\bar{p}$ than a large sample size.

Suppose a lot of 100 is represented by ten samples, each made up of ten units. If the lot is 10 per cent defective, there are ten defective units in all.

In order to have each of the ten samples representative of the lot, there would have to be nine effective and one defective unit in each sample. However, if samples of ten are selected at random from the 100 lot, the natural odds are that a sample or samples with more than one defective will be selected. If one sample is selected with two defectives, the sample is 20 per cent defective; more significantly, there must be one perfect (no defective) sample left in the balance of the other nine samples, for there are only ten defective units in all. Now, if one sample with three defectives is selected, which is a normal expectation, then there must remain two perfect samples with the other seven samples the average of 10 per cent defective. Thus, there are seven samples the same as the lot, one poorer, and two better. Obviously, if one sample is selected with four defectives, there must be three perfect samples which are better than the lot for the one defective sample poorer than the lot. The other six samples would be the same as the lot. This explains the fact that small samples appear better than the lot more often than they appear poorer.

The basic idea in constructing control charts where the sample size, n, varies more than the 1.2 (n min.) is to resort to re-establishment of the control limits for each lot. The calculation is identical to the one given previously, with the difference that new three sigma limits are calculated. The application of this idea is illustrated in the next example, the data for which are in Table 6.2.

$$\bar{p} = \frac{1608}{32{,}000} = 0.05 \text{ or } 5.0\ \% \text{ for center line}$$

For $n = 800$

$$0.05 \pm 3\sqrt{\frac{(0.05)(1 - 0.05)}{800}}$$

$0.05 \pm 3(0.0077)$

0.05 ± 0.023

$\text{UCL} = 0.073$ or $7.3\ \%$

$\text{CL} = 0.05$ or $5.0\ \%$

$\text{LCL} = 0.027$ or $2.7\ \%$

For $n = 400$

$$0.05 \pm 3\sqrt{\frac{(0.05)(1 - 0.05)}{400}}$$

$0.05 \pm 3(0.0109)$

0.05 ± 0.0327

$\text{UCL} = 0.0827$ or $8.3\ \%$

$\text{CL} = 0.05$ or $5.0\ \%$

$\text{LCL} = 0.017$ or $1.7\ \%$

In this example note that two sets of limits are calculated, one for the data where $n = 800$, and one for the data where $n = 400$. Note also that the center line, $\bar{p} = 5.0$ per cent, is the same over the entire production period. Finally, note that fractional values of $\bar{p}$, rather than per cent values, are used in the actual cal-

TABLE 6.2
Inspection Results for Viscose Rayon Cones

Cones inspected per day	Second quality cones	% Second quality	Cones inspected per day	Second quality cones	% Second quality
800	36	4.5	800	42	5.2
800	42	5.2	800	38	4.8
800	38	4.8	800	40	5.0
800	40	5.0	800	39	4.9
800	45	5.6	800	44	5.5
800	39	4.9	400	18	4.5
800	36	4.5	400	22	5.5
800	38	4.8	400	19	4.8
800	43	5.4	400	24	6.0
800	40	5.0	400	18	4.5
800	42	5.2	400	17	4.2
800	39	4.9	400	16	4.0
800	47	5.9	400	19	4.8
800	43	5.4	400	17	4.2
800	46	5.8	400	21	5.2
800	42	5.2	400	23	5.8
800	36	4.5	400	18	4.5
800	39	4.9	400	17	4.2
800	40	5.0	400	20	5.0
800	38	4.8	400	22	5.5
800	42	5.2	400	18	4.5
800	40	5.0	400	20	5.0
800	38	4.8	400	19	4.8
800	44	5.5	400	22	5.5
800	41	5.1	400	21	5.2
			$\Sigma =$ 32,000	1608	

culation; per cent values (whole numbers) can be used, but care should be taken in the setting up of the work.

For example, the formula for limits for $n = 800$ above would be:

$$5.0 + 3\sqrt{\frac{5(100 - 5)}{800}}$$

The control chart for the example is shown in Fig. 6.2.

If the sample size varies from day to day to the extent that it is necessary to adjust the control limits very frequently, it is usually more convenient to calculate the standard deviation based on

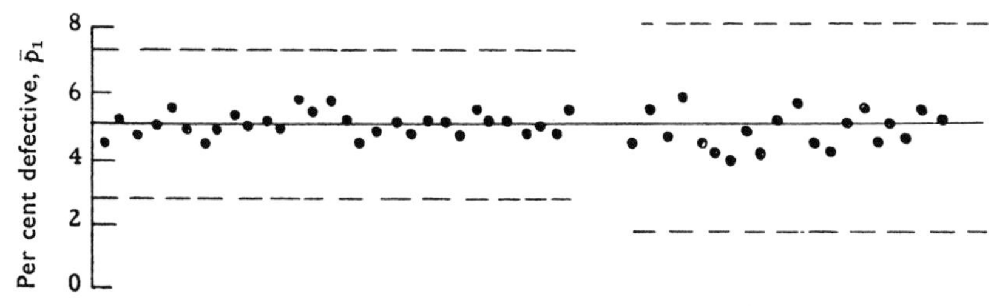

Fig. 6.2. Control chart for data from Table 6.2.

an $n = 1$ and then divide the standard deviation by the square root of the daily sample size to establish the limits.

Number Defective ($\bar{p}n$ charts)

In those cases where the sample size remains constant, it is possible to analyze the quality of production by using the actual number of defectives rather than the per cent or fraction of the total that these defectives represent. But, the sample size must remain constant. Control charts based on the number of defectives can be established in much the same way as for fraction defective. These charts are termed $\bar{p}n$ charts, the $\bar{p}n$ representing the number of defectives (for $\bar{p}$ is the fraction defective and n is the number of units in the sample).

The choice of whether to use the $\bar{p}n$ analysis and chart or the $\bar{p}$ approach seems to settle down to a question primarily of convenience. For example, if yarn is wound on dye tubes on a 100-spindle package winder, and each doff is checked for some quality (such as package density, diameter, or weight), then each 100 packages represents a sample unit of $n = 100$. This value of n would remain constant, so that it would be quite logical to analyze the data on the number defective rather than per cent defective.

The reason for the limitation that the $\bar{p}n$ method requires a constant sample size can be explained quite simply. The number of defective units is plotted directly on the control chart; the number of defectives is a direct function of the sample size, for with the larger sample size, the defectives increase, and with the smaller sample size, the defectives decrease. For example, if the normal sample size used for inspecting cones of yarn is 1,000 cones and the average defective is 50 (or 5 per cent), the center line for the $\bar{p}n$ chart would be 50; if the sample size were changed to 500 cones, the average number of defective cones would probably be 25 (or 5 per cent), and the control limits and center line for the $\bar{p}n$ chart would be changed to the new set of conditions. As pointed out in the discussion of the $\bar{p}$ chart, the only change usually required on that type of chart when the sample size changes is in the control limits. However, it should be remembered that *both* the center line and control limits have to be adjusted for a significant change in sample size when plotting the $\bar{p}n$ chart.

The standard deviation for this type of distribution is calculated by the formula

$$\sigma_{\bar{p}n} = \sqrt{\bar{p}n(1 - \bar{p})}$$

and the three sigma limits are established by

$$\bar{p}n \pm 3\sqrt{\bar{p}n(1 - \bar{p})}$$

where $\bar{p}n$ is the average number of defective pieces or units per sample, and $\bar{p}$ is the average fraction defective. In actual practice, where $\bar{p}$ is about 0.05 (or 5 per cent) or less, the value $(1 - \bar{p})$ may be omitted from the preceding equations and the formula for control limits becomes

$$\bar{p}n \pm 3\sqrt{\bar{p}n}$$

This is satisfactory since a $\bar{p}$ of 5 per cent or less will have little effect on the final answer. This is shown by the following calculations, in which $\bar{p}n = 20$ and $\bar{p} = 5$:

Full calculation for control limits

$$\bar{p}n \pm 3\sqrt{\bar{p}n(1-\bar{p})}$$
$$20 \pm 3\sqrt{20(1-0.05)}$$
$$20 \pm 3\sqrt{19.0}$$
$$20 \pm 3(4.36)$$
$$20 \pm 13.1$$
$$\text{UCL} = 33.1$$
$$\text{LCL} = 6.9$$

Short cut method ($\bar{p} \leqq 5\%$) for control limits

$$\bar{p}n \pm 3\sqrt{\bar{p}n}$$
$$20 \pm 3\sqrt{20}$$
$$20 \pm 3(4.47)$$
$$20 \pm 13.4$$
$$\text{UCL} = 33.4$$
$$\text{LCL} = 6.6$$

EXAMPLE. In Table 6.2 two sets of data are listed, one for a sample size of 800 and one for 400. The calculations to set up a control chart for these two conditions using the number defective $\bar{p}n$ approach is given herewith.

For $n = 800$, the average number of second-quality cones is

$$\frac{1217}{30} = 40.6 \text{ cones}$$

This is the $\bar{p}n$ value. Note that there are 30 sets of $n = 800$.

For $n = 400$, the average number of second-quality cones is

$$\frac{391}{20} = 19.6 \text{ cones}$$

This is the $\bar{p}n$ value. Note that there are 20 sets of $n = 400$.

The control limits are based on the formula $\bar{p}n \pm 3\sqrt{\bar{p}n}$, for the $\bar{p}$ values in each case are:

$$\bar{p} = \frac{40.6}{800} = 5.08 \qquad \bar{p} = \frac{19.6}{400} = 4.9$$

For $\bar{p}n \pm 3\sqrt{\bar{p}n}$

$$40.6 \pm 3\sqrt{40.6}$$
$$40.6 \pm 3(6.37)$$
$$40.6 \pm 19.1$$
$$\text{UCL} = 59.7$$
$$\text{CL} = 40.6$$
$$\text{LCL} = 21.5$$

$$19.6 \pm 3\sqrt{19.6}$$
$$19.6 \pm 3(4.43)$$
$$19.6 \pm 13.29$$
$$\text{UCL} = 32.9$$
$$\text{CL} = 19.6$$
$$\text{LCL} = 6.3$$

The control chart for this example is Fig. 6.3.

As another example, suppose that it has been found that the average number of misgraded cuts of fabric per 100-cut bale is 2.84, which is equivalent to 2.84 per cent misgraded. The control

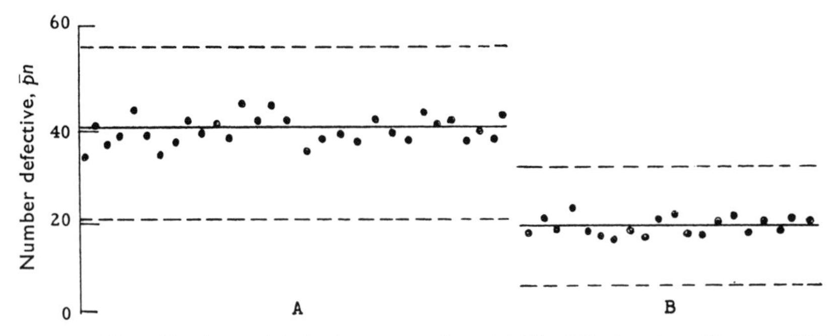

Fig. 6.3. Number of defective cones from Table 6.2. A: chart for $n = 800$; B: chart for $n = 400$.

limits for this type of chart would be calculated in the same manner as described above. The fraction defective is less than 5 per cent, so the formula $\bar{p}n \pm 3\sqrt{\bar{p}n}$ can be used.

$$2.84 \pm 3\sqrt{2.84}$$
$$2.84 \pm 5.07$$
$$\text{UCL} = 7.91$$
$$\text{CL} = 2.84$$
$$\text{LCL} = 0$$

Number of Defects ($\bar{c}$ *charts*)

When interest centers around the number of defects in an individual sample, as for example the defects per unit of area, per cut of fabric, per cone, and the like, control charts are constructed based on the number of defects. This is known as the $\bar{c}$ chart.

The same conditions relative to sample size that were described for the $\bar{p}n$ chart hold also for the $\bar{c}$ chart. That is, the sample size should remain constant since there is no way of adjusting the control limits on the $\bar{c}$ chart without knowing the value of the new $\bar{c}$, and of course this must be calculated for each significant change in sample size. In actual practice it is usually not considered necessary to establish new control limits for small short-term variation in the sample size unless the variation exceeds 5 per cent.

There is one advantage that the $\bar{c}$ chart has over the other types of control charts, and that is that the units can be so chosen that the lower limit on the chart will not be zero. For example, if it is found that the average number of defects per cut are of such a magnitude that the lower limit for the $\bar{c}$ chart is zero, it is possible to express the defects as the average number per roll or some other convenient unit of length or area. There is a definite advantage in not having the lower limit of the chart equal to zero; with a chart having a lower limit of zero, it is impossible to see any significant improvement in the number of defects since there is no chance of having a number of defects less than zero. However, with the lower limit greater than zero, it is possible to detect any significant improvement in quality.

The standard deviation for this type of distribution is calculated by the formula

$$\sigma_{\bar{c}} = \sqrt{\bar{c}}$$

and the control limits are represented by

$$\bar{c} \pm 3\sqrt{\bar{c}}$$

where $\bar{c}$ is the average number of defects per unit.

For example, an investigation into the number of end breaks per cake when rewinding from viscose cakes revealed 216 breaks for 120 cakes, which is equivalent to 216 ÷ 120 or 1.8 breaks per cake. This is equal to $\bar{c}$.

To make a control chart of the breaks per cake, the control limits would be calculated as follows:

$$\bar{c} \pm 3\sqrt{\bar{c}}$$

$$1.8 \pm 3\sqrt{1.8}$$

$$1.8 \pm 3(1.34)$$

$$\text{UCL} = 5.8$$

$$\text{CL} = 1.8$$

$$\text{LCL} = 0$$

In this example, individual cakes are being studied, so that the breaks in individual cakes are being considered. As a result, each entry on the chart represents the breaks per cake.

For a plant processing several thousand cakes per day, the amount of time involved in plotting the breaks per cake would be unreasonable.

For this reason, it would be more satisfactory to plot the average breaks per cake per side, or per machine, or per thousand pounds of production, or some other convenient unit of measure. This reflects in closer control limits and in a somewhat lesser degree of knowledge of the process, for such large units are used.

As an example, assume that in the foregoing example 150 positions (or cakes) are taken as the sample unit. The calculation for control limits follows the equation

$$\bar{c} \pm 3\sqrt{\frac{\bar{c}}{n}}$$

where n is the number of cakes.

$$1.8 \pm 3\sqrt{\frac{1.8}{150}}$$

$$1.8 \pm 3\left(\frac{1.34}{12.25}\right)$$

$$1.8 \pm 0.328$$

$$\text{UCL} = 2.13$$

$$\text{CL} = 1.8$$

$$\text{LCL} = 1.47$$

Suppose it is desired to base the limits on 1,000-pound lots. With each cake averaging 2 pounds, the limits would be based on 500-cake lots ($1.8 \pm 3\sqrt{1.8 \div 500}$) and the values plotted at each interval would be the actual average breaks per 1000 pounds (or the 500 cakes).

Another method which is even more simple than the two described above would be to plot the number of breaks per machine. The average number of breaks which could be expected for a 150-cake machine would be (1.8)(150) or 270 breaks. The control limits for this chart would be:

$$270 \pm 3\sqrt{270}$$
$$270 \pm 49.29$$
$$\text{UCL} = 319$$
$$\text{CL} = 270$$
$$\text{LCL} = 221$$

Some other common applications for this type of chart are defects per piece or cut of fabric, ends down at spinning, defects per cone, etc.

The calculations involved in establishing $\bar{c}$ control limits can be reversed for another use, and this is to determine the average number of defects (the $\bar{c}$ value) that can be tolerated to meet some maximum condition (the upper control limit). Thus, the mill decides on a maximum number of defects that can be allowed in some process; the question arises as to what the average level of number of defects should be in order to give some assurance that the entire product will fall below the upper control limit. The procedure involves first, the decision as to the maximum number of defects (UCL), and secondly, the calculations solving the equation $\text{UCL} = \bar{c} + 3\sqrt{\bar{c}}$ in order to find the value of $\bar{c}$.

Example. In grading fabric, it is common practice to assign point values to each defect, these values varying with the intensity of the defect, and being added to give a total point value for each piece graded. (The distribution of point values does not follow exactly the $\bar{c}$ distribution, however it is a reasonably good approximation.) Assume that a mill decides after investigation and establishment of this system that the maximum number of points acceptable for first-quality fabric is 80; beyond an 80-point total, the fabric is down-graded to second quality. The question that must be answered is: What must be the average number of points per piece of first-quality fabric in order that the mill can be assured that practically all the fabric will be first quality? What the mill is actually saying is that the upper control limit for points is 80, so what is the average value of $\bar{c}$?

The mathematics of the problem are:

$$\text{UCL} = \bar{c} + 3\sqrt{\bar{c}}$$

Substituting $$80 = \bar{c} + 3\sqrt{\bar{c}}$$

or $$\bar{c} + 3\sqrt{\bar{c}} - 80 = 0$$

To solve for $\bar{c}$, let $\bar{c} = x^2$, so that

$$x^2 + 3x - 80 = 0$$

This follows the equation:

$$ax^2 + bx + c = 0, \text{ with } a = 1,\ b = 3, \text{ and } c = -80$$

for which

$$x = \frac{-b \pm \sqrt{(b)^2 - 4ac}}{2a}$$

Substituting:

$$x = \frac{-3 \pm \sqrt{(3)^2 - 4(-80)}}{2(1)}$$

$$= \frac{-3 \pm \sqrt{329}}{2}$$

$$= \frac{-3 \pm 18.14}{2}$$

Using plus values:

$$x = \frac{15.14}{2} = 7.57$$

Therefore

$$\bar{c} = x^2 = (7.57)^2 = 57.30$$

Thus, the average points representing defects that should warrant approximately all first-quality cuts are 57.30.

If one more step is taken, the upper and lower limits can be calculated, and the control chart constructed. The upper control limit should check with the original specified value of 80.

$$57.30 \pm 3\sqrt{57.30}$$
$$57.30 \pm 3(7.57)$$
$$57.30 \pm 22.71$$
$$\text{UCL} = 80.0$$
$$\text{CL} = 57.3$$
$$\text{LCL} = 34.6$$

Guide for Installation. The mathematics and mechanics have been covered for the construction of five types of statistical quality control charts, and with a little practice and technical ability and plenty of common sense, the construction and use of these charts should not be too difficult. As a matter of fact, this is usually the simplest part of the job. When the time arrives for the installation of a statistical quality control system, the diplomacy involved usually overshadows by far the technical skill necessary. Listed below are a few suggestions which may help when installing and using statistical quality control charts and systems.

1. Find the process where excessive trouble occurs.
2. Find the quality characteristics of this process that are most troublesome, or that seem to be out of control.
3. Study the manner of testing.
4. Obtain past records.
5. Decide what possible charts you can make.

TABLE 6.3

Guide for Selection of Method of Analysis

Problem	Type of chart				
	$\bar{X}$	R	$\bar{p}$	$\bar{p}n$	$\bar{c}$
Per cent bad quality			×		
Number of bad pieces or units				×	
Number of defects or bad places per piece or per unit					×
Number of bad units with a reasonably constant sample size				×	
Number of bad units with a sample size that varies considerably from lot to lot, day to day, or week to week			×		
The strength of yarn, cord, fabric, etc.	×				
The amount of variation in strength within the breaks or number from a bobbin, cone, or spool of yarn.		×			
Yarn number, roving size, sliver size, lap weight	×				
Fiber properties, such as strength, length, fineness, etc.	×				

6. Decide what type chart you want to make. See Table 6.3. $\bar{p}n$ and $\bar{p}$ require more calculations; $\bar{c}$ requires less calculations; $\bar{X}$ and R require less sampling.
7. List possible causes of trouble.
8. Decide on a common sense method of sampling.
9. Select a sampling schedule.
10. Analyze both past and present data.
11. Make one chart at a time when beginning, and be sure that this chart is a success before moving on to new charts.
12. Advise all supervisory and technical personnel involved with the process upon which the chart is to be based. This is just as important, if not more important, than any of the points listed above.

Problem Section

TABLE 6.4

Per Cent Second-Quality Fabric Produced Per Day

Cuts inspected per day	% seconds	Cuts inspected per day	% seconds
600	3.4	598	4.3
594	3.8	600	3.5
605	3.0	602	3.9
610	3.2	598	3.3
598	3.2	601	3.4
603	3.6	420	3.7
595	3.0	425	3.6
600	3.7	418	3.5
606	3.5	422	4.4
598	3.1	419	3.8
597	3.9	422	3.8
599	3.4	421	4.3
595	3.2	425	4.2
597	3.5	418	3.7
604	3.7	419	3.6
602	3.5	421	3.7
600	4.3	420	3.8
603	3.7	421	3.5
601	3.6	421	3.8
603	4.0	420	4.0

Practice Problem 1. a. Construct and plot a statistical control chart using data in Table 6.4. b. Is the process in a state of statistical control?

Hints for Solution. Would variation in sample size cause trouble in using a $\bar{p}n$ chart? If so, can you get $\bar{p}n$? Would it be preferable to use $\bar{p}$ analysis, and would n be within the 20-per cent limits? Does the data lend itself to using $\bar{c}$ chart?

TABLE 6.5

Inspection Results for Gingham Fabric

Cuts inspected per day	No. of seconds	Cuts inspected per day	No. of seconds
850	49	848	46
860	42	860	47
840	40	570	25
855	41	577	28
845	42	584	27
870	42	590	30
855	49	585	32
830	43	584	29
850	45	577	33
843	42	579	29
854	49	588	26
860	46	584	31
858	43	585	31
858	47	590	33
862	42	575	29
848	43	582	27
854	42	580	31
862	44		

Practice Problem 2. Construct and plot a control chart for per cent second quality using the data in Table 6.5.

Practice Problem 3. a. Construct and plot a statistical control chart using the data in Table 6.6 below. b. Does the process exhibit a state of statistical control? c. The company producing this material has been offered a contract for several thousand yards, but the contract states that a first-quality piece of the material shall contain not more than 16 defect points. Should the company accept the contract? If so, what per cent

TABLE 6.6

Number of Defect Points Per 60-Yard Cut of Chambray Fabric

Cut no.	No. of points	Cut no.	No. of points
1	3	23	4
2	4	24	9
3	7	25	4
4	4	26	8
5	9	27	9
6	7	28	4
7	7	29	6
8	4	30	7
9	2	31	8
10	1	32	2
11	6	33	8
12	7	34	5
13	6	35	1
14	6	36	7
15	2	37	4
16	5	38	1
17	6	39	6
18	8	40	2
19	5	41	2
20	9	42	7
21	5	43	8
22	6	44	4

of their production would you expect to meet the requirements? Why?

Practice Problem 4. a. Construct and plot a statistical quality control chart for the data in Table 6.7. b. Are any of the inspectors consistently better or worse than the others? If so, which inspectors?

Practice Problem 5. a. Construct and plot $\bar{p}$ and $\bar{p}n$ charts for the data in Table 6.8. b. For this particular application, which type of chart would you recommend?

TABLE 6.7

Per Cent of Towels Misgraded by Inspectors Per 144-Towel Bundle

Insp. no.	% misgraded	Insp. no.	% misgraded
21	3.4	22	3.6
20	4.0	21	3.3
18	2.5	24	3.4
19	4.5	20	3.7
20	3.8	19	4.8
21	2.9	17	3.5
18	1.9	18	2.9
18	2.2	17	3.4
20	2.7	21	3.4
19	5.2	20	3.7
17	3.6	19	5.2
23	3.4	22	4.1
22	3.6	18	3.0
21	3.9	23	3.6
20	3.4	22	3.8
18	2.4	21	3.0
17	3.7	18	2.9
19	4.3	19	5.8
20	3.6	18	2.2
22	3.8	17	3.0
23	3.5	20	3.3
17	3.3	19	5.6
19	4.7	21	2.9
20	3.0	20	3.1

Practice Problem 6. a. Construct and plot a control chart for the data in Table 6.9. b. Are any of the rolls out of control?

TABLE 6.8
Number of Unsatisfactory Laps Produced by Picker Per Day

Laps per day	No. of rejects	Laps per day	No. of rejects
855	5	888	3
894	8	890	1
882	7	646	2
884	9	638	1
860	6	639	3
881	2	629	2
869	8	632	1
872	6	642	0
896	3	632	2
886	1	640	2
876	2	635	1
884	5	642	3
879	3	649	2
877	4	635	4
890	3	642	2
885	2	629	1
873	7	630	1

TABLE 6.9
Defect Points for a Cotton Fabric

Roll no.	Points	Roll no.	Points
1482	52	1582	73
1490	66	1586	50
1493	66	1588	78
1494	55	1589	50
1502	58	1592	64
1512	67	1598	62
1516	62	1605	54
1529	56	1612	68
1540	61	1614	54
1541	58	1622	67
1550	64	1629	64
1570	56	1631	53

CHAPTER VII

Sample Size

When the subject of sampling is being discussed, the question, "How many samples should we test?" usually arises. This question asked in this manner gives no indication as to what set of conditions the sample is to satisfy, and the quality control engineer can be of little or no service when it comes to establishing sample size unless he knows what is required of the sampling plan. He must know something of the process involved, the characteristic being measured, and what degree of protection is demanded of the sample selected. For the question to be sensible, it should have been, "How many samples should we test in order that our test results will be within a given per cent of the process average a certain per cent of the time?" For example, it might be, "How many spinning bobbins should we take at one time so that the skein test results will be within $\pm$ 0.25 pounds of the true process average 95 per cent of the time?" The question now sets forth definite conditions which the sampling program and sample size must meet.

With this set of conditions suggested, it then becomes necessary to establish the most sensible and in some cases the most economical limits. For example, is it necessary to be within a certain amount of the true average 80, 85, 90, or 95 per cent of the time? This value must be set by some competent person and should be given some careful consideration before any final decision is made. The same is true of the allowable sampling error; is it necessary to be within one per cent of the true average, within $\frac{1}{2}$ per cent, within 2 per cent, or within some other given range?

As stated above, the selection of the random sampling error and probability factor should be made only after careful consideration of all variables involved.

Sample Size for the Normal Distribution

The calculation for the appropriate sample size for averages is a fairly simple and easy calculation, and at the same time is an accurate method for determining the number of tests necessary to satisfy a given set of circumstances.

For example, suppose that a certain yarn number has an average skein strength of 90 pounds and a standard deviation of 7 pounds. The question can now be asked, "How many tests must be made so that the average of the test results will be within 1.5 per cent of the population average about 95 per cent of the time?" Summarizing the information available: $\bar{X} = 90$ pounds; $\sigma = 7$ pounds; $E = 1.5$ per cent, which is the allowable sampling error; $P = 95$ per cent, which is the probability that the average of the test results falls within the given limits, i.e., the equivalent area under the frequency distribution curve from $-t\sigma$ to $t\sigma$, where t in this case is 1.96, shown by the shaded area in Fig. 7.1.

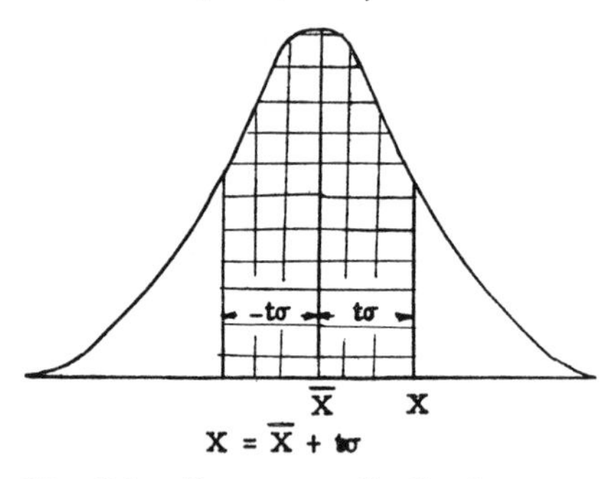

Fig. 7.1. Frequency distribution curve.

Note that t is a ratio; for each value of t, there is a definite probability. The areas under the curve (as in Table 1 of Appendix I) are from central ordinate to $\bar{X} + t\sigma$, and the values given must be multiplied by two to get the total area. For calculation of n number of tests in the formula below, the values of t in Table 7.1 will suffice in most cases. In textile work, a probability of 95 per cent is most frequently used.

By using the following formula, which is presented in the *A.S.T.M. Standards on Textile Materials*, it is possible to calculate the number of tests necessary to satisfy the above set of conditions:

$$n = \frac{t^2V^2}{E^2}$$

$$= \frac{(1.96)^2\,[(\frac{7}{90})(100)]^2}{(1.5)^2} = 103$$

This means that if a total of 103 tests are made, the average of these will be within 1.5 per cent of the population average approximately 95 per cent of the time. This method is only as

TABLE 7.1

Values of t

Probability, P %	t
90	1.645
95	1.960
99	2.576

reliable as the estimate of the coefficient of variation. It follows that if a large number of tests were made when calculating the original standard deviation, then this calculation gives more reliable results.

Quite frequently it is desirable to say that the average of the test results will be within plus or minus a given amount, such as plus or minus 2 pounds or 1.5 pounds, etc., rather than expressing it as a per cent. When this is the case, the following formula is used:

$$n = \left(\frac{t\sigma}{E}\right)^2$$

To illustrate this, use the same data: $\bar{X} = 90$ pounds; $\sigma = 7$ pounds; $E = 1.5$ per cent (1.5 per cent of $90 = 1.35$ pounds); $P = 95$ per cent, for which $t = 1.96$.

$$n = \left(\frac{t\sigma}{E}\right)^2$$

$$= \left[\frac{(1.96)(7)}{1.35}\right]^2 = 103$$

It should be noted that this method gives the same value for n as the previous method.

The preceding formulas may be used for purposes other than determining the number of measurements. Suppose that it is impossible for the laboratory to perform 103 tests of the material discussed above, and it can test only 65 samples. In that case, what is the probability that the average of the test results will be within 1.5 per cent of the population average?

Given: $\bar{X} = 90$ pounds; $\sigma = 7$ pounds; $V = 7.8$ per cent; $E = 1.5$ per cent; $n = 65$.

Find: P, the probability. Using the same equation as in the first example, solve for t.

$$n = \frac{t^2 V^2}{E^2}$$

$$65 = \frac{t^2(7.8)^2}{(1.5)^2}$$

$$t^2 = \frac{(65)(2.25)}{60.84}$$

$$t^2 = 2.40$$

$$t = 1.55$$

In referring to Table 1.2 in Appendix I, it is found that a t value of 1.55 is equivalent to approximately 86 per cent (significance of 0.14). (To obtain the value of significance from Table I.2, use the value $n = 120$, and interpolate between the value $t = 1.041$ at 0.3 significance or 70 per cent probability and $t = 1.658$ at 0.1 level or 90 per cent probability to get a value of 86 per cent for $t = 1.55$. Similar results would be obtained using $n = 60$, but it is customary to use the next larger value of n.) With this set of circumstances, it can be said that the average of 65 tests will be within 1.5 per cent of the population average approximately 86 per cent of the time.

In actual practice, it is usually not customary to designate the probabilities numerically, but rather by limits or ranges. That is, in place of saying a probability of 95 per cent or 97 per cent, the range from 95 to 100 per cent is usually referred to as having an extremely high probability, and the range from 90 to 95 is classed simply as having a high probability.

A little can be said now as to the intelligent selection of the probability, P, and the sampling error, E. As shown by two of the examples above, only a slight increase in the probability will cause a comparatively large increase in the number of tests, and by the same token, a small decrease in the allowable sampling error will also result in a large increase in the number of tests. This means that the quality control engineer must decide on the most economical set of conditions and at the same time afford satisfactory protection. Generally speaking, the sampling errors selected for textile work should range from 2 to 5 per cent and the probability from 90 to 95 per cent.

Sample Size for the Poisson Distribution

The calculations necessary for determining the sample size for the Poisson distribution can be very similar to those for normal distribution if the interest centers around a certain type of inspection, that is, if the problem centers around the average number of defects per unit of area or length, such as the number of defects per yard in a piece, cut, or roll of fabric. When the fraction defective, p' (where p' is the fraction defective of the population), is small, for example less than 0.05, the binomial approximates the Poisson distribution, which is designated by $p'n$ or $\bar{p}n$, the mean number of defectives per sample. Similarly, $\bar{c}$, the number of defects per unit such as area, follows the Poisson distribution, with the standard deviation equal to $\sqrt{\bar{c}}$.

It is possible to determine the number of samples to test in order to catch a shift in quality for the purpose of control. It is generally acceptable in deciding the size of sample to use to get one that would yield a 0.50 probability (or more) of showing on any single sample the shift decided on. To do this, $\bar{c}$ must be known, as well as the extent of the shift to be caught.

Now, the control limits for the $\bar{c}$ chart are $\text{UCL} = \bar{c} + 3\sqrt{\bar{c}}$ and $\text{LCL} = \bar{c} - 3\sqrt{\bar{c}}$ where the samples are all the same size. If the size of the samples varies, then the limits would vary on the control chart (except that LCL may equal zero in all cases, which indicates the need for larger samples). For example, if the units inspected were pieces of broadcloth, based on 60-yard length units, and some lengths dropped to 50 yards, and others were double-cut lengths of 120 yards, then the deviations would have to be corrected for the number of units of length in each sample. If the unit length in the above is 60 yards, then the number of defects per 60 yards would be the basic unit for the $\bar{c}$ chart. If the length of a second sample is 50 yards, then the units in this sample would be $\frac{5}{6}$; for a 120-yard length sample the unit length would be 2. The inspection length units are termed k. Therefore, for a sample of k units, the limits are changed to

$$\text{UCL} = \bar{c} + 3\sqrt{\frac{\bar{c}}{k}}$$

$$\text{LCL} = \bar{c} - 3\sqrt{\frac{\bar{c}}{k}}$$

It is known that the 0.50 probability point comes close to the upper and lower control limits of the $\bar{c}$ chart. Also, when the $\bar{c} = \text{UCL}$, there will be a 0.50 chance of a value being above the limit with normal distribution, for half the frequency distribution curve would be above UCL. If the value of k inspection units is known, then the size of sample is known, for k can be chosen so that the shift in quality is equal approximately to $3\sqrt{\bar{c}/k}$. The shift in quality can be called d.

$$d = 3\sqrt{\frac{\bar{c}}{k}} \text{ or } 3\sqrt{\frac{\bar{c}}{n}}$$

where n is the same as k and is number of inspection units in the sample.

For example, let the inspection be one of fabric, with rolls each containing 720 yards. Past tests have determined that the average number of defects per yard is 0.20. The question relates to the number of rolls (or yards) of fabric that would have to be inspected to give a 0.50 protection against a shift in the average number of defects per yard of 0.015 (i.e., in order that the average number of defects in the sample will not differ from the process average by more than 0.015 defects per yard).

$$\bar{c} = 0.20$$

$$d = 0.015$$

$$0.015 = 3\sqrt{\frac{0.20}{k}}$$

$$\text{or } k = \frac{9(0.20)}{(0.015)^2} = 8000$$

Thus, the yards to inspect are 8000, or 8000 ÷ 720 = 11.1 rolls. In practice, 11 or 12 rolls would be inspected. Instead of defects per yard, the same results will apply if defects per roll of 720 yards are used.

If a probability of detecting a difference is desired, then a different approach must be used. It is recommended that reference be made to statistical textbooks for these methods, for the design of a test becomes somewhat complicated.

As with the $\bar{c}$ analysis, the size of the sample can be determined for fraction defectives. With fraction defectives, sample sizes should be large enough to get some defective articles in the sample; this is particularly true with small values of $\bar{p}$ average fraction defectives. Also, the normal curve can be used to compute probabilities if $\bar{p}n \geqq 5$.

It has been previously shown that for fraction defectives

$$\sigma = \sqrt{\frac{\bar{p}(1 - \bar{p})}{n}}$$

where n is number of units inspected.

Under the above conditions, the control limits are set at 3σ, or

$$3\sqrt{\frac{\bar{p}(1 - \bar{p})}{n}}$$

It can be reasoned that a new process fraction defective $\bar{p}_2$ could be set; it is desired that there be a 0.50 chance or more of spotting this shift in a single sample.

Suppose the present $\bar{p}_1$ fraction defective is 0.025; we desire the 0.50 chance of spotting a shift of $d = 0.010$ to a new level of $\bar{p}_2 = 0.035$ fraction defective. Therefore, a value of n must be determined that will satisfy the conditions. The UCL is set at a level equal to $\bar{p}_1 = 0.025$ plus the increase of $d = 0.010$, or at $\bar{p}_2 = 0.035$, where the increase $d = 0.010$ is that for which a 0.50 chance of detection in a single sample is desired. At the UCL equal to $\bar{p}_2$, with normal distribution, 50 per cent of the points may be above the UCL.

With $\bar{p}_1$ at present level, and $\bar{p}_2$ at the desired level for detecting a shift, then

$$\text{UCL}_{\bar{p}_1} = \bar{p}_1 + 3\sigma_{\bar{p}_1}$$

and

$$\text{UCL}_{\bar{p}_1} = \bar{p}_2$$

Therefore,

$$\bar{p}_2 = \bar{p}_1 + 3\sqrt{\frac{\bar{p}_1(1 - \bar{p}_1)}{n}}$$

$$\bar{p}_2 - \bar{p}_1 = d = 3\sqrt{\frac{\bar{p}_1(1 - \bar{p}_1)}{n}}$$

and

$$n = \frac{9(\bar{p}_1)(1 - \bar{p}_1)}{d^2}$$

In the example,

$$n = \frac{9(0.025)(1 - 0.025)}{(0.010)^2} = 2194$$

Therefore, a sample size of 2194 (or 2200) yards would have to be inspected to have a probability of 0.50 of detecting a change from 2.5 per cent to 3.5 per cent defective.

Using the same set of conditions as above, what sample size is needed to detect a change to 4.0 per cent defective?

$$\bar{p} = 0.025$$

$$d = 0.04 - 0.025 = 0.015$$

$$n = \frac{9\bar{p}(1 - \bar{p})}{d^2}$$

$$= \frac{9(0.025)(0.975)}{(0.015)^2} = 975$$

This gives some indication of the large change in sample size for a comparatively small change in the difference in quality which is to be detected.

The methods given for determining the sample size for the binomial and Poisson distributions are intended only for cases which are represented by the examples given. There are a number of different methods which can be used for different circumstances pertaining to the inspection of fabrics and materials in the textile field.

Selected Bibliography

There are several excellent reference books which give inspection tables and sampling plans beyond the scope of this book. For anyone interested in designing an elaborate inspection plan, the books listed below treat the subject in more detail.

Peach, *Industrial Statistics and Quality Control*, Edwards and Broughton, Raleigh.

Simon, *An Engineer's Manual of Statistical Methods*, John Wiley, New York.

Dodge and Romig, *Sampling Inspection Tables*, John Wiley, New York.

Schrock, *Quality Control and Statistical Methods*, Reinhold, New York.

Duncan, *Quality Control and Industrial Statistics*, Richard D. Irwin, Chicago.

CHAPTER VIII

Differences between Averages

The statistical techniques discussed in this chapter are designed to be helpful in making decisions relative to differences that occur in processes or materials. One fundamental point that must be remembered when studying these techniques is that the test results, even though the test specimens may be drawn from the same sample, are not likely to be equal. For example, if a piece of fabric is cut into two equal parts and several strength tests made on each piece, it is extremely unlikely that the averages of the results would be numerically equal. The difference between averages will always vary, and the magnitude of the variation can be from a very small difference to one very large. The problem usually involved is: Should this difference between the averages be disregarded, or should it be given consideration? In other words, is the difference significant or is it insignificant?

One important use of these techniques is to measure the results of any change that has been made in a process or material. If a change is made in a process, does the change really make a difference in the quality, and if so, what is the magnitude of the real change? Very often after a change in processing is made, it is rather difficult to make a satisfactory recommendation without some analysis of the data, and even then, unless the analysis is complete, the recommendations may be erroneous.

The techniques to be studied are powerful and useful tools; however, they are not intended to replace engineering ability, to differentiate between types of errors, nor are they intended to replace the value of sound jugdment relative to cost control and expenditures.

Distribution of Differences. Since this problem centers around differences, the type of frequency distribution involved is the distribution of differences. Suppose, for example, that a series of averages, $\bar{X}$, are tabulated and then the average, $\bar{\bar{X}}$, of these numbers obtained. For a perfectly symmetrical distribution, the average, median, and mode are identical. With this true, then it can be said that the value of $\bar{X}$ occurring most frequently in the distribution will be of the same numerical value as the average $\bar{\bar{X}}$, and that this value $\bar{X} - \bar{\bar{X}}$ will equal zero and also will occur

TABLE 8.1

Example of Distribution of Differences

$\bar{X}$	$(\bar{X} - \bar{\bar{X}})$	$\bar{X}$	$(\bar{X} - \bar{\bar{X}})$
49	−1	50	0
51	+1	48	−2
50	0	51	+1
51	+1	49	−1
50	0	47	−3
48	−2	50	0
49	−1	51	+1
51	+1	51	+1
50	0	49	−1
50	0	50	0
52	+2	52	+2
49	−1	50	0
53	+3	49	−1
$\Sigma\bar{X} = 1300$	$n = 26$	$\bar{\bar{X}} = \dfrac{1300}{26} = 50$	

most frequently. As shown in Table 8.1, the value 0 occurs most frequently, the value 1 the next most frequently, and so on. With the differences plotted in the form of a frequency distribution as shown in Fig. 8.1, it can be seen what is meant by the term *distribution of differences.*

Difference of Means where n is Comparatively Large ($n > 30$). It is known that the means tend to follow normal distribution; it follows that the differences between means tend to follow a normal distribution, as shown under idealized conditions in Fig. 8.1. Thus $d = \bar{X}_1 - \bar{X}_2$, the difference between two means, tends to be normally distributed. In calculating for differences, the value of d is first obtained. Next, the standard deviation of the difference between means, σ_d (or as it is often termed, the standard error of the difference), is calculated from

$$\sigma_{\bar{X}_1-\bar{X}_2} = \sigma_d = \sqrt{\frac{\sigma_1^2}{n_1} + \frac{\sigma_2^2}{n_2}}$$

where σ_1 and σ_2 are the standard deviations of the two samples respectively.

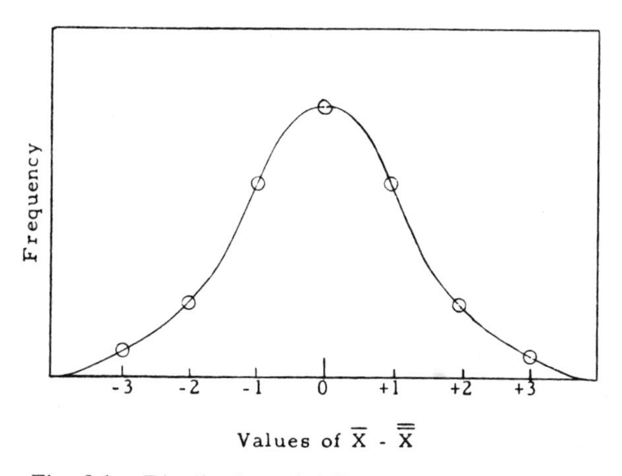

Fig. 8.1. Distribution of differences from Table 8.1.

The hypothesis under which the test operates is that $\bar{X}_1 = \bar{X}_2$; then there is a risk of rejecting $\bar{X}_1 = \bar{X}_2$ if they happen to be equal. This risk must be small, generally 0.05 or less (5 in 100) for significance. The risk is obtained by first finding the ratio of d/σ_d where d is $\bar{X}_1 - \bar{X}_2$, and then finding the area under the normal probability curve (Table 1 in Appendix I). The area gives the probability of chance causing the difference.

The following data were taken from an experiment where the strength of an experimental fabric was compared with the strength of a standard fabric.

Experimental: $\bar{X}_1 = 97.0$; $\sigma_1 = 5$; $n_1 = 64$.

Control: $\bar{X}_2 = 94.4$; $\sigma_2 = 4.5$; $n_2 = 64$; $(\bar{X}_1 - \bar{X}_2) = d = 2.6$

The experimental fabric shows a greater strength by 2.6 pounds than the standard fabric. Is this difference of 2.6 due to the construction of the fabric, or is it due to chance variation?

To answer this question, the standard deviation of the distribution of the differences, σ_d, must be calculated since this particular distribution is the one involved.

$$\sigma_d = \sqrt{\frac{\sigma_1^2}{n_1} + \frac{\sigma_2^2}{n_2}} = \sqrt{\frac{(5)^2}{64} + \frac{(4.5)^2}{64}} = \sqrt{0.3906 + 0.3164} = 0.84$$

The next step is to find the ratio of the difference, d, to the standard deviation of the difference, σ_d, or ratio of d/σ_d.

$$d/\sigma_d = \frac{d}{\sigma_d} = \frac{|\bar{X}_1 - \bar{X}_2|}{\sigma_d} = \frac{97.0 - 94.4}{0.84} = \frac{2.6}{0.84} = 3.1$$

By looking up this ratio of 3.1 in Table I.1, Areas Under the Normal Probability Curve, it is found to have a value of 0.49903. This means that 49.903 per cent of the area of the curve is included from the center line to the positive and negative side respectively, or that 49.903 sample differences in 100 would fall between $\bar{X}_1 - \bar{X}_2 = 0$ and $\bar{X}_1 - \bar{X}_2 = +2.6$. Consequently there are $50 - 49.903$ or 0.1 in 100, or 1 in 1000 chances that a difference might occur due to chance as great or greater than $\bar{X}_1 - \bar{X}_2 = +2.6$. Therefore, the difference is significant.

Looking at it another way, (49.903)(2) or 99.806 sample differences would fall between $\bar{X}_1 - \bar{X}_2 = +2.6$ and -2.6; thus, $100 - 99.806$ or 0.194 sample differences in 100 would be as great or greater than $+2.6$ and -2.6. In other words, there is a probability of 0.00194 of an occurrence of a deviation as great or greater than 2.6 due to chance. With this probability, it clearly indicates that a difference is not due to chance; therefore, it is considered significant. This is shown graphically in Figs. 8.2 and 8.3. In Fig. 8.2,

$$\sigma_{\bar{X}_2} = \text{standard error of mean} = \frac{\sigma_2}{\sqrt{n}} = \frac{4.5}{\sqrt{64}} = 0.562$$

The distribution curve in Fig. 8.2 has been drawn using a center line, $\bar{X}_2$, as an estimate of the population average, $\bar{X}_p = 94.4$, and the ± 1, 2, and $3\sigma_{\bar{X}_2}$ limits are indicated with the corresponding numerical value for $\bar{X}_2$. At the $+3\sigma$ limit, the value of $\bar{X}_2$ is 96.1; compare this to the value of the average $\bar{X}$ of 97.0. The area outside of 3σ (or 3 sigma limits) is 0.0013, or a probability of only 1.3 in 1000 that any value of $\bar{X}_2$ would be above 96.1 as a

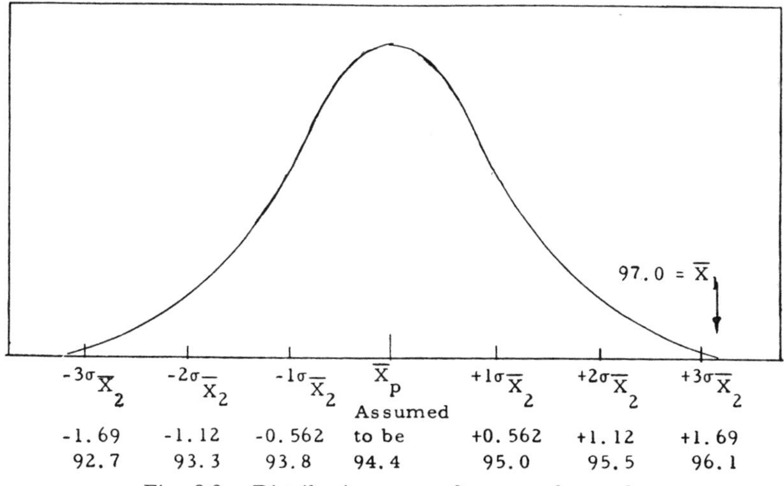

Fig. 8.2. Distribution curve for control sample.

result of chance (or 2.6 in 1000 that any value would be above 96.1 or below 92.7). Therefore, something other than chance is causing the difference between the two means.

Fig. 8.3 may be analyzed in much the same way, but in this case the frequency distribution of the difference is shown. The center line is taken at $\bar{X}_1 - \bar{X}_2 = d = 0$, and then the 1, 2, and $3\sigma_d$ values shown. The actual value $d = 2.6$ is greater than the value of d at the 3 sigma limit of 2.52; the probability of chance causing this difference is less than 1.3 in 1000.

Another approach to the analysis of probability of chance causing the difference is to select a critical value of probability, and then to see how the actual ratio of d/σ_d compares to this. If the critical value is 1.96, which gives a single tail probability of 0.025 (or two-tail probability of 0.05) for large samples, then with a ratio of $d/\sigma_d = 3.1$, we can say that the difference is significant. If a critical value of 2.58 is used, then the single tail probability is 0.005. With an actual ratio of d/σ_d of 3.1 as compared to 2.58, we can say that the differences are highly significant. However, we have already shown by other methods what the probabilities were of chance causing the difference. This latter approach, which is actually a confidence limit method, can be used if desired.

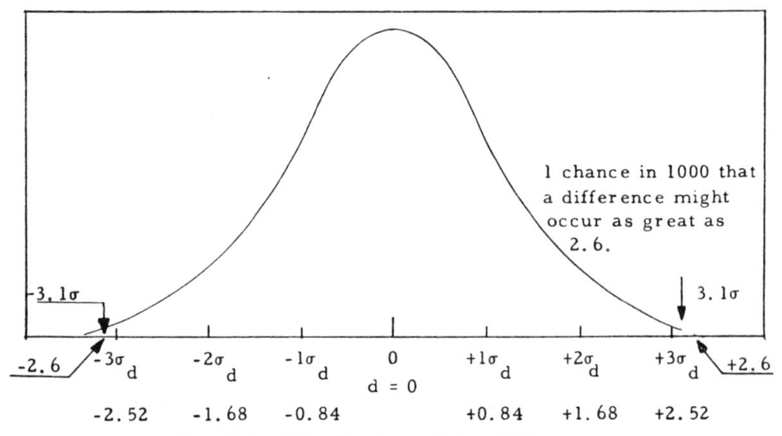

Fig. 8.3. Distribution of the differences.

To summarize the question as to when a difference is considered significant and when it is not considered significant, the following general rule may be used as a practical guide. If the probability of occurrence is less than 1 in 100, the difference can be considered as real or significant; if the probability of occurrence is between 1 and 5 in 100, then it can be said that there is some evidence of a difference, but that it is not conclusive; and if the probability is greater than 5 in 100, then the difference is usually considered as not being significant. Exactly which significance limit to use depends upon the particular situation.

Difference of Means where n is Comparatively Small ($n \leq 30$). With small values of n, the estimate of the universe σ is not known or is not good enough, and so the t distribution is resorted to. The distribution should be normal. As previously explained, the term t is a ratio and has no units. In the frequency distribution sketch, Fig. 7.1, the area under the curve is taken as unity. Then, for example, any area to the right of the ordinate X at a distance $t\sigma$ from the center line is representative of the frequency of observations having values greater than $\bar{X} + t\sigma$. Likewise, the area to the left of the other ordinate at $-t\sigma$ gives frequencies of less than $\bar{X} - t\sigma$. For a large population, the areas are those shown in Table I.1. When the differences between two means are being studied, the use of the t relationship is used:

$$t = \frac{d}{\sigma_d}$$

where d is the difference $\bar{X}_1 - \bar{X}_2$, and σ_d is the standard error of the difference. There is a different probability curve for t, which depends on the number of degrees of freedom involved (see below). If the degrees of freedom are infinite, then the probabilities in the t tables are the same as for the normal probability tables, for the estimated standard deviation approaches the true value.

In comparing the means of two small ($n \leq 30$) samples, calculations are made to determine if chance would cause a difference in means that is greater than the observed difference. As for large samples, the distribution of d for many experiments would give a normal distribution, which would be the distribution of the difference between two means, with the mean difference equal to zero.

In making the calculations, first an estimate of a common σ is made:

$$\sigma = \sqrt{\frac{\Sigma(X_1 - \bar{X}_1)^2 + \Sigma(X_2 - \bar{X}_2)^2}{n_1 + n_2 - 2}} = \sqrt{\frac{\text{Sum of squares}}{n_1 + n_2 - 2}}$$

or,

$$\sigma = \sqrt{\frac{n_1 \sigma_1^2 + n_2 \sigma_2^2}{n_1 + n_2 - 2}}$$

where Σ is summation of all values; $\Sigma(X_1 - \bar{X}_1)^2$ is the sum of all the differences squared between individuals and average for the first series and $\Sigma(X_2 - \bar{X}_2)^2$ the same for the second series; $n_1 + n_2 - 2$ is the degrees of freedom and is used when n_1 and

n_2 are small. (The total value of $n_1 + n_2 - 2$ is the n value used in t-tables of probability.) In the second alternate equation, σ_1^2 and σ_2^2 are the variances for the two series.

Then the standard error of the mean difference is calculated:

$$SE_d = \sigma_d = \sigma\sqrt{\frac{1}{n_1} + \frac{1}{n_2}}$$

Finally,

$$t = \frac{d}{\sigma_d}$$

By reference to t tables for the calculated value of t at the degrees of freedom used, the probability is obtained. The interpretation of the probability in relation to d is the same as described above for large samples.

The application of this approach is illustrated in the following examples. The example for Table 8.2 follows the formulas given above; Table 8.3 gives the same data, but handles these data in a slightly different way. The example for Table 8.4 shows a difference that is significant but not to as high a degree as for the other examples.

TABLE 8.2

Strength of a Standard and Experimental Cord

Standard			Experiment		
X	$(X - 45)$	$(X - 45)^2$	X	$(X - 48)$	$(X - 48)^2$
45.3	0.3	0.09	49.2	1.2	1.44
47.4	2.4	5.76	48.5	0.5	0.25
44.8	−0.2	0.04	47.6	−0.4	0.16
46.7	1.7	2.89	49.8	1.8	3.24
44.6	−0.4	0.16	48.4	0.4	0.16
$\bar{X} = 45.76$	$\Sigma = 3.8$	$\Sigma = 8.94$	$\bar{X} = 48.70$	$\Sigma = 3.5$	5.25

Corrected sum of squares $= 8.94 - \frac{(3.8)^2}{5} = 6.06$

Corrected sum of squares $= 5.25 - \frac{(3.5)^2}{5} = 2.80$

Sum of squares $= 6.06 + 2.80 = 8.86$

$$\sigma = \sqrt{\frac{\text{Sum of squares}}{n_1 + n_2 - 2}}$$

or,

$$\text{Variance} = \sigma^2 = \frac{\text{Sum of squares}}{n_1 + n_2 - 2}$$

$$\sigma = \sqrt{\frac{8.86}{8}} = \sqrt{1.11} = 1.05$$

$$t = \frac{d}{\sigma_d} = \frac{d}{\sigma\sqrt{\frac{1}{n_1} + \frac{1}{n_2}}}$$

$$t = \frac{|48.70 - 45.76|}{1.05\sqrt{\frac{1}{5} + \frac{1}{5}}} = \frac{2.94}{0.66} = 4.45$$

Degrees of freedom $= n_1 + n_2 - 2 = 8$

By locating the t values of 4.45 with 8 degrees of freedom in Table 1.2 in Appendix I, it is found to have a level of significance of less than 1 per cent, so the difference of 2.94 pounds must be considered significant; that is, there is less than 1 per cent probability that chance would cause a difference of $d = 2.94$ pounds.

TABLE 8.3

Strength of a Standard and Experimental Cord

Standard		Experiment	
X	X^2	X	X^2
45.3	2052.09	49.2	2420.64
47.4	2246.76	48.5	2352.25
44.8	2007.04	47.6	2265.76
46.7	2180.89	49.8	2480.04
44.6	1989.16	48.4	2342.56
$\Sigma = 228.8$	10475.94	$\Sigma = 243.5$	11861.25
$\bar{X} = \frac{228.8}{5} = 45.76$		$\bar{X} = \frac{243.5}{5} = 48.70$	

$$\sigma = \sqrt{\frac{\Sigma X^2}{n-1} - \frac{(\Sigma X)^2}{n(n-1)}}$$

$$\sigma_1 = \sqrt{\frac{10475.94}{4} - \frac{(228.8)^2}{(5)(4)}} = 1.23$$

$$\sigma_{1\bar{X}(n=5)} = \frac{1.23}{\sqrt{5}} = 0.550$$

$$\sigma_2 = \sqrt{\frac{11861.25}{4} - \frac{(243.5)^2}{(5)(4)}} = 0.837$$

$$\sigma_{2\bar{X}(n=5)} = \frac{0.837}{\sqrt{5}} = 0.374$$

$$\sigma_d = \sqrt{(0.550)^2 + (0.374)^2} = 0.66$$

$$t = \frac{|48.70 - 45.76|}{0.66} = 4.45$$

The results are the same as with those under Table 8.2 and are analyzed in the same manner.

TABLE 8.4
Skein Strength of 60s Yarn from Two Different Suppliers

Supplier II			Supplier I		
X	$(X-60)$	$(X-60)^2$	X	$(X-60)$	$(X-60)^2$
62	2	4	64	4	16
64	4	16	62	2	4
58	−2	4	61	1	1
60	0	0	65	5	25
61	1	1	63	3	9
57	−3	9	64	4	16
59	−1	1	63	3	9
64	4	16	64	4	16
62	2	4	64	4	16
58	−2	4	60	0	0
$\Sigma = 605$	5	59.0	$\Sigma = 630$	30	112.0
$\bar{X} = 60.5$			$\bar{X} = 63.0$		
Corrected sum of squares $= 59 - \frac{(5)^2}{10} = 56.5$			Corrected sum of squares $= 112 - \frac{(30)^2}{10} = 22$		

Sum of squares $= 56.5 + 22 = 78.5$

$$\sigma = \sqrt{\frac{\text{Sum of squares}}{n_1 + n_2 - 2}}$$

(Note that variance σ^2 could be used here.)

$$\sigma = \sqrt{\frac{78.5}{18}} = \sqrt{4.36} = 2.09$$

$$t = \frac{d}{\sigma\sqrt{\dfrac{1}{n_1} + \dfrac{1}{n_2}}}$$

$$= \frac{|63 - 60.5|}{2.09\sqrt{\dfrac{1}{10} + \dfrac{1}{10}}} = \frac{2.5}{(2.09)(0.447)} = 2.66$$

For $n = 18$ degrees of freedom and $t = 2.66$, the probability of occurrence of a difference of 2.5 as a result of chance is slightly greater than 1 per cent, as shown in Table I.1.

For a probability of 0.05 (5 times in 100) and with a two-tail test for plus or minus a critical value, the critical value of t at 18 degrees of freedom is ± 2.101. As the calculated t is 2.66, which is outside the critical value 2.101, it can be assumed that a difference d exists. However, the level of significance is not high, and evidence is not conclusive.

Difference of Means where $n_1 \neq n_2$. The calculations for significant differences under these conditions is slightly more complicated due to the fact that the variances must be weighted according to the degrees of freedom of each set of data.

Under these conditions,

$$t = \frac{d}{\sqrt{\dfrac{(n_1 + n_2)(\Sigma d_1^2 + \Sigma d_2^2)}{n_1 n_2[(n_1 - 1) + (n_2 - 1)]}}}$$

In connection with the weighting of variances, it is well to remember the following points. (a) It is incorrect to take the average of a series of standard deviations. The standard deviations should be squared, which will give the variance, and the variances averaged; i.e. variances are additive quantities. (b) It is incorrect to average variances from different size samples unless the variances have been weighted according to the proper degrees of freedom. (c) Where the average variance is to be calculated, it is incorrect to pool all of the individual measurements and calculate only the one variance.

Significance of Variance. In general terms, the study of variance (σ^2) is important for two reasons. One application is the study of change in variation in a product or process, resulting from some change in the process. For example, if the twist is changed in a yarn, it would be important to know not only the fact that it had a significant effect on strength but also if it had a significant effect on the variation in strength. A similar common application of variance studies is the comparison between operators and machines performing the same function. The second important influence of variance is in connection with the t test for significant differences. One of the underlying assumptions of the t test is that the two standard deviations belong to the same distribution or that they are both estimates of the same standard deviation. If this is not true, then the standard t test loses most of its effectiveness. To test the difference between variances, the technique known as the F test is used.

The F test consists of calculating the ratio between two variances. Which of the two variances is used as the numerator and which as the denominator depends upon the previous experience with the process.

With no previous experience with the variances (two-tail test),

$$F = \frac{\text{Larger variance}}{\text{Smaller variance}}$$

For instance, if two methods of testing fabric strength give variances of 12.6 for $n = 25$ and 8.9 for $n = 30$, is the variation between the two methods significant?

$$F = \frac{12.6}{8.9} = 1.42$$

The numerator has 25 — 1 or 24 degrees of freedom; therefore, locate 24 on the n_1 (top) column of the variance ratio table. The denominator has 29 degrees of freedom, which is located in the vertical (n_2) column. By converging these two points, the ratio at the 5 per cent level is found to be 1.9. Since this is larger than 1.42, there appears to be no significant difference in the variances. However, since this is the two-tailed test, the levels are 10 per cent and 2 per cent, rather than 5 per cent and 1 per cent. (Table 4 in Appendix I).

With some previous experience with the variances, the experimenter will know which of the two variances is expected to be larger. Under these cricumstances,

$$F = \frac{\text{Variance expected to be larger}}{\text{Variance expected to be smaller}}$$

If the standard deviation for one yarn size has been 5.3 and has been consistently higher than the other yarns, and the process has been changed so that the standard deviation of 150 tests is now 4.7, is the new yarn really better than the old yarn?

$$F = \frac{(5.3)^2}{(4.7)^2} = 1.27$$

With $n_1 = \infty$ and $n_2 = 149$, the ratio is 1.0. Therefore, the new yarn has exhibited a significant decrease in variation.

Difference of Means Between Percentages. The handling of problems involving percentages is similar to the methods previously discussed. The method in this case is illustrated by the following problem, the data for which are given in Table 8.5.

$$\sigma_{\bar{p}} = \sqrt{\frac{\bar{p}(1 - \bar{p})}{n}}$$

$$\sigma_d = \sqrt{\sigma_{\bar{p}_1}{}^2 + \sigma_{\bar{p}_2}{}^2}$$

From Table 8.5, $\bar{p}_1$ is 0.051, and $\bar{p}_2$ is 0.071. The calculations for significance follow.

$$\sigma_{\bar{p}_1} = \sqrt{\frac{(0.051)(1 - 0.051)}{17648}}$$
$$= 0.00165 \text{ or } 0.165\,\%$$

TABLE 8.5
Inspection Results of Two Fabrics: A Standard Pattern *vs.* An Experimental Pattern

Standard pattern			Experimental pattern		
Cuts [a]	No. of seconds	% seconds	Cuts [a]	No. of seconds	% seconds
500	25	5.0	100	7	7.0
510	27	5.3	115	10	8.7
513	30	5.8	97	9	9.3
493	22	4.5	110	11	10.0
497	26	5.2	107	8	7.5
503	25	5.0	109	12	11.0
505	27	5.3	100	9	9.0
498	30	6.0	98	11	11.2
509	32	6.3	94	7	7.4
512	19	3.7	99	10	10.1
505	22	4.4	105	6	5.7
498	25	5.0	103	5	4.9
500	20	4.0	100	6	6.0
500	29	5.8	98	4	4.1
515	24	4.7	95	6	6.3
513	30	5.8	99	5	5.1
509	33	6.5	103	6	5.8
511	20	3.9	105	4	3.8
510	35	6.9	100	3	3.0
508	32	6.3	114	7	6.1
500	22	4.4			
500	29	5.8			
490	30	6.1			
497	20	4.0			
505	22	4.4			
500	28	5.6			
505	19	3.8			
500	24	4.8			
495	22	4.4			
515	29	5.6			
510	30	5.9			
508	21	4.1			
506	25	4.9			
508	23	4.5			
500	20	4.0			
$\Sigma = 17648$	$\Sigma = 897$		$\Sigma = 2051$	$\Sigma = 146$	

$$\bar{p}_1 = \frac{897}{17648} = 0.051 \text{ (or 5.1 \%)}$$

$$\bar{p}_2 = \frac{146}{2051} = 0.071 \text{ (or 7.1 \%)}$$

[a] Cuts of fabric inspected per day.

$$\sigma_{\bar{p}_2} = \sqrt{\frac{(0.071)(1 - 0.071)}{2051}}$$

$$= 0.00568 \text{ or } 0.568\ \%$$

$$\sigma_d = \sqrt{(0.165)^2 + (0.568)^2}$$

$$= 0.590\ \%$$

$$\frac{X}{\sigma_d} = \frac{|7.1 - 5.1|}{0.590} = 3.39$$

By referring to either Table I.1 or I.2, the value 3.39 is found to be considerably below the 0.001 level of significance. From this, it can be concluded that if the true difference between $\bar{p}_1$ and $\bar{p}_2$ is zero, a difference of 2.0 per cent might occur less than 0.1 per cent of the time in favor of the standard fabrics. In other words, the hypothesis that $\bar{p}_1 = \bar{p}_2$ (or $\bar{p}_1 - \bar{p}_2 = 0$) is rejected, for 2 per cent difference would be caused by chance less than 1 time in 1000. Therefore, a real difference is present between the fabrics, and with all other things being equal, the increase in the per cent of seconds must be attributed to the pattern change.

Confidence Intervals. Confidence intervals (also referred to as reliability of means) measure the limits within which a population mean would be expected to lie for a given probability. The data in Table 8.6 represent the skein strength for an 18/1 cotton yarn, and as shown, the average is 129.3 pounds and the standard deviation is 5.2 pounds. The confidence intervals or the reliability of the sample mean at the 95 per cent level are calculated as follows:

$$\bar{X} \pm t\frac{\sigma}{\sqrt{n}}$$

where $\bar{X}$ = sample mean of 129.3; $n = 10$; $t = 2.262$ for $n-1=9$ at the 5 per cent level; $\sigma = 5.2$.
Confidence intervals are:

$$129.3 \pm (2.262)\frac{(5.2)}{\sqrt{10}} = 129.3 \pm 3.75$$

or

$$125.6 \text{ to } 133.0$$

From this it is safe to assume (95 times in 100) that the population mean falls between 125.6 and 133.0 pounds. In other words, if all the yarn produced had been tested, the chances are very good that the average would have been between 125.6 and 133.0 pounds. If the one per cent level had been used with

TABLE 8.6

Skein Strength for 18/1 Cotton Yarn

Skein strength	Strength-129	(Strength-129)^2
133	4	16
132	3	9
132	3	9
137	8	64
131	2	4
118	−11	121
127	− 2	4
128	− 1	1
130	1	1
125	− 4	16
$\Sigma = 1293$	3	245
$\bar{X} = 129.3$		

$$\text{Corrected sum of squares} = 245 - \frac{(3)^2}{10} = 244.1$$

$$\sigma = \sqrt{\frac{244.1}{n-1}} = \sqrt{\frac{244.1}{9}} = 5.2$$

$t = 3.250$, then the population mean would be between the resulting limits 99 times out of 100. Assume that instead of ten tests, twenty-five tests were made with an average $\bar{X} = 129.3$ and a standard deviation $\sigma = 5.2$. What effect would the increase in sample size have on the confidence limits? The calculation would be the same as before except that with $n = 25$ and $n - 1 = 24$, t at the 0.05 level becomes 2.064.

$$129.3 \pm (2.064) \frac{(5.2)}{\sqrt{25}}$$

$$129.3 \pm 2.15$$

$$127.15 \text{ to } 131.45$$

This shows that by increasing the sample size, the confidence limits become closer or the reliability of the mean is greater. This is nothing new to any mill man or quality control engineer; however, these calculations will show the decreasing rate at which reliability is obtained by continuing to increase the sample size and as a result, increasing the cost of testing. The question to be answered is, what is the maximum number of tests to be made? Obviously, all testing laboratories wish to produce test results that are as reliable as possible; however, a balance must be reached between the reliability of the results and the economical factors. To be assured of maximum protection at a minimum of cost, the number of tests necessary to give adequate protection should be calculated as outlined previously in this book and in most other books dealing with quality control and statistical methods.

Confidence Limits for Differences Between Means. When problems arise involving the difference between two means, it is frequently desirable to know confidence intervals for the difference. This is calculated in a manner somewhat similar to the one outlined previously. Using the example discussed for the difference of skein strength in Table 8.4, the problem is as follows for the 95 per cent level.

$$|\bar{X}_1 - \bar{X}_2| \pm t\sigma\sqrt{\frac{1}{n_1} + \frac{1}{n_2}}$$

where $|\bar{X}_1 - \bar{X}_2| = 2.5$; $n_1 = n_2 = 10$; $t = 2.101$ for $n_1 + n_2 - 2$ or 18 at the 5 per cent level, and

$$\sigma = \sqrt{\frac{78.5}{18}} = 2.09$$

Confidence limits are:

$$|63.0 - 60.5| \pm (2.101)(2.09)\sqrt{\frac{1}{10} + \frac{1}{10}}$$

$$= 2.5 \pm 1.95$$

or,

$$0.55 \text{ to } 4.45$$

This shows that the difference in yarn strength from the two suppliers will be between 0.55 and 4.45 pounds, 95 per cent of the time.

Problem Section

Practice Problem 1. Is the difference between the two yarns in Table 8.7 to be considered as significant? Why?

TABLE 8.7

Tenacity of Viscose Rayon as Produced By Different Spinning Machines

Machine 1	Machine 2
2.10	2.18
2.11	2.17
2.16	2.19
2.09	2.12
2.10	2.15
2.11	2.16
2.08	2.18
2.11	2.19
2.12	2.14
2.11	2.13

Practice Problem 2. Is the difference in the yarn strengths to be considered as significant? Why?

TABLE 8.8

Skein Strength of 60/1 Cotton Yarn Produced From Different Fiber Mixes

Mix	Skein strength ave.	Standard deviation	No. of skeins tested
Mix 1	34.1 lbs.	4.7 lbs.	105
Mix 2	32.8 lbs.	4.2 lbs.	72

Practice Problem 3. If all of the defective quills listed in Table 8.9 can be classed as machine defects, can the difference in defective quills produced by the two machines be considered significant? Why?

TABLE 8.9
Inspection Results of Quills Wound On Two Different Winders

Machine	Quills inspected	No. defective
Machine 1	1230	26
Machine 2	885	17

Practice Problem 4. From the data in Table 8.10 determine the following: a. If the difference in strength and elongation is significant. b. If the variation in the strength and elongation is

TABLE 8.10
Test Results of One Type Fiber Using Two Methods of Testing

Fiber property		Old method	New method
Strength:	$\bar{X}$	14.3 gms.	13.6 gms.
	σ	4.47 gms.	3.41 gms.
	n	100	42
Elongation:	$\bar{X}$	44.8 %	56.42 %
	σ	13.82 %	21.0 %
	n	100	42

significant. c. Confidence limits for the strengths and elongations. d. The most difference likely to be found between the two methods. e. The method you would recommend and why.

CHAPTER IX

Correlation

Correlation may be defined as an expression of relationship, with the relationship being between two or more variables. The use of correlation and the fitting of a regression line to data has several applications in the textile industry. There are several types of correlation, such as simple, non-linear, multiple, and partial; however, only simple correlation will be discussed. The application of the other types of correlation should be applied only by a mathematician or statistician with a competent knowledge of the theories and techniques involved.

For correlation, which is the process of establishing the extent of a mutual or reciprocal relationship, the values of Y in terms of X are wanted, so that Y can be predicted in terms of X. The method known as least squares is used to determine the extent of the relationship. In this approach, two conditions are satisfied for the regression line: the algebraic sum of distances of points from the line must be zero, and the sum of the squares of these deviations from the line must be a minimum (smaller than from any other line, which in effect determines the line). The line has the equation of the general form $Y = a + bX$, assuming the relationship is linear.

The formulas for the trend line are:

$$\Sigma Y = an + b\Sigma X \tag{1}$$

$$\Sigma XY = a\Sigma X + b\Sigma X^2 \tag{2}$$

$$Yc = a + bX \tag{3}$$

where a is the intercept; b is the slope; and n is the number of pairs of observations; Yc is the calculated value of Y for any real value of X.

It will be found that equations (1) and (2) can usually be

solved simultaneously. In setting up for calculation, it is wise to list in columns the values of X and Y with their totals ΣX and ΣY; XY with total ΣXY, and X^2 and Y^2 with totals ΣX^2 and ΣY^2.

These alternate equations may be better in some cases:

$$a = \frac{\Sigma X^2 \Sigma Y - \Sigma X \Sigma XY}{n\Sigma X^2 - (\Sigma X)^2} \tag{4}$$

$$b = \frac{n\Sigma XY - \Sigma X \Sigma Y}{n\Sigma X^2 - (\Sigma X)^2} \tag{5}$$

Use either one or the other for a or b, then solve for b or a as follows:

$$a = Y - bX$$

$$b = \frac{Y - a}{X}$$

The regression line can be plotted from any two pairs of values of Y_c and X. The computed values of Y are estimates, with the extent or degree of correlation determining how close to actual values these are.

Coefficient of Correlation. The degree of relationship, i.e., the degree of correlation, is designated by the letter r. Perfect correlation exists when the dependent variable is entirely dependent on the independent variable, so that changes in the independent variable cause the variations in the other. Such a correlation is signified as a correlation coefficient of 1.00. Also, if the dependent variable increases (or decreases) as the independent variable increases (or decreases), then there is a positive relationship. On the other hand, if one increases and the other decreases (or *vice-versa*), then the correlation is negative. Thus, values of r vary from 1.0 to -1.0. For industrial data, a coefficient of correlation, whether plus or minus, between .80 and 1.0 is considered as excellent, between .60 and .80 as good, between .30 and .60 as fair, and from 0 to .30 as of doubtful value. The coefficient of correlation should be used with caution. It is occasionally very easy to draw erroneous conclusions when using only this measure as a means of determining the relationship between variables. For this and other reasons, some of the present day statisticians discourage the use of this particular technique without further investigation of the data.

To calculate the coefficient of correlation, the following formulas can be used:

$$r = \sqrt{\frac{[a\Sigma Y + b\Sigma XY] - \bar{Y}\Sigma Y}{\Sigma Y^2 - \bar{Y}\Sigma Y}} \tag{6}$$

(The sign of r is the same as that of b, which is the slope of the line represented by $Y = a + bX$).

$$r = \frac{\overline{XY} - \bar{X}\bar{Y}}{\sigma_x \sigma_y} \tag{7}$$

Also,

$$r = \frac{n\Sigma XY - (\Sigma X)(\Sigma Y)}{\sqrt{[n\Sigma X^2 - (\Sigma X)^2][n\Sigma Y^2 - (\Sigma Y)^2]}} \tag{8}$$

or,

$$r = \frac{\Sigma(X - \bar{X})(Y - \bar{Y})}{\sqrt{\Sigma(X - \bar{X})^2 \Sigma(Y - \bar{Y})^2}} \tag{9}$$

The data shown in Table 9.1 represent the lengths of cotton fibers as measured by a cotton classer and the Fibrograph. To calculate the coefficient of correlation, use the formula (6) above.

The values for a and b come from the formula for the regression equation discussed on the following pages.

$$r = \sqrt{\frac{[(-0.00528)(112.17) + (1.03048)(117.1224)] - (1.06828)(112.17)}{120.177 - (1.06828)(112.17)}}$$

$$= 0.88$$

Since r takes the same sign as b, the positive value of the square root is taken and $r = +0.88$.

Another method for calculating the coefficient of correlation is by use of the relationship (7) above. Using the same data and formula (7), the calculation is as follows:

$$r = \frac{1.11545 - (1.0418)(1.06828)}{(0.0494)(0.0575)} = +\,0.89$$

since

$$\sigma_x = \sqrt{\frac{\Sigma X^2}{n} - \left(\frac{\Sigma X}{n}\right)^2} = \sqrt{1.08779 - (1.0418)^2} = 0.0494$$

and

$$\sigma_y = \sqrt{\frac{\Sigma Y^2}{n} - \left(\frac{\Sigma Y}{n}\right)^2} = \sqrt{1.1445 - (1.06828)^2} = 0.0575$$

TABLE 9.1

Cotton Fiber Lengths in Inches

Classer X	Fibrograph Y	X^2	Y^2	XY
1.06	1.07	1.1236	1.1449	1.1342
1.03	1.07	1.0609	1.1449	1.1021
1.06	1.06	1.1236	1.1236	1.1236
1.06	1.13	1.1236	1.2769	1.1978
1.06	1.13	1.1236	1.2769	1.1978
1.06	1.08	1.1236	1.1664	1.1448
1.06	1.12	1.1236	1.2544	1.1872
1.09	1.16	1.1881	1.3456	1.2644
1.09	1.13	1.1881	1.2769	1.2317
1.06	1.09	1.1236	1.1881	1.1554
1.06	1.11	1.1236	1.2321	1.1766
1.06	1.09	1.1236	1.1881	1.1554
1.03	1.09	1.0609	1.1881	1.1227
1.03	1.04	1.0609	1.0816	1.0712
1.03	1.08	1.0609	1.1664	1.1124
1.00	1.04	1.0000	1.0816	1.0400
1.06	1.05	1.1236	1.1025	1.1130
1.03	1.06	1.0609	1.1236	1.0918
1.06	1.09	1.1236	1.1881	1.1554
1.06	1.12	1.1236	1.2544	1.1872
⋮	⋮	⋮	⋮	⋮
Σ = 109.39	112.170	114.2184	120.177	117.1224
$\bar{X}$ = 1.0418	1.06828	1.08779	1.1445	1.11545
n = 105				

Regression Line. The regression line is the line of best fit with respect to the data under consideration.

One very common way of placing a line through a scatter

diagram (see Fig. 9.1) is to draw a line through the points using only the eye and judgment in deciding approximately where the line should be drawn. A more exact and accurate method is to

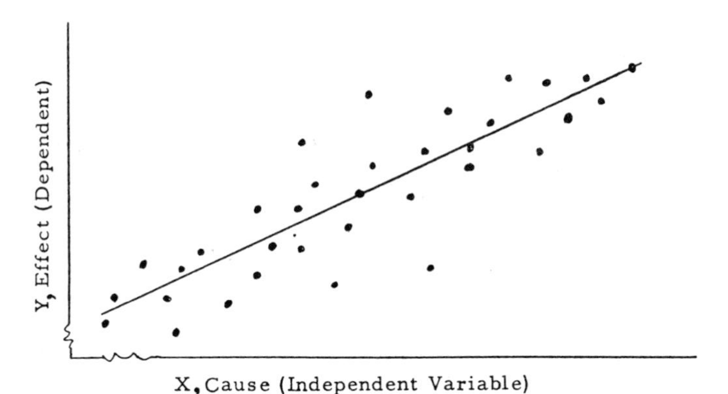

Fig. 9.1. Scatter diagram with regression line (line of best fit or trend line)

fit a line to the data by the least squares method, which means that the line is so placed that the sum of the squares of the distances of the points from the line are at a minimum.

To calculate the line, it must be decided first which of the variables is to be considered as dependent upon the other. For example, if the effect of moisture on yarn strength were being studied, it is obvious that the yarn strength, with all other things being equal, is dependent upon the amount of moisture present. Once the dependent variable has been established, the scatter diagram is plotted with the dependent variable always being plotted on the Y or vertical axis, as shown in Fig. 9.1.

With Y as the dependent variable, the equations (1), (2), and (3) given above are used to calculate the regression line, provided the relationship is linear.

Using again the data from Table 9.1 and the two equations $\Sigma Y = an + b\Sigma X$ and $\Sigma XY = a\Sigma X + b\Sigma X^2$, the substitution becomes:

$$112.170 = 105a + b109.39$$
$$117.1224 = 109.39a + b114.2184$$

To solve simultaneously,

$$\frac{109.39}{105} = 1.0418$$

$$\begin{aligned}
(1.0418)(112.170) &= (105)(1.0418)a + b(109.39)(1.0418) \\
(-)116.8587 &= (-)109.39a + (-)b113.9625 \\
117.1224 &= 109.39a + b114.2184 \\
\\
0.2637 &= 0.2559b \\
b &= 1.03048
\end{aligned}$$

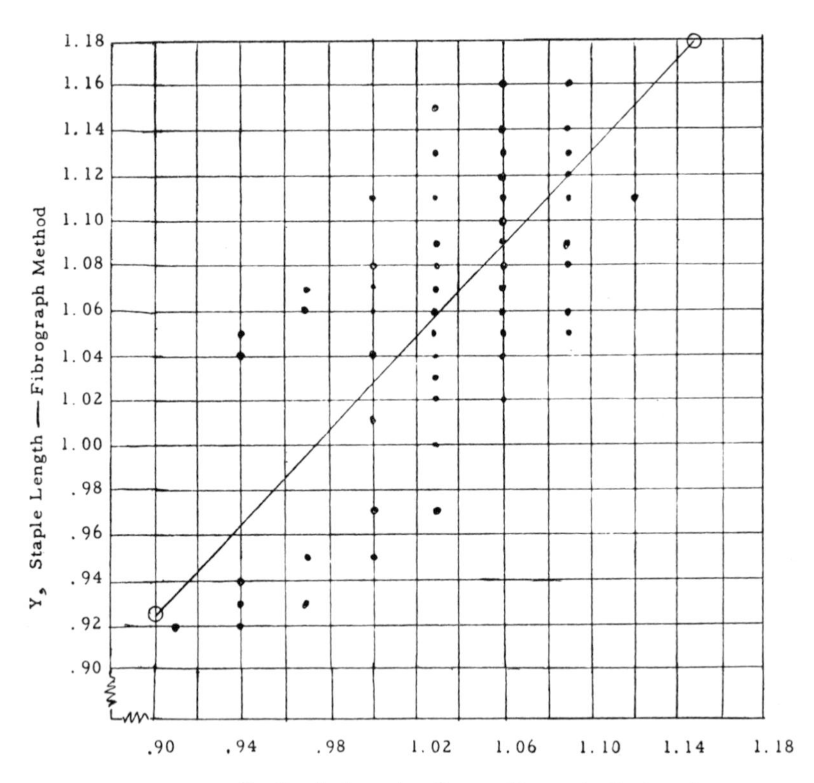

Fig. 9.2. Correlation of cotton classer's designation with fibrograph determination of cotton staple length.

$$Y = -0.00528 + 1.03048X$$
$$r = 0.89$$

To solve for *a*:

$$112.170 = 105a + (1.03048)(109.39)$$
$$105a = 112.170 - 112.7240$$
$$a = -0.00528$$

The equation for the line then becomes

$$Y_c = -0.00528 + 1.03048X$$

where Y_c is the calculated value of Y in terms of X.

To apply the line to the data, it is necessary first to plot the data in the form of a scatter diagram as shown on Fig. 9.2. After the scatter diagram has been plotted, the line can then be added. To do this, values must be established for both X and Y which satisfy the equation. For an X of 0.90, the corresponding value for Y_c would be

$$Y_c = -0.00528 + (1.03048)(0.90) = 0.922$$

And for an X of 1.15, the corresponding value of Y would be

$$Y_c = -0.00528 + (1.03048)(1.15) = 1.18$$

With these two points known, they are entered on the chart and a straight line drawn between the points.

Control Limits with Regression Line. At certain times it is advantageous to use three sigma limits in connection with the regression line when the process being investigated changes with respect to time or some other variable.

The values in Table 9.2 represent the neps formed by Deltapine 14, $1\frac{3}{32}$-inch Middling cotton with respect to time after stripping the card.

The first steps for this type of problem are identical to the previous example. Plot the scatter diagram, Fig. 9.3, for the data in Table 9.2 and make the necessary calculations.

Regression line:

$$\Sigma Y = na + b\Sigma X$$
$$\Sigma XY = a\Sigma X + b\Sigma X^2$$
$$1572.4 = 11a + b3150$$
$$554{,}271.0 = 3150a + b1{,}354{,}500$$
$$\frac{3150}{11} = 286.36$$

Multiply the first equation by 286.36, and the equations become

$$(-)450{,}278.12 = (-)3150a + (-)902{,}045.34b$$
$$554{,}271.0 = 3150a + 1{,}354{,}500b$$

$$10{,}399.29 = 452{,}454.66b$$
$$b = 0.2298$$

TABLE 9.2

Formation of Neps by Deltapine # 14 Cotton With Respect to Time After Stripping Card [a]

Mins. after stripping, X	Neps/gm Y	X^2	Y^2	XY
0	66.4	0	4,408.96	0
30	89.3	900	7,974.49	2,679.0
60	82.9	3,600	6,872.41	4,974.0
120	105.1	14,400	11,046.01	12,612.0
240	148.1	57,600	21,933.61	35,544.0
300	153.9	90,000	23,685.21	46,170.0
360	154.5	129,600	23,870.25	55,620,0
420	173.9	176,400	30,241.21	73,038.0
480	200.3	230,400	40,120.09	96,144.0
540	188.5	291,600	35,532.25	101,790.0
600	209.5	360,000	43,890.25	125,700.0
$\Sigma = 3{,}150$	1,572.4	1,354,500	249,574.24	554,271.0
$\bar{X} = 286.36$	142.95	123,136.36	22,688.57	50,388.27
$n = 11$				

[a] Data courtesy of Research Department, School of Textiles, N. C. State College.

Then, solving for a

$$1572.4 = 11a + (3150)(0.2298)$$
$$a = \frac{1572.4 - 723{,}87}{11}$$
$$a = 77.14$$

With $Y_c = a + bX$, the line represented by $Y_c = 77.14 + 0.2298X$ can be fitted to the scatter diagram.

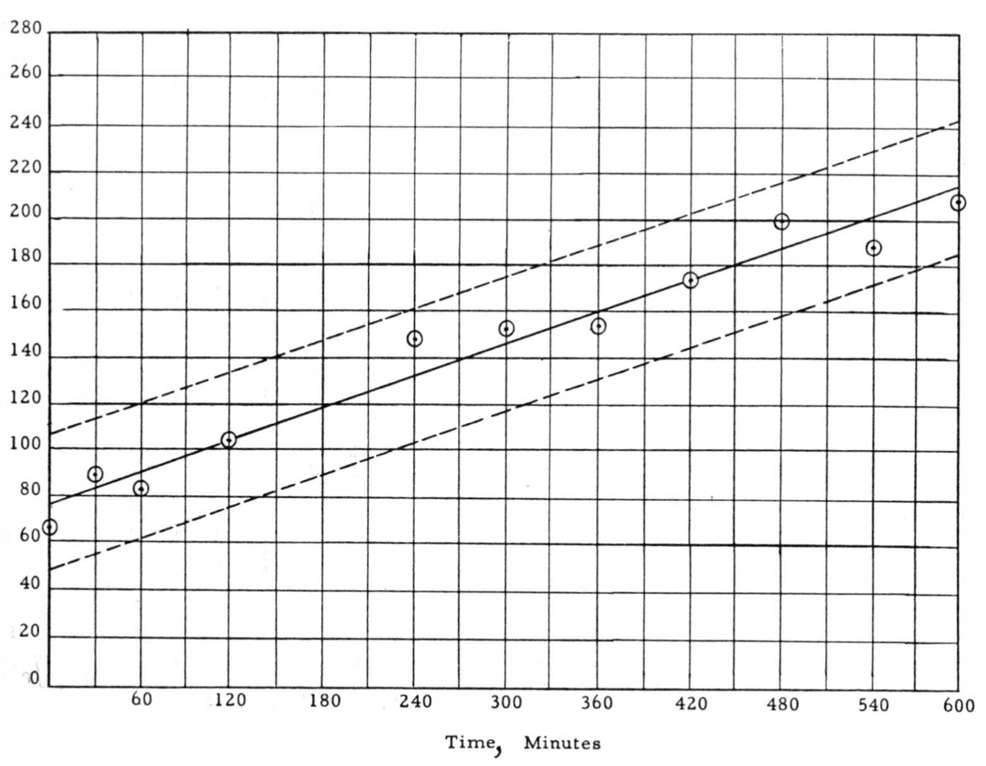

Fig. 9.3. Correlation of neps formed at carding with time after stripping for Deltapine No. 14 cotton.

$$Y = 77.14 + 0.2298X$$
$$r = 0.98$$

Coefficient of correlation:

$$r = \sqrt{\frac{(a\Sigma Y + b\Sigma XY) - \bar{Y}\Sigma Y}{\Sigma Y^2 - \bar{Y}\Sigma Y}}$$

$$= \sqrt{\frac{[(77.14)(1572.4)+(0.2298)(554{,}271)]-(142.95)(1572.4)}{249{,}574.74 - (142.95)(1572.4)}}$$

$$= \sqrt{0.9633} = 0.98$$

or,

$$r = \frac{\overline{XY} - \bar{X}\bar{Y}}{\sigma_x \sigma_y}$$

$$= \frac{50{,}388.27 - 40{,}935.16}{(202.8)(47.5)} = 0.98$$

With

$$\sigma_x = \sqrt{\frac{\Sigma X^2}{n} - \left(\frac{\Sigma X}{n}\right)^2}$$

$$= \sqrt{\frac{1{,}354{,}500}{11} - \left(\frac{3150}{11}\right)^2}$$

$$= 202.8$$

and

$$\sigma_y = \sqrt{\frac{\Sigma Y^2}{n} - \left(\frac{\Sigma Y}{n}\right)^2}$$

$$= \sqrt{\frac{249{,}574.47}{11} - \left(\frac{1572.4}{11}\right)^2}$$

$$= 47.5$$

The standard deviation, σ_{ys} used in establishing the control limits on the line of regression is calculated as follows:

$$\sigma_{ys} = \sqrt{\frac{\Sigma Y^2 - (a\Sigma Y + b\Sigma XY)}{n}}$$

$$= \sqrt{\frac{249{,}574.74 - [(77.14)(1572.4) + (0.2298)(554{,}271)]}{11}}$$

$$= \sqrt{82.57} = 9.1$$

or,

$$\sigma_{ys} = \sigma_y \sqrt{1 - r^2}$$

$$= 47.5 \sqrt{1 - (0.98)^2}$$

$$= 9.1$$

With the value σ_{ys} equal to 9.1, the three sigma limits, 3(9.1) or 27.3, may then be placed with the regression line as shown on Fig. 9.3, and the result is a statistical quality control chart.

TABLE 9.3

Neps Formed At Carding With Respect to Time After Stripping

Mins. after stripping	Neps/gm.
0	65.5
30	69.2
60	69.1
120	81.1
180	77.2
240	81.0
300	99.8
360	95.0
420	93.4
480	113.1
540	99.1
600	101.7

Problem Section

Practice Problem 1. From the data in Table 9.3, find: a. formula for regression line; b. coefficient of correlation (is the correlation excellent, good, or fair?); c. control limits. d. Plot data and enter regression line and limits.

CHAPTER X

Humidity and Moisture

Most textile fibers are hygroscopic; that is, they have the ability to absorb or give up moisture. This moisture is picked up or absorbed by the hygroscopic material from the atmosphere if the relative amount of moisture in the air is greater than that in the material. Conversely, the moisture will be given up by the material if the relative amount of moisture in the air is less than that in the material. These changes are due to differences between the vapor pressure of the atmosphere and of that within the material. When these vapor pressures approach equilibrium, as indicated by the equalization of the rate of exchange, then for practical purposes hygroscopic equilibrium has been attained.

Under natural conditions the amount of moisture in the atmosphere is continually changing; this results in varying the amount of moisture contained by a hygroscopic material exposed to this atmosphere. The result of this is a change in the physical characteristics of the material. For instance, cotton absorbs moisture rapidly when exposed to high humidity, and as a result, the weight of the material as well as its strength increases and other properties change. Linen shows substantial increases in strength as the moisture content is increased. Man-made fibers generally show reductions in strength with corresponding increases in elongation as their moisture contents are increased. These changes are high for viscose rayon and low on nylon and Dacron. Animal fibers show slight decreases in strength with increases in moisture content. Practically all textile materials show increased pliability and a greater immunity to static electrical influence with increases in moisture content.

Textile manufacturing operations are conducted to a large extent in a humidified atmosphere. Under ideal humidity conditions, the following advantages are realized in processing:

(*1*) Reduction in generation of static electricity. (*2*) Materials are more easily workable due to increased pliability. (*3*) Reduction in amount of dust and fly. (*4*) Allows for the retention of moisture already within the material. (*5*) Permits greater bodily comfort for personnel in cool weather.

Humidity

Humidity is the term used to describe the moisture existing in the atmosphere. Absolute humidity is the actual density of water vapor in the atmosphere; it is independent of temperature or degree of saturation of the air. It can be expressed as the weight in grains of water vapor in one pound of dry air or it may be expressed as grains of water vapor per cubic foot of air.

Relative humidity is the ratio of the weight of the existing vapor tc that of the maximum weight of vapor which the space could contain at the same temperature, expressed as a percentage. Air that is saturated with water vapor would have a relative humidity of 100 per cent. Air that is perfectly dry would have a relative humidity of 0 per cent. The ability of air to hold moisture is dependent upon its temperature. The higher its temperature, the more moisture it can hold. For example, a pound of dry air at 32°F can hold a maximum of 26 grains of water vapor; when it contains this amount, the atmosphere is saturated and it has a relative humidity of 100 per cent if it contains this amount of moisture. At 70°F a pound of dry air can hold 110 grains, and at 90°F it can hold 218 grains; in each case the relative humidity is 100 per cent if the air contains this amount of moisture.

If we assume a condition in which warm air has a high relative humidity, then by chilling that air a temperature is reached at which the air becomes saturated. In other words, further reduction in temperature would result in condensation of the water vapor from the air. This temperature, the dew point, can be defined as the temperature at which condensation of moisture would occur if the existing atmosphere were cooled without change in vapor pressure.

Measurement of Humidity

To measure the relative humidity in the atmosphere, a sling

psychrometer can be used. This is an instrument made up of two thermometers properly mounted on a metal frame attached to a handle. One of these thermometers is known as the dry bulb thermometer and records the dry bulb temperature. The second of the matched thermometers has its bulb covered with a wick which can be wet before use. This records the wet bulb temperature which is the temperature at which the atmosphere would become saturated by the evaporation of water.

The sling psychrometer is accepted as the most accurate instrument for measuring relative humidity. When used for calibrating other types of hygrometers, 12-inch or 15-inch psychrometers are preferred.

To make a determination of relative humidity with the use of the sling psychrometer, the first step is to moisten the clean wick of the wet bulb thoroughly. The wet bulb extends beyond the dry bulb so the dry bulb will be maintained dry and so the wet bulb can be wet easily by dipping it in a small bottle of distilled water at room temperature. The wick must be free from any foreign matter and should not be handled, and water should be distilled because of the cooling effect of salts and impurities in tap water. The instrument is swung preferably in a vertical plane by holding onto the handle and allowing the frame to swing through the iar. It should be noted that the thermometers are mounted so that they act independently and both face the air current. The speed of the whirling should be such as to give a speed of 15 ± 5 feet per second to the wet bulb.

The wet bulb should be read after swinging for about 30 seconds. Swinging should be repeated for approximately 20 seconds and a second reading taken. This should be repeated until equilibrium of the wet bulb is reached, which will be indicated by two successive similar readings. The wet bulb should be read first and promptly, and then the dry bulb without contact of the hands with the bulbs.

Charts have been prepared for converting the readings of wet and dry bulb temperatures to relative humidity. One type of chart shows the dry bulb temperature readings in a vertical column and the difference between the dry and wet bulb temperatures at the top across vertical columns which show the relative humidity. See Appendix I.

Hygrometers. A hygrometer is any properly calibrated instrument used for determining the moisture condition of the air. The sling and recording psychrometers are the two most common textile types of hygrometers.

For a continuous record of humidity, various types of recording hygrometers or psychrometers are used. Some of these have provision for maintaining a definite movement of air across the sensitive elements of the thermometer and are equipped with automatic clocking and recording devices capable of making a continuous chart of time, wet and dry bulb temperatures, or relative humidity.

Standards. The standard atmospheric test condition for nearly all textile materials is:

Dry bulb	70°F ± 2°F
Relative humidity	65 % ± 2 %

For testing under mill operating conditions, it is more important to keep the relative humidity at 65 per cent than it is to maintain standard dry bulb temperature, inasmuch as materials are affected

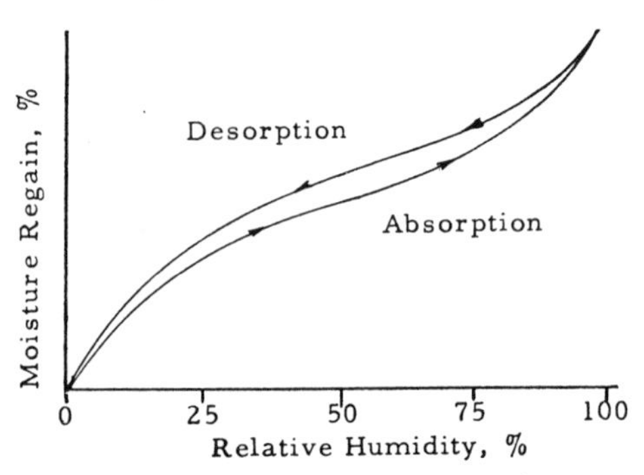

Fig. 10.1. Typical moisture hysteresis curves.

to a much greater degree by humidity than by temperature. Materials to be tested are conditioned in an atmosphere of standard relative humidity and temperature until equilibrium is reached. Equilibrium may be considered to have been obtained when there is no progressive change in weight with successive weighings. On heavy woven cotton fabrics equilibrium should be reached in four to five hours. There are some variations in the moisture equilibrium specifications for different types of textiles, but the above statement will cover most conditions. It is generally accepted that equilibrium should be approached from the dry side. The reason for this is that most textile materials will pick up less moisture at any given atmospheric condition if it is in a drier state initially. This is illustrated in Fig. 10.1. The absorption curve shows the trend of the amount of moisture that will be picked up by a material if originally dry. The desorption curve shows the amount of moisture remaining in the material if originally in a wet state. For any given humidity, the moisture regain will be higher for desorption than absorption.

Moisture in Materials

There are two ways of describing the amount of moisture in a textile material. The moisture content is the per cent weight of moisture based on its original *As-Is* state.

Moisture regain is the per cent weight of moisture based on the oven-dry weight of the material. The regain of a material is considered to be the amount of moisture at the time of test.

$$\text{Per cent of moisture content} = \frac{\text{Original weight} - \text{dry weight}}{\text{Original weight}}(100)$$

$$\text{Per cent of moisture regain} = \frac{\text{Original weight} - \text{dry weight}}{\text{Dry weight}}(100)$$

The following example illustrates the two calculations:

Weight of specimen in room *As Is* = 105 grains

Oven-dry weight = 100 grains

$$\text{Per cent moisture content} = \frac{105-100}{105}(100) = 4.75\%$$

$$\text{Per cent regain} = \frac{105-100}{100}(100) = 5.0\%$$

"Reference regain is an arbitrary regain decided upon by the parties to an agreement" (A.S.T.M. Committee D-13)

Commercial regain is a standard adopted for business purposes and is generally very close to the regain of a material exposed to standard test conditions.

TABLE 10.1

Regain Values [a]

Fiber	Regain %	
	Approx. at standard conditions	Accepted commercial regains
Cotton, raw	8.5	
Cotton, yarn (common basis)	7.0–7.5	
Silk	11.0	11.0
Wool	16.0	
Wool tops (with oil)	19.0	
Woolen yarn, carded	17.0	13.0
Worsted yarn, Bradford wet spun		13.0
Worsted yarn, French dry spun		15.0
Wool and worsted fabrics	16.0	
Linen	12.0	
Jute	13.8	
Viscose rayon	13.0	11.0
Cuprammonium rayon	11–12.5	11.0
Acetate rayon	6.0	6.5
Nylon	3.8–4.0	4.5
Glass	less than 0.5	
Vicara (Azlon)		10.0
Acrilan (polyacrilic)	1.7	1.5
Orlon (polyacrilic)	1.0–2.0	1.5
Dacron (polyester)	0.4	0.0
Dynel	less than 1	0.0
Vinyon	0.0–0.3	0.0

[a] From various sources

Standard regain is the regain of a material obtained under standard test conditions when approached from the dry side.

Some of the man-made textile fibers have relatively high regains, whereas others resist the absorption of water. Fibers that accept moisture readily are classified as hydrophilic materials, and those that do not are classified as hydrophobic materials.

In Fig. 10.1 no values have been entered for the regain per cent, as relative values are hypothetical. For cotton at room temperature the regain at 100 per cent relative humidity would be approximately 23 per cent and for wool, approximately 33 per cent.

Many fabrics and yarns are made from mixtures of more than one type of fiber with the result that there is a change in the standard or commercial regain of the mixture. The commercial regain of such a mixture can be calculated by knowing, first, the percentage of each type of fiber in the mixture and, second, the commercial regain of each type of fiber (see Table 10.1):

$$\text{Regain \% of mixture} = \frac{(\%A)(R_a)+(\%B)(R_b)+\ldots}{\%A+\%B+\ldots}$$

where $\%A$ = per cent of first type of fiber; $\%B$ = per cent of second type of fiber; R_a = commercial regain per cent of first type of fiber; R_b = commercial regain per cent of second type of fiber.

For example, a yarn is made up of 40 per cent viscose and 60 per cent acetate. What is its regain per cent?

$$\text{Mixture regain} = \frac{(40)(11.0)+(60)(6.5)}{40+60} = 8.30\ \%$$

In those cases where there is no established commercial regain, a suitable value based on actual regain should be accepted by mutual consent.

The regain of a yarn composed of two different fibers, such as viscose and acetate rayons, can be obtained in another way by the use of Fig. 10.2. As shown in the diagram, for any percentage of two fibers (equalling a total of 100 per cent), the regain per cent for the composite material can be found easily. In the example shown, point A is the 11 per cent commercial regain for viscose rayon, and B is the $6\frac{1}{2}$ per cent value for acetate type of rayon. For a blend containing 40 per cent of viscose and 60 per cent of acetate, follow the vertical line from the 40 per cent value to *C*, and opposite this on the vertical axis can be read the average regain value, 8.3 per cent in this case.

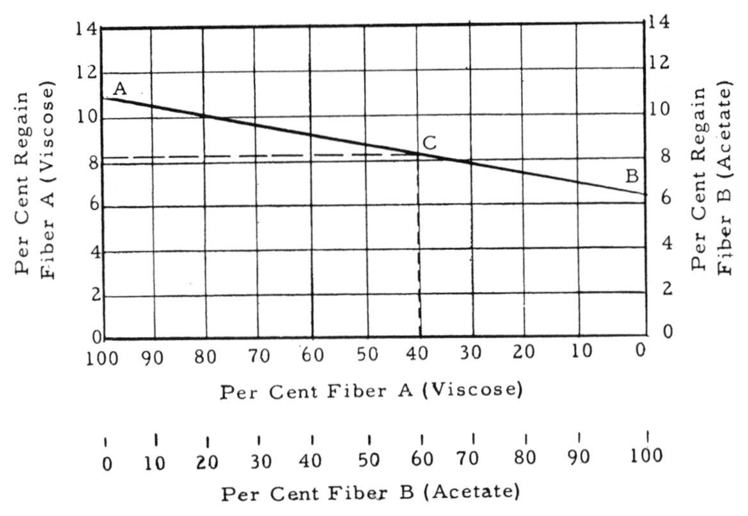

Fig. 10.2. Mixture regain of a mixture.

Drying Ovens

The laboratory determination of the amount of moisture in a specimen is accomplished by drying in any one of the specially designed laboratory ovens. The Emerson oven is a semi-automatic electrical unit provided with thermostats to control the heat generated by the electrical heating units; a small fan exhausts air to control temperature. Specimens can be placed in small cages calibrated for weight and which can be suspended from rotatable crosspieces which are under external control. A fine chemical balance is attached permanently on top of the oven and is equipped with connecting wire links so that the cages within the oven can be weighed without their removal. The rotatable crosspieces facilitate the positioning of the cages for weighing. Thus, the moisture determinations can be made without removing the samples from the oven.

In testing for the moisture in textiles, the Brabender oven has also proven very satisfactory because of its semi-automatic features.

The equipment consists of two units: First, a special weighing-in scale for weighing the 10-gram samples used (the semi-automatic feature depends on an initial sample of 10 grams); and secondly, the oven with sample dishes.

The weighing-in scale is accurate to plus/minus 0.002 grams. It is calibrated so that the weight indicator points to zero with the 10 grams of sample on the scale pan; deviations of sample weight above or below the standard show as plus or minus.

The oven itself has ten sample stations, so that ten samples can be tested simultaneously. For each station there is a small perforated basket $1\frac{1}{4}$ inch high capable of holding the 10-gram sample; these baskets are all of the same weight. These baskets can be inserted through a small door in the tester; as each sample is inserted, a round platform within the tester is rotated a notch by an external handwheel.

Electric drying elements supply the heat for drying out samples; a forced draft with vents provides the circulation of heated air to dry samples quickly. The temperature within the oven can be controlled accurately over a range of from 115°F to 320°F (by an electrical contact thermo-regulator which has a sliding contact point movable from the outside by a magnet). Accuracy of temperature is about $\frac{1}{4}$°F. In starting a test, i.e., in first heating the oven, it takes about ten to fifteen minutes to reach the controlled temperature. After the oven is once heated, open textile materials can be dried out in twenty-five to thirty minutes.

After the samples are dried, the turntable is rotated to the desired sample position for test, and by movement of a lever outside the oven, three prongs lift the selected basket for weighing. The built-in weighing device is enclosed in a glass chamber. The accuracy is 0.002 gram; the balance is oil-dampened. The weighing shows on an illuminated target type dial, and shows the loss of weight, not in grams, but in per cent. Thus, the scale reads in per cent moisture content. For example, for a 10-gram sample losing 1 gram of weight due to moisture, the scale would read 10 per cent; if 2 grams loss, it would show 20 per cent, etc. When turning on the switch to illuminate the dial, the forced draft is cut out, thus preventing any distortion due to air currents.

Divisions on the scale are marked at each 0.2 per cent, but readings can be taken between calibration lines.

The recalibration of the weighing device, if once it is out of calibration, is a fussy job requiring painstaking care and experimentation. However, the instruments should be calibrated over the entire range of moisture contents being tested. Thus, the weighing device might be in perfect balance at 2 per cent moisture, and out substantially at 6 per cent moisture. Calibration, if attempted, should be done with accurate dead weights.

Inasmuch as most textile moisture determinations are expressed as moisture regain rather than moisture content, the values from the Brabender should be corrected to regain. The formula for this is:

$$\text{Moisture regain, }\% = \frac{(100)(\text{moisture content, }\%)}{100 - \text{moisture content, }\%}$$

In testing for textiles, oven temperatures should be set at 220° – 230°F as recommended by A.S.T.M. It will be found that the rate of drying is very rapid at the outset; however, drying should be continued until equilibrium is obtained.

Corrections for Regain. With changes in the actual regain of textiles, there are corresponding changes in weight. Knowing what the actual regain of a specimen is, it is possible to calculate the weight corrected to a standard regain.

In the case of cotton yarn, the correction formula is as follows:

$$N = \frac{n(100 + R_a)}{100 + R_s}$$

in which N = the yarn number corrected to standard (or reference regain); n = the yarn number at regain R_a; R_a = actual regain at time of test; R_s = standard (or reference) regain.

NOTE. The above formula applies to the indirect system of numbering, in which yarn number is a function of length/weight. For the direct system of numbering, the formula would take the form:

$$N = \frac{n(100 + R_s)}{100 + R_a}$$

For example, if the average size of a cotton yarn is 42.0, and if a moisture determination test reveals that the regain of the yarn was 4 per cent at the time of testing, then the yarn size corrected to commercial standard regain of 7 per cent is:

$$\text{Corrected size} = (42.0)\frac{(100 + 4)}{(100 + 7)} = 42.0\left(\frac{104}{107}\right) = 40.8$$

Another standardized procedure is the calculation of corrected commercial weights of various textile yarns and threads. The method, somewhat similar to that shown above, is shown by the following derivation:

Let A = the net weight of a test specimen; D = the oven-dry weight of the same specimen; R_c = the commercial regain of the fiber tested; R_a = the actual regain of the specimen; W = the gross weight of the material including the container, as received; t = the tare weight of the container, paper, spools, tubes, etc.

$$\text{Corrected weight} = \text{Oven-dry weight}\left(\frac{100 + \text{commercial regain}}{100}\right)$$

$$= D\left(\frac{100 + R_c}{100}\right)$$

The commercial regain would be set according to the tabulation given in Table 10.1. The actual regain is

$$R_a = \frac{100(A - D)}{D}$$

For an entire lot, the commercial or corrected weight is calculated as follows:

$$\text{Commercial weight} = \frac{(W - t)(100 + R_c)}{(100 + R_a)}$$

The equation can be simplified as shown herewith:

$$\text{Commercial weight} = \frac{(W - t)(100 + R_c)}{100 + \dfrac{100(A - D)}{D}}$$

$$= \frac{(W - t)(100 + R_c)D}{100D + 100A - 100D}$$

$$= \frac{(W - t)(100 + R_c)D}{100A}$$

(The last formula follows A.S.T.M. specifications for correction of commercial weights.)

If skeins of yarn are to be corrected to a common basis of commercial regain, the following formula is used:

$$\text{Weight corrected to commercial regain} = \frac{(\text{Actual skein weight})(100 + R_c)}{100 + R_a}$$

where R_c is the commercial regain.

Corrections can be made, also, for the strength of yarns or fabrics where the conditions under which the tests are performed are not standard. It is necessary, of course, to determine the actual regain of the specimens at the time of test, inasmuch as the corrections depend on regain. For cotton yarns and fibers, the formula for correcting the observed strength at any regain to the strength at standard (or commercial) regain is:

$$S_c = (Sm)\frac{(100 + fR_s)}{(100 + fR_a)}$$

where S_c = corrected strength; S_m = observed machine reading or average; R_s = standard regain; R_a = actual regain; f = correction factor (approximately 6 for cotton yarns).

It will be noted that the observed strength is corrected by a ratio, which in turn depends upon regains and the correction factor. This correction factor can be established accurately for any yarn or fabric by the following method.

Several series of strength tests are made of the material in question over a wide range of moisture regains. The results of these tests are plotted as shown, for example, by the curve *XY* in Fig. 10.3. A straight line, *AB*, is drawn tangent to the curve *XY* at the point where the rate of change is constant. Ordinarily it will be found that the rate of change is constant between the limits of 3 per cent and 9 per cent regain. The slope of the line *AB* is such that for a change from 0 to 10 per cent regain, the strength increases from 70 to 112 pounds, a change of 42 pounds, or a change of 4.2 pounds for each per cent change in regain. At 0 per cent regain, the assumed strength is 70 pounds. Therefore, based on this oven-dry assumed strength, the correction factor becomes 4.2×100 divided by 70, or 6 per cent. This factor is the per cent change in strength for each per cent change in regain, in other words, a 6 per cent change in strength for each per cent change in moisture regain. However, this should only be used in the normal range of regains and not to be used for extremely low or high humidities.

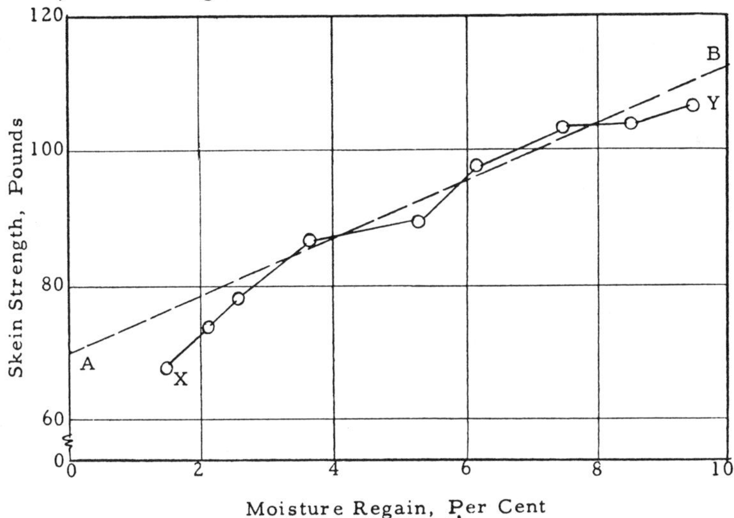

Fig. 10.3. Effect of regain on the skein strength of cotton yarn.

As an example, suppose skeins of yarn are tested in a mill laboratory that is humidified but not air conditioned. On a particularly damp day the relative humidity would be high, and consequently the yarn size heavy and strength high if compared to results taken under standard conditions. On this particular day, skeins of 20/1 yarn are tested for strength, yarn size, and then oven-dried. The data obtained are as follows:

Skein strength, S_m	110 lbs.
Yarn number, n	19.4/1
Yarn sample weight, *As Is*, A	41.23 grains
Yarn sample weight, dry, D	37.48 grains
Standard reference regain, R_c	7–1/2 %
Strength correction factor, f	6 %

Find the actual regain (R_a), the corrected yarn number (N), and strength corrected to standard reference regain (S_c).

$$\text{Actual regain, } R_a = \frac{41.23 - 37.48}{37.48}(100) = 10.0\,\%$$

$$\text{Corrected yarn number, } N = 19.4\left(\frac{100 + 10.0}{100 + 7.5}\right) = 19.85/1$$

$$\text{Corrected skein strength, } S_c = \frac{110(100+6\times 7\text{–}1/2)}{(100+6\times 10)} = 99.7 \text{ pounds}$$

CHAPTER XI

Cotton Staple Length and Grade

General Aspects

The raw material used by the cotton textile industry, the cotton fiber, is one of the most variable materials used by any industry. The properties of cotton fibers vary for all different varieties of cotton, for different growth areas, for different climatic conditions, and from year to year. In the strictest sense, only a semblance of control can be maintained over these variables during the growing of the fibers even under ideal conditions. With all of these as well as other variables introduced into any one crop, the problem of selecting the proper cotton for textile processing can be complex.

To aid the mill in the selection of a particular cotton to be used for a specific purpose and also to aid in predicting the performance of the fibers during processing as well as in the finished product, a series of tests have been developed which are in more or less common use throughout the cotton spinning industry.

Some of these tests are performed by the United States Department of Agriculture. The results of these tests are available for general use.(1) They are published as a series of supplements and then in final summary form for each crop year. The purpose of these series of test results is to keep the industry informed as to the quality, fiber properties, and processing characteristics of some varieties of cotton grown in different areas and harvested at different times. By use of these reports, it is possible to decide with reasonable success what particular variety or varieties of fiber are needed to meet a given set of conditions, and this knowledge can be of great assistance to the individual responsible for selecting the cotton to be purchased.

Once the proper cotton has been selected, it is then the responsibility of the cotton classer and the quality control laboratory to measure the quality of the cotton purchased. To be assured that the quality level is maintained, there are several tests which are available, the most important of which are: staple, grade, and character, which are measured by the cotton classer, and fiber length, fineness, maturity, and strength, which are measured in the quality control laboratory. In addition, greater emphasis is being placed now on waste analysis, nep potential, and color.

The frequency with which any or all of the laboratory tests is made depends entirely upon the particular mill and the manner in which the results are to be used. Some mills use the test results as a means of selecting cotton to be purchased, whereas others use the results to measure the quality of cotton already purchased so that blends and mixtures can be more intelligently established and processing difficulties anticipated before the fibers are actually processed.

For mills to use fiber testing as a basis for purchasing cotton, it is first necessary to determine exactly what type of fibers are needed to manufacture a given end product. This can be done first by consulting the U. S. Department of Agriculture reports and purchasing those cottons which most nearly meet the requirements, and secondly, by keeping a complete set of records as to fiber properties, and processing performance for the varieties used. With this information, it is then possible to establish the minimum requirements for the cotton. Once these standards have been established, they can thereafter be used as a guide or set of specifications by which varieties may be selected and purchased.

With sufficient background of fiber test results, it is possible to predict processing difficulties which may be expected with a particular type or group of fibers, as well as the anticipated quality level of yarn and fabric. For the production department this type of information and assistance can be of great value. It is also possible to predict within normal limits the yarn strength which can be expected from a shipment or batch of cotton if the fiber properties are known and sufficient data are available on the strength and performance of previous tests.

The measuring of fiber properties should be done at standard conditions [1] or in an atmosphere where the temperature and relative humidity are kept at a constant level. Variation of moisture content in the cotton fiber will affect the test results. For example, cotton tends to staple shorter when dry than when wet; it appears to be coarser when wet than when dry; and where the fibers must be weighed, the weight will vary with the moisture content. Moisture also affects the strength of cotton. The higher the moisture content, the stronger the fibers will appear to be. Therefore, for these reasons, it is much more satisfactory to test in a controlled atmosphere whenever possible.

The definition of staple as quoted from *Handbook for Licensed Classifiers* is, "The length by measurement of a selected portion of the fibers, which, although every sample contains fibers of many different lengths, by custom is assigned to a sample or bale as a whole."

Practically every bale of cotton used by the textile industry is sampled several times: by the gin for their use in selling and for state agricultural evaluation; by the merchant for his use, or for his customer's evaluation; by the mill after purchase to verify the quality. The evaluation of the bale is made from the small samples taken from the bale where each sample is a composite one made up of one half from the top and the bottom of the bale.

The determination of the staple length, grade, and character of the bale is made from the samples by the cotton classer, whether he represents the gin, the merchant, or the mill. These same samples are used by the merchant or the spinner for his laboratory analysis of the cotton.

The determination of staple by a cotton classer is more of an art than a science. However, science has yet to replace this method or technique of measurement, and it is a universal language understood throughout the world and used as one of the bases for establishing the price of the fiber. To aid the classer in establishing staple and grade of cotton, official standards are available

[1] Where standard conditions are specified, they refer to a dry bulb temperature of 70°F and a relative humidity of 65 per cent.

from the U. S. Department of Agriculture. There are official standards for both staple and grade, and in addition to being used as an aid to the classer, they are also used in case of arbitration.

The accurate determination of staple length is important not only from a financial point of view, but also from the standpoint of processing and quality. With all other things being equal, the longer the staple length, the stronger the yarn. Therefore, it is necessary that an accurate estimate of the staple length be made so as to assure a finished product of sufficient strength and uniform quality. If a staple of insufficient length is used, it will result in a weak, uneven yarn. On the other hand, if a staple length is used which is longer than the standard, it will mean that the cost of the raw material is greater than necessary to produce a given quality, and the margin of profit on the material will be reduced or completely eliminated. In addition to this, fiber breakage will occur during drafting and the quality of the yarn is likely to be reduced.

Influence of Staple and Yarn Number on Skein Strength

Some of the relationships of staple length to skein strength are shown in Figs. 11.1, 11.2, 11.3, and 11.4. The data used in plotting these charts were taken from the U. S. Department of Agriculture report, "Summary of Fiber and Spinning Test Results for Some Varieties of Cotton Grown by Selected Cotton Improvement Groups, Crop of 1951." Since the data are not representative of all cottons grown and are for only a period of one year, the actual values should be used with caution since they may not be directly applicable to any particular situation. However, the relationships discussed are generally true, and most cottons can be expected to perform in the same general pattern as shown in these charts.

Fig. 11.1 shows the relationship of skein strength to yarn number for some of the common staple lengths. The skein strength is that of a 120-yard length with 80 $1\frac{1}{2}$ yard wraps; thus, 160 ends are under test simultaneously on a pendulum-type tester. Curves of this shape are characteristic of this relationship.

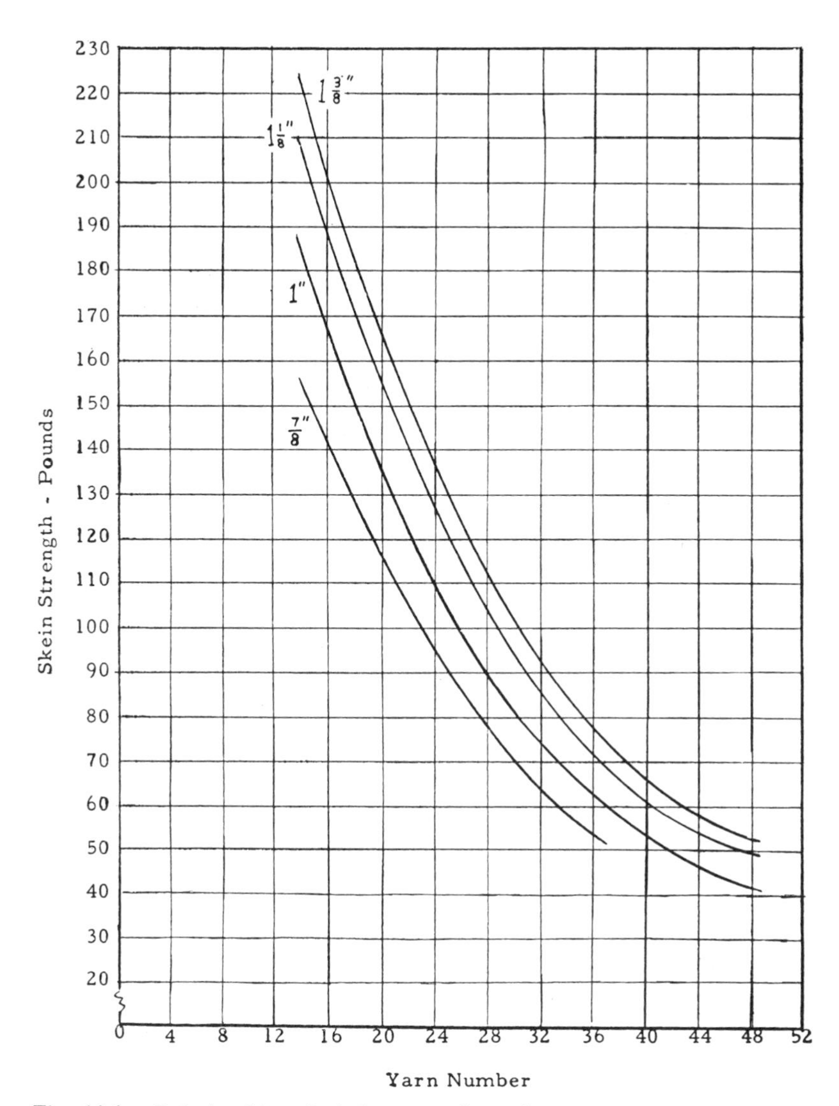

Fig. 11.1. Relationship of skein strength and yarn number for carded yarns made from different staple length cottons.

Fig. 11.2 shows the relationship of count-strength product and yarn number for different staple lengths. The slope of these lines indicates some interesting relationships. As can be seen, the

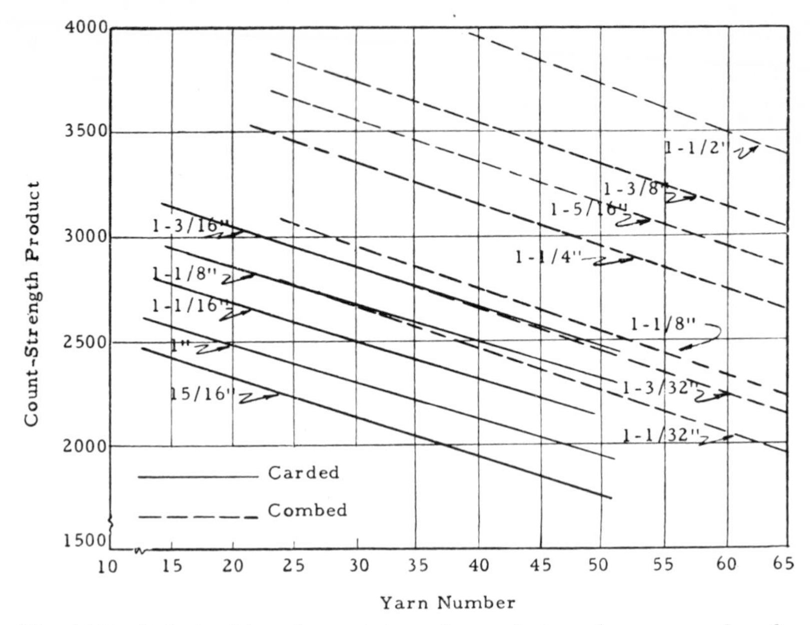

Fig. 11.2. Relationship of count-strength product and yarn number for different staple lengths.

TABLE 11.1
Amount of Change in Count-Strength Units for a Change of One Yarn Number for Carded Yarns [a]

Staple length (ins.)	Count-strength units (slope)
$\frac{13}{16}$	25.1
$\frac{7}{8}$	23.1
$\frac{15}{16}$	23.3
1	21.2
$1\frac{1}{16}$	21.3
$1\frac{1}{8}$	17.4
$1\frac{3}{16}$	17.3
$1\frac{1}{4}$	15.9
Average	21.7

[a] The data in this table are based on more than 70,000 individual skein tests representing 425 different cottons.

slope for the carded yarns is approximately the same (18.3) regardless of the staple length used. As a matter of fact, the slopes are probably somewhat closer than would be found if a large number of tests were made. Nevertheless, they would be in the approximate range shown in the figure. The relationships of staple length, yarn number, and count-strength product are discussed in a U. S. Department of Agriculture Circular (2). One of the results of this extensive study was the slope of the lines for different staple cotton. These are shown in Table 11.1.

These results show a variation from 15.9 to 25.1 count-strength units for staple varying from $1\frac{1}{4}$ to $\frac{13}{16}$; however, as pointed out by Campbell in the following example, this only makes a difference of 0.5 pounds in the skein strength which is well within the experimental error normally expected in skein testing.

$$S_2 = \frac{C_1 S_1 - (C_2 - C_1)21.7}{C_2}$$

where C_1 = Observed yarn count (English Cotton System) 20.75/1

C_2 = Specified or "nominal" count 22.0/1

S_1 = Observed skein strength . . . . 110.0 lbs.

S_2 = Corrected skein strength . . . . Unknown

$$S_2 = \frac{(20.75)(110) - (22 - 20.75)(25.1)}{22}$$

$$= 102.3 \text{ lbs. (using highest value of slope} = 25.1)$$

$$S_2 = \frac{(20.75)(110) - (22 - 20.75)(15.9)}{22}$$

$$= 102.8 \text{ lbs. (using lowest value of slope} = 15.9)$$

The formula used above for correcting skein strength to a specified number or for estimating the strength which could be expected from another yarn number of the same construction and from the same cotton was developed by Campbell as a result of the study reported in the circular (2). The explanation of the formula is rather simple. If there is an average change of 21.7 count-strength units for each change of one yarn number, this relationship can be expressed as follows:

$$C_2 S_2 = C_1 S_1 - (C_2 - C_1)(21.7)$$

For example, if the count-strength product for a skein of 20/1 yarn is 2000, the count-strength product of a 21/1 yarn would be

$$2000 - (21 - 20)(21.7) = 2000 - (1)(21.7) = 1978.3$$

If the count-strength product for a 19/1 is desired, the substitution would be the same, and the problem would then become

$$2000 - (19 - 20)(21.7) = 2000 - (-1)(21.7) = 2000 + 21.7 = 2021.7$$

Since the skein strength is usually the value desired it can be calculated by rearranging the formula as follows:

$$S_2 = \frac{C_1 S_1 - (C_2 - C_1)(21.7)}{C_2}$$

It will be found that the count-strength correction value of 21.7, although yet recommended by A.S.T.M., is somewhat high, and the value 18.3 is suggested for average use; the reason for this lies in differences in processing methods, increased information, and new varieties of cotton available since the method was first established. Another value for this correction factor is 17.64(3).

The preceding discussion and formulas are all based on the assumption that the relationships are linear. Actually, the relationships are very slightly curva-linear; however, the curve is slight and for all practical purposes it can be considered as linear over a short segment of the line. The above formula should only be used to correct skein strengths over a small range of yarn numbers, but it may be used to estimate skein strengths over a wide range of numbers, provided the corrections are within the spinning limits of the fibers involved.

Figs. 11.3 and 11.4 show the relationship between staple length and strength for some common yarn numbers. Fig. 11.3 shows the effect of staple length on the yarn skein strength. By examining Fig. 11.4, it can be seen that the combed yarns are more sensitive to staple length change than carded yarns. The lines representing carded yarns have a slope of approximately 91, whereas the combed yarns have a slop of 100: i.e., the count-strength product

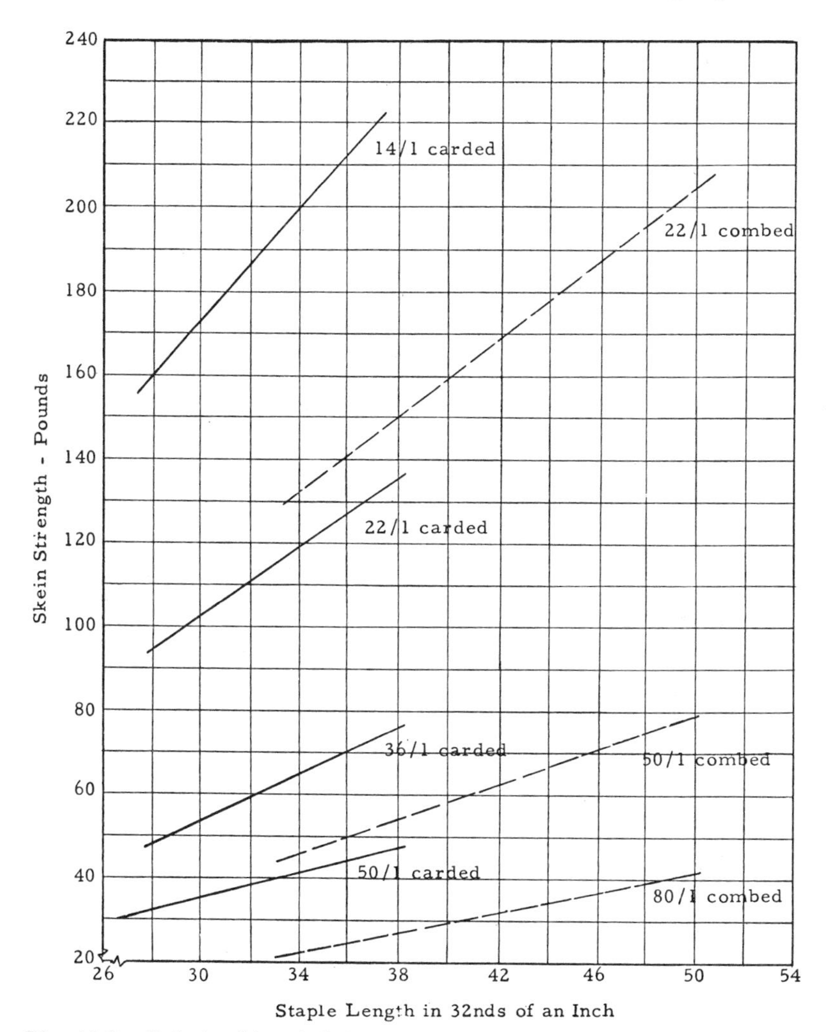

Fig. 11.3. Relationship of skein strength to staple length for carded and combed cotton yarns.

for carded yarn changes 91 units for each 32nd change in staple length, and for combed yarn the change is 100.

To calculate the estimated skein strength for a carded yarn pro-

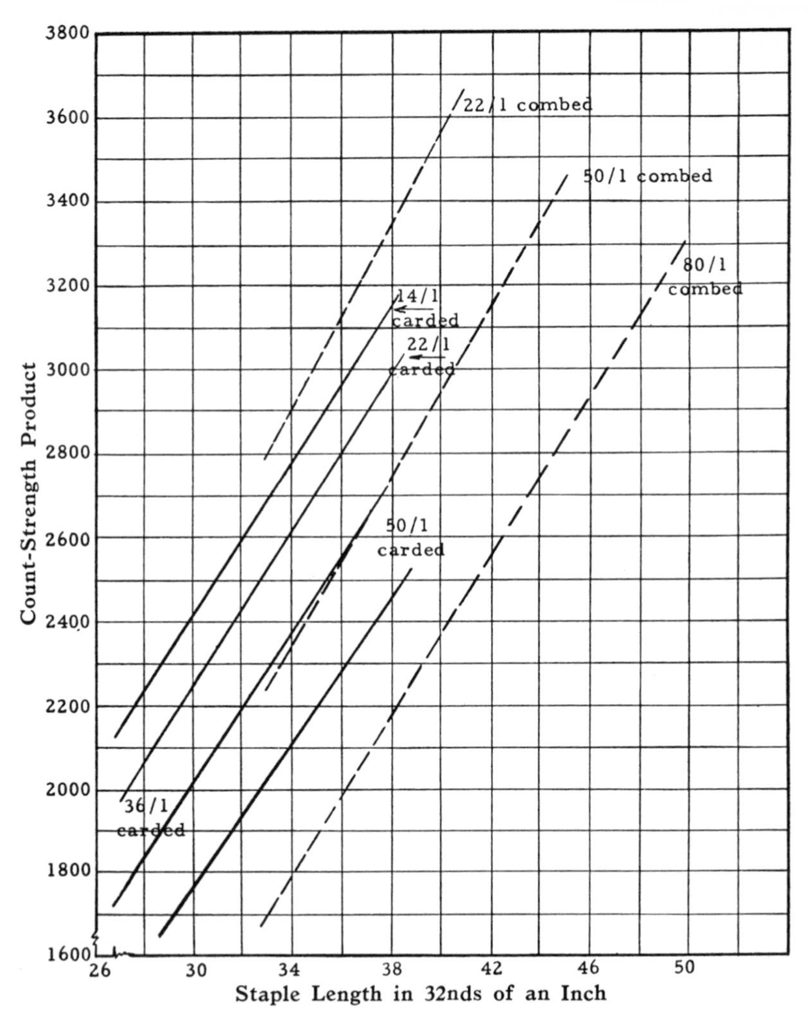

Fig. 11.4. Relationship of count-strength product and staple length for some common yarn numbers.

cessed on the long draft system, the following formula may be used:

$$S = \frac{91.49(L) - 18.27(C) - 70.3}{C}$$

where L = staple length in 32nds of an inch; C = yarn number; S = estimated skein strength.

For example, if the skein strength is to be estimated for a 20/1 yarn made from $1\frac{1}{8}$-inch cotton, the calculation would be:

$$\text{Estimated skein strength} = \frac{91.49(36) - 18.27(20) - 70.3}{20} = 143$$

(In the formula, 18.27 represents the count-strength correction value.)

If it is desired to determine the strength of a particular variety relative to the strength that should be obtained on an average as a result of years of testing experience, then the comparison is made on the basis of the above formula, and a *yarn strength index* is obtained. The following example illustrates this application.

A Rowden cotton of $\frac{15}{16}$-inch staple length ($= \frac{30}{32}$nds) has the following skein strengths:

14s	154 pounds
22s	91 pounds

Determine the yarn strength index. (Use preceding formula.)

$$\text{Est. standard strength, } 14s = \frac{91.49(30) - 18.27(14) - 70.3}{14} = 172.8$$

$$\text{Est. standard strength, } 22s = \frac{91.49(30) - 18.27(22) - 70.3}{22} = 103.3$$

The strength index is equal to

$$\text{Index \%, } 14s = \frac{154}{172.8}(100) = 89.1$$

$$\text{Index \%, } 22s = \frac{91}{103.3}(100) = 88.1$$

$$\text{Average strength index} = 88.6 \text{ (Use 89)}$$

In this example, 14/1 and 22/1 yarns have been used, for these are two of the standard yarns used in the variety test reports prepared by the U. S. Department of Agriculture. (Other yarns are 50/1 and 80/1.) Furthermore, in these reports a yarn strength index is listed for each variety tested, and this index is similar to the one calculated above.

A variation of the above offers a method of determining the "equivalent staple length" which is defined as the performance of a cotton of given staple compared with the performance of an average cotton of the same staple.[1] For example, if a $1\frac{1}{4}$-inch cotton when spun into a 20/1 yarn gives a count-strength product of 2450, to what staple length is that equivalent in an average cotton?

$$L = \frac{SC + 18.27(C) + 70.3}{91.49}$$

$$= \frac{2450 + 18.27(20) + 70.3}{91.49} = 31.5$$

Expressed in 32nds of an inch, this gives a skein strength equivalent to that of a $\frac{31}{32}$ to a 1-inch cotton of average quality, processed on the long draft system.

The estimated skein strength for combed yarns processed on the long draft system is calculated by increasing the carded yarn strength by 10 per cent.

Cotton Staple Length

The method of stapling recommended by the U. S. Department of Agriculture (4) is as follows:

Grasp in the two hands a tuft of cotton of a size convenient for the purpose (about $\frac{1}{4}$ of an ounce), holding it firmly between the thumb and forefinger of each hand with the thumbs placed together, the fingers being turned in toward the palms of the hands, and the middle joints of the second, third, and fourth fingers of each hand touching the corresponding joints of the fingers of the other hand, so as to give a good leverage for breaking the cotton.

Pull the cotton slowly with about the same leverage of each hand on the joints of the fingers, separating the tuft of cotton into two parts.

Discard the part remaining in the right hand.

Grasp with the thumb and forefinger of the right hand the end of the tuft of the cotton retained in the left hand. The point of pressure on the cotton

[1] This value was formerly reported in variety tests but was discontinued because of abuse.

in the left hand is just below the joint of the thumb and at the nail joint of the forefinger.

With the right hand, draw a layer of fibers from the cotton held in the left hand. Retain in the right hand the layer so drawn. Repeat this operation four or five times, placing each successive layer directly over the fibers previously drawn, using care to see that the ends of all the layers are even with each other between the thumb and forefinger of the right hand.

After discarding the cotton in the left hand, hold the fibers thus obtained between the thumb and forefinger of the right hand and smooth them with the thumb and forefinger of the left hand.

Place these fibers on a flat horizontal surface with a black background, preferably black velvet.

Block off the ends of the fibers with a cotton-stapling rule so as to indicate the length of the bulk of the fibers, and measure the distance between the blocked-off ends.

If this method is correctly and consistently followed, the length obtained should agree with that obtained by comparison with pulls from the official staple types.

Cotton Grade

The grade of cotton falls within one of five groups: gray, white, spotted, tinged, and yellow-stained. However, within these groups there are different grades, depending upon the amount of foreign

TABLE 11.2

Official Grades of American Upland Cotton

Gray	White	Spotted	Tinged	Yellow Stained
GMG	Good Middling [a]	GM Sp	GMT	GMYS
SMG	Strict Middling [a]	SM Sp	SMT [a]	SMYS
MG	Middling [a]			
SLMG	Strict Low Middling [a]	M Sp	MT [a]	MYS
	Low Middling [a]	SLM Sp	SLMT [a]	
		LM Sp	LMT [a]	
	Strict Good Ordinary [a]			
	Good Ordinary [a]			

[a] Indicates grades for which grade boxes of the standards are available. Grades above the line are deliverable on futures contracts; those below the line are not deliverable on futures contracts.

matter in the cotton. There are a total of twenty-four different grades of upland cotton which are listed in Table 11.2. (For the 1960 revision, see Appendix V.)

To determine the grade of a given unknown sample of cotton, a "loaf" is shaped by hand to match those in the grade boxes, and a visual comparison is made. The unknown sample is thereby fitted to some particular grade.

The color of the cotton has been, and yet is in many cases, evaluated by this same visual comparison method. However, an electronic instrument, the Nickerson-Hunter cotton colorimeter, has been developed which determines the color accurately and quickly. The instrument measures reflectance and the degree of yellowness. The color values that are measured by the instrument

TABLE 11.3

Relationship of Grade of Cotton to Picker and Card Waste and Non-lint Content

Grade	Picker and card waste [a] % by weight	Non-lint content [b] % by weight
Good Middling	6.3	2.4
Strict Middling	7.2	2.9
Middling	8.1	3.7
Strict Low Middling	9.3	5.1
Low Middling	12.5	7.6
Strict Good Ordinary	15.6	11.0
Good Ordinary	18.3	17.0

[a] Carded at 9½ pounds per hour.
[b] By use of the Shirley Analyzer.

can be plotted on the color diagram supplied with the instrument, and thus compared with the color diagrams made from the official grade standards. The Department of Agriculture has available for purchase actual color diagrams of the standards, thus simplifying the evaluation procedure for the cotton technologist.

As stated previously, one criteria for the grading of cotton is the amount of waste or foreign matter present. This relationship is shown in Table 11.3. The data in this table were extracted from the United States Department of Agriculture report on Cotton Testing Service for February, 1955.

In comparing the data in the preceding table with individual samples, it should be remembered that the rate of carding, nature of foreign matter present in the cotton, and any other reason for placing the cotton in any particular grade classification will influence the relationship of grade and waste. If the cotton, for example, were graded down due to poor color, the results would likely be different with respect to total waste and non-lint content.

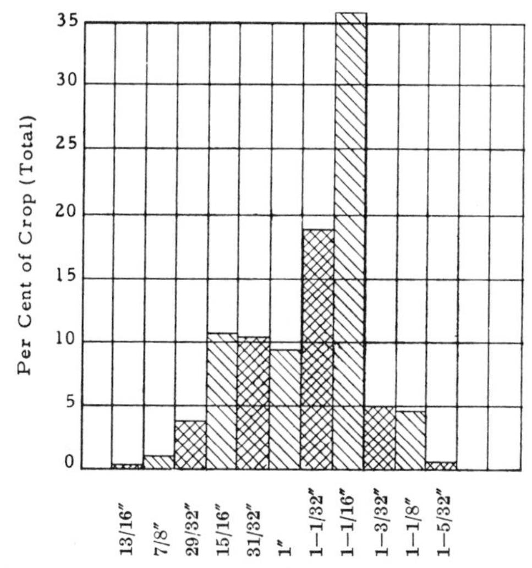

Fig. 11.5. Distribution of grade and staple lengths for the 1958 crop (U. S. Uplands cotton).

The term preparation refers to the treatment to which the cotton has been subjected during ginning, that is, whether cotton as it leaves the gin is smooth and free from neps, naps, gin-cut fibers, etc., or whether it has an excessive amount of these defects.

Standards have been issued which aid the classer in determining the quality of the preparation. The standards are listed as A, B, and C. For the preparation to be classed as A, the cotton must be free from the defects listed above. Preparation B indicates a slight amount of neps, naps, and rough cotton in general, whereas preparation C indicates considerable damage with neps, naps, stringy cotton, etc., predominating throughout the sample.

The staple length standards and grade boxes came into being

as a result of the need of having standards that could be used for reference, not only in widely scattered areas but also from season to season. These standards formed the first basis of stabilizing technical knowledge concerning cotton; they formed the yardsticks for the determination of relative values in commercial transactions and are yet the background and basis for the purchase and sale of cotton and cotton futures.

However, since the first cotton was grown for commercial use, there always has been another term used to describe cotton: character. The statement sometimes quoted that character is "those elements of cotton quality which are not included in grade or staple length" conveys the general idea of the all-inclusiveness of the term, but the description does not give the full importance or conciseness that is deserved. To the old-time cotton classer, character meant just what the word implied; the distinctive quality belonging to a particular batch of cotton. Actually, he was evaluating the cotton as he classed it for such distinguishing characteristics as fiber fineness (or "silkiness" and "smoothness"), strength (by breaking fibers), maturity, and amount of waste.

The synthetic fiber gave the final impetus to the need for expressing "character" in technical terms, for these new fibers are completely described by their manufacturers for all technical qualities. The character of cotton, as for the manmade fibers, then emerges as a broad array of technical measurements, each of which contributes to the full knowledge of the material under study. Some of the tests included in these measurements are given on the following pages.

References

1. Fiber and Spinning Test Results for Some Varieties of Cotton Grown by Selected Cotton Improvement Groups, Crop of 19—, U. S. Dept. Agr., Agr. Marketing Service, Cotton Div.
2. Campbell, M. E., *U. S. Dept. Agr. Circ. No.* 413.
3. Webb, R. W., "Improved Equations for Predicting Skein Strength of Carded Yarn with Special Reference to Commercial Production of American Cotton," U. S. Dept. Agr., 1955.
4. "The Classification of Cotton," *U. S. Dept. Agr., Misc. Publ. No.* 310 (1938).

CHAPTER XII

Cotton Fiber Length Analysis

Among all the tests that are made to evaluate the physical properties of cotton, probably none are more frequently used and in most cases more important than the length. Staple length is associated with not only spinning performance but also yarn properties and characteristics. Some factors that are influenced by length are spinning limit (how fine a yarn may be spun), yarn appearance, evenness, and strength.

Because American cotton of long staple is limited in quantity, a price differential must be paid by the spinner to get added staple length. The cost increase for extra staple length is very impressive to the spinner. For instance, in July of 1955, the average premium of 1 inch staple over the market standard of $\frac{15}{16}$ inch was nearly $1\frac{1}{2}$ cents per pound, for $1\frac{1}{16}$-inch staple the premium was more than 3 cents, and for $1\frac{1}{8}$-inch staple it was more than 6 cents per pound.

Staple length and its effect on yarn characteristics has been discussed; in this chapter the measurement of fiber length is to be covered. The difference between the two terms, staple length and fiber length, may be somewhat confusing. The term staple length is used to describe the length of cotton as measured by the hand techniques described in the preceding chapter. The results of this stapling by hand give an apparent length which is neither the shortest, the longest, nor the average length of the fibers in the sample. It will be shown below that this length is that of the fibers where about 25 per cent of the fibers by weight are longer, and 75 per cent shorter than the length obtained. In comparison, fiber length refers to the length and length distribution of the individual fibers in a sample of cotton as opposed to this apparent length of a discrete group of many fibers as a unit. It is the length obtained by laboratory techniques. As a result, the data are more accurate and more suited to scientific study than those obtained by the classer. Furthermore, the data from these laboratory tests provide much more information concerning the sample tested. At the same time, there should be no inference drawn to the effect that hand classing of cotton gives erroneous or worthless results. As a matter of fact, because of the speed of hand classing and reproducibility of results, practically every bale of the multi-million bale crop in the U. S. is hand classed and bought and sold on the basis of the results of this hand classing. Up to the present time the use of laboratory instruments and techniques has not been aimed at supplanting, but rather at supplementing hand classing. It is a means for verifying the results of hand classing. It provides a background of information for predicting yarn and fabric properties. It can be used to give the data for in-process control of machines. In this latter case, for example, a composite sample made from a blend representing several bales of cotton from a single source can be analyzed for fiber length; the results can be used to determine roll settings. Thus, the evaluation of fiber length made up of its various elements (mean length, upper quartile or upper-half mean length, and uniformity) becomes a very valuable source of information for quality control, both in the purchase of the raw material and in its processing.

There are two methods for laboratory analysis used to measure fiber length. The first of these, by the Fibrograph, is the more rapid of the two, but results are not as detailed, nor are they considered as accurate. However, speed of testing is very high. The second method, the fiber array, by use of the Suter-Webb Sorter, is much slower, but results are more accurate. The use of this technique has been relegated in recent years almost entirely to that of research, whereas the Fibrograph technique has found its place in the routine area of testing and quality control. Both methods will be explained.

Fibrograph

Although the Fibrograph does not give all of the details pertaining to fiber length measurements that the fiber array method does, it is, nevertheless, a valuable instrument for measurements which are helpful to the quality control department. The instrument is used chiefly to estimate the average or mean length and the upper half mean length of fibers approximately $\frac{1}{4}$ inch and longer; from these, other values such as fiber length uniformity may be calculated.

Theory of Machine.[1] The Fibrograph is an optical instrument with light-sensitive cells for scanning parallel cotton fibers and simultaneously drawing a length-frequency curve. Due to the method of holding the sample in the comb, the instrument does not include in its measurement those fibers of approximately $\frac{1}{4}$ inch and shorter. Therefore, the use of the instrument for determining the length of machine waste results in entirely erroneous and misleading measurements, inasmuch as a large proportion of the fibers in waste is made up of very short fibers.

Procedure for Preparation of Test Sample

A. Preliminary

1. Select at least three pinches or small tufts of fibers at random from laboratory sample. The weight of the composite test sample should be approximately 400 mg.
2. The small tufts when selected are placed on top of each other, and the entire bundle of fibers is separated or pulled apart by gripping the ends of the mass with the thumbs and forefingers and pulling the sample into two parts. The two parts of the sample are then superimposed upon each other with the two broken edges being directly in line with each other.

B. Combining and Transferring

1. The prepared sample, as described above, is combed onto one of the combs at random. To accomplish this, hold the comb with the teeth pointing upward, and comb the broken edges of the sample first. Continue combing until all of the sample is uniformly distributed over the full length of the comb.
2. The comb with the sample is held in one hand with the teeth

[1] A new semi-automatic model is now available. Its use eliminates some of the steps in the procedure given.

upward. A second comb, with the teeth pointing downward, is gently combed through the fibers, with the penetration of the teeth increasing gradually into the fibers with each stroke. After this combing, all of the sample is transferred to one comb. To transfer the sample: (a) Hold both combs with the teeth pointing upward. (b) Place one comb in a position so that the teeth almost touch the base of the teeth of the other comb. (c) Deposit the fibers onto the teeth of the bottom comb by pushing straight down with the top comb.

3. Repeat combing and transferring two additional times. On the last transfer, the fibers are distributed evenly over the lengths of both combs, with an equal amount being on each comb. The combing strokes given to the cotton are approximately as follows: twenty for the original combing, ten for the cotton after the first transfer, and ten strokes after the cotton has been transferred the second time.

Procedure for the Use of the Fibrograph (See Fig. 12.1)

A. Preliminary

1. Turn switch marked "Line" (1) to *ON* position.
2. Press button (2) at left end of fluorescent light holder to turn on light.
3. Wait twenty minutes before using.

B. Machine Adjustment

1. Turn galvanometer switch (3) to *OFF* position.
2. Light image pointer should be at central mark of index (4).
3. If not, adjust with knob (5) on right side until image is centered.
4. Turn switch marked "Galv" (3) to *ON* position. (Leave line switch (1) in *ON* position.)
5. Turn "Bal" switch (6) to "Adj."
6. If light image is off, adjust image to index line (4) with balance knob (7).
7. Turn balance switch (6) to "TEST."
8. Insert card.

9. Turn right handwheel (8) to line up index line (9) on card carrier with index line on plate (10).
10. Adjust light image to central mark of index (4) with zero knob (11).

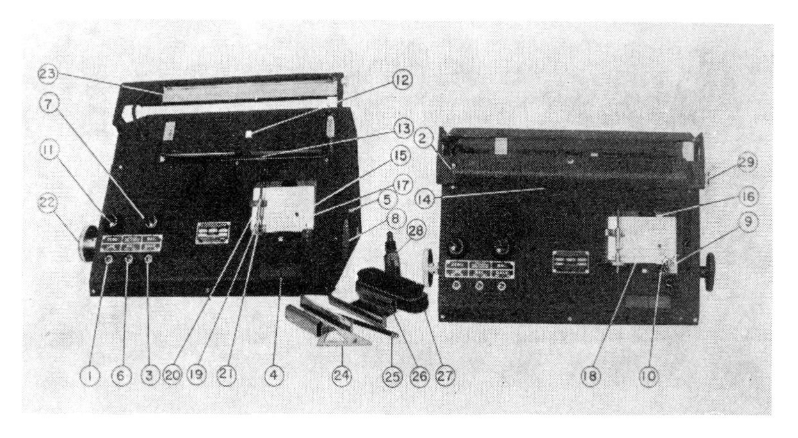

Fig. 12.1. Fibrograph.

C. Test

1. Lift fluorescent light holder and insert combs in the slots with the fibers pointing downward.
2. Brush the fibers down and replace the light holder. This will throw the light image off.
3. With the right handwheel, line up the light image with the center mark.
4. Let the pen rest on the card. Be sure the pen is in starting position.
5. The two handwheels, one of which controls the comb holding the fibers and the other the card position are turned forward simultaneously, keeping the light image at the center mark at all times; the pen draws the curve (See Figs. 12.2 and 12.3).
6. Turn the left handwheel in reverse as far as possible (without force) to make the vertical line.
7. Release the pen and allow it to slide down to complete the vertical line *BC* in Fig. 12.3.

8. Turn the right handwheel in reverse as far as possible (without force) to make the horizontal line *CD* perpendicular to *BC*.

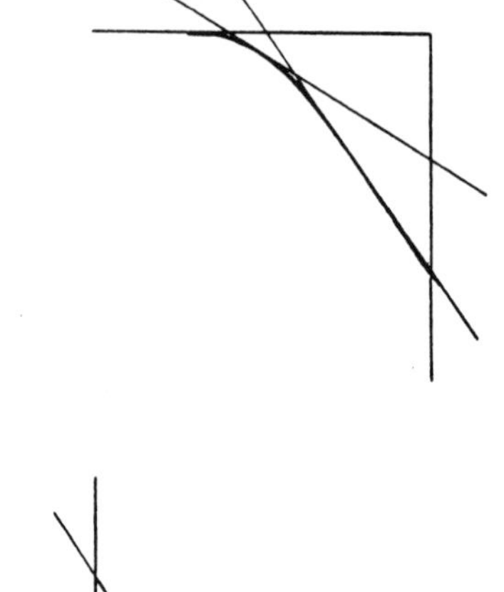

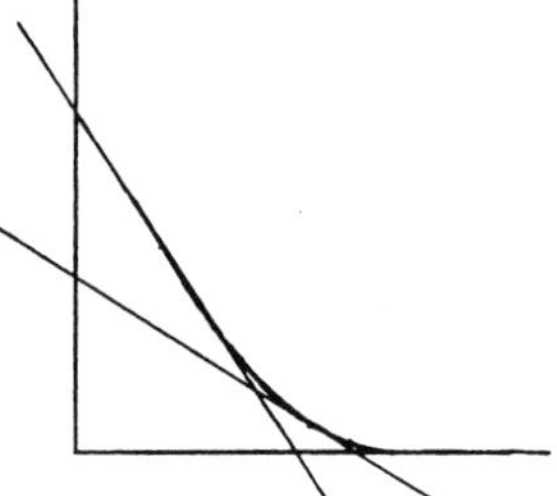

Fig. 12.2. Curves as drawn by fibrograph.

D. Calculations. Curve *A*, axis *BC*, and axis *CD* in Fig. 12.3 are produced on the Fibrograph. With the use of the fibroscale, the fibrogram in Fig. 12.3 is completed as follows:

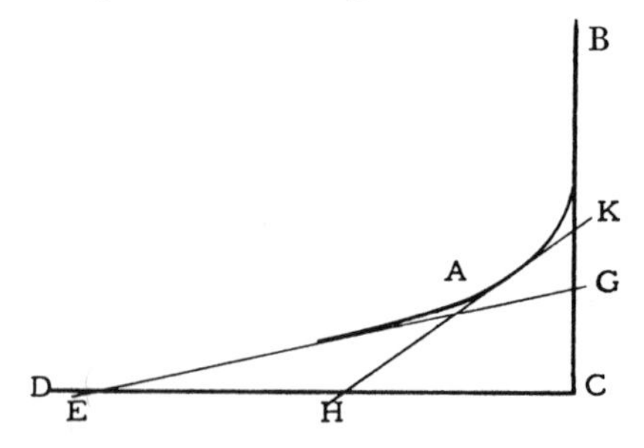

Fig. 12.3. Analysis curve.

1. Draw tangent *EG* through the starting point of curve *A* and extend it until it intercepts *DC* at *E* and *BC* at *G*.

2. With fibrogram, bisect *EC* at *H* and draw tangent *HK* from *H* to curve *A* and extend it until it intercepts *BC* at *K*.
3. Measure *CK* in inches with the fibroscale Fig. 12.4. This is the upper-half mean length of the fibers.

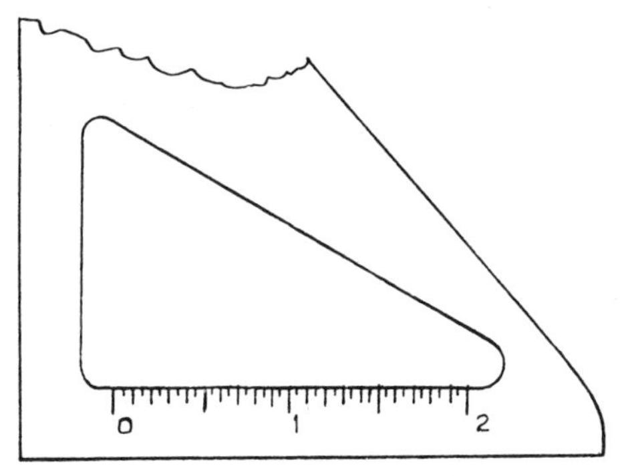

Fig. 12.4. Fibroscale.

4. Measure *CG* in inches with the fibroscale. This is the mean length of the fibers.
5. Divide the mean length by the upper-half mean length and multiply by 100 to obtain the per cent uniformity ratio.

$$\frac{\text{Mean length}}{\text{Upper-half mean length}}(100) = \%\text{ uniformity ratio}$$

Mean Length. The mean length as measured by the Fibrograph is an estimate of the average length of the cotton fibers longer

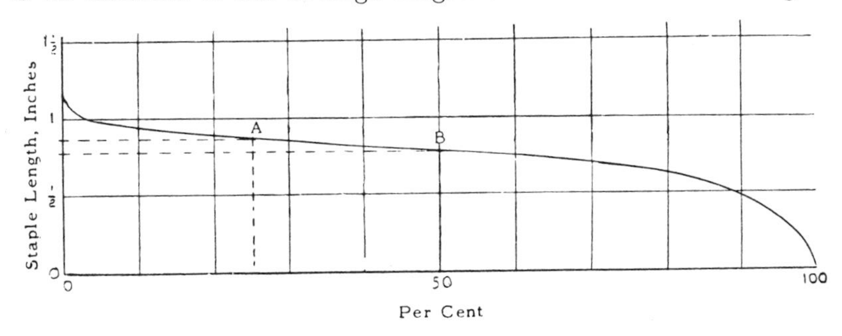

Fig. 12.5. Typical array diagram. A: approximate upper-half mean length; B: approximate mean length. Upper 25 % = 1.110″; coefficient of variation = 28.2 %; mean length = 0.939″.

than approximately $\frac{1}{4}$ inch in the test specimen. The upper-half mean length is a value approximating the staple length as determined by the cotton classer. The correlation between the upper-half mean length and the classer's length is excellent, the difference between the two values usually being within $\frac{1}{32}$ of an inch. See Fig. 12.5.

Relation to Twist. The United States Department of Agriculture published in "Summary of Fiber and Spinning Test Results for Cotton Varieties Grown by Selected Cotton Improvement

TABLE 12.1

Relationship of Upper-Half Mean Lengths and Twist Multipliers for Maximum Yarn Strength [a]

Upper half mean length, in. (Fibrograph)	Twist multiplier	Upper half mean length, in. (Fibrograph)	Twist multiplier
0.62 and shorter	5.35	0.98–1.01	4.10
0.63–0.66	5.15	1.02–1.05	4.05
0.67–0.70	5.00	1.06–1.09	3.95
0.71–0.74	4.85	1.10–1.13	3.90
0.75–0.78	4.70	1.14–1.16	3.85
0.79–0.82	4.60	1.17–1.20	3.80
0.83–0.86	4.45	1.21–1.24	3.75
0.87–0.89	4.35	1.25–1.28	3.70
0.90–0.93	4.25	1.29–1.32	3.65
0.94–0.97	4.20	1.33–1.36	3.60

[a] Source: U. S. Department of Agriculture.

Groups, Crop of 1951" a table of values giving the relationship of upper-half mean lengths and twist multipliers which were suggested for maximum yarn skein strength. The twist multipliers, being based on past experience and previous tests, should be used only as a guide since several factors such as processing techniques, year to year variation in cotton, etc., are likely to influence the results. See Table 12.1.

Uniformity Ratio. The uniformity ratio is a measure of the uniformity of fiber lengths in the sample, expressed as a per cent. As shown in the section describing calculations, the uniformity ratio is the ratio of

$$\frac{\text{Mean length}}{\text{Upper-half mean length}}(100)$$

where the mean length is the average length of all fibers, approximately $\frac{1}{4}$ inch and longer, and the upper-half mean length is the estimate of the length of the fibers in the upper half portion as measured by the fibrograph.

Thus, this uniformity value is the ratio of the average length of all fibers measured by the instrument to the length of the choice long fibers. As the difference between these two lengths increases, either with a decrease in the mean length or an increase in the upper-half mean length, the degree of uniformity decreases. This reflects as a lower uniformity ratio.

For example,

Mean length, in.	Upper-half mean, in.	Diff., in.	Uniformity ratio, %
0.80	1.00	0.20	80
0.75	1.00	0.25	75
0.80	1.06	0.26	75

A sample with all fibers of the same length would have a uniformity ratio of 100 per cent, whereas a sample with large variations in fiber length would have a low ratio. Table 12.2 gives the rating of fibers with respect to uniformity ratio.

TABLE 12.2

Ratings of Cotton Fibers with Respect to Uniformity Ratio [a]

Uniformity ratio	Rating
Above 80	Uniform
76 to 80	Average uniformity
71 to 75	Slightly irregular
70 and below	Irregular

[a] Source: United States Department of Agriculture

Number of Tests. The American Society for Testing Materials, Committee D-13, recommends "five test specimens and five fibrograms shall be made from each laboratory sample." However, the number of test specimens and fibrograms may vary depending upon individual laboratory equipment, personnel, and volume of testing. In addition, the number of samples will vary with the amount of variation found in the test results if the sample sizes are determined statistically.

Frequency of Tests. The frequency of the tests depends entirely on the individual mill and the method in which the results are to be used. For example, if the tests are performed merely as a check test on material being used, the frequency may be daily, weekly, or some other convenient schedule. However, if the cotton is being bought against specifications and the test results are to be used to determine if the fibers are meeting specifications, the frequency and volume would be increased to guarantee satisfactory protection. In any event, frequent check tests on samples of known length are a necessity. Daily check tests are recommended as a minimum, and additional tests whenever an operator changes from one length cotton to another during any one day.

Fiber Array Method, Suter-Webb Sorter

The Suter-Webb method of determining fiber length measurements, as compared to the Fibrograph method, is much more time-consuming, inasmuch as the sorting of fibers is largely a hand operation. However, the additional amount of information resulting from the use of this method will in some cases justify its use in preference to the Fibrograph. This is especially true in research or advanced quality control work.

Theory of Test. The Suter-Webb method is based on the principle of separating a sample of cotton fibers into carefully segregated small groups of fibers of known length. The groups of fibers are formed with respect to length in increments of $\frac{1}{8}$ inch. These groups are then weighed and the data evaluated in such a manner as to give an array or diagram of fiber length distribution with pertinent measurements.

Procedure for Preparation and Sorting of Test Sample

A. Preliminary

1. Prepare master sample by spreading cotton into a thin layer and taking several small pinches from the top and bottom. (A.S.T.M. recommends a total of 48 pinches, 24 from the top and 24 from the bottom, each pinch weighing approximately 25 milligrams.) Continue to pull and lap the sample with the fingers without the loss of the cotton until the fibers are parallel and free from leaf, stem, and other foreign matter.
2. Select a 75 $\pm$ 0.4-milligram representative test sample from the master sample.

B. Sorting

1. Place test sample upon one set of combs and carefully insert fibers into combs by use of finger and depressor.
2. With forceps, grip a few of the protruding fibers through the combs, place on other set of combs, and press into the combs by use of the depressor. Continue this process, dropping the combs as necessary to grip the fibers until all fibers have been transferred to the second set of combs.
3. Revolve the sorting apparatus 180 degrees and proceed to transfer fibers, as described above, to empty combs. Care should be taken to align and straighten the fibers.
4. Revolve the sorting apparatus 180 degrees. Draw small groups of fibers from the combs and place successively as straight as possible upon a velvet-covered plate. (A.S.T.M. recommends that the size of groups or pinches be such that not less than 50 or more than 100 groups be formed for short staple cotton [less than $1\frac{1}{8}$ inches in length] and not less than 80 nor more than 125 for long-staple cotton [$1\frac{1}{8}$ inches and over in length]. The U.S.D.A. laboratories have found that the number of pulls for an array should be allowed to vary over a narrow range for uniform and consistent results. Generally, the more pulls used for an array, the longer the cotton will be. They recommend 65 to 75 for short cottons.)

Grouping and Weighing of Fibers

A. Grouping. Arrange fibers into groups by measuring the length of each small group of fibers and grouping together all fibers that fall with a range of $\frac{1}{8}$ inch. For example, all small groups ranging in length from 18/16 to 20/16 would be placed

FIBER LENGTH CALCULATIONS

TEST Number 1 SAMPLE IDENTIFICATION 7626

COTTON Deltapine 14

Mid-Point 1/16" (L)	Groups	Total	Sum of Array Groups (W)	Length Squared (L^2)	Weight x Length (WL)	Weight x Length Squared (WL^2)
33				1089		
31				961		
29				841		
27				729		
25				625		
23				529		
21	6	6	5.00	441	105.00	2205.00
19	5	5	10.55	361	200.45	3808.55
17	10+4	14	23.80	289	404.60	6878.20
15	7+2	9	11.20	225	168.00	2520.00
13	7	7	9.85	169	128.05	1664.65
11	3+3	6	3.65	121	40.15	441.65
9	4	4	4.45	81	40.05	360.45
7	3	3	1.00	49	7.00	49.00
5	2	2	2.05	25	10.25	51.25
3	2	2	1.88	9	5.64	16.92
1	1	1	.45	1	.45	.45
Totals			73.88		1109.64	17996.12

CALCULATIONS

39.35 − 18.47 = 20.88

To Calculate 25% Point:

$\frac{16}{16}$ = 1.000 Inches

$\frac{20.88}{23.80}$ (⅛) = 0.110 Inches

Σ = 1.110 Length in Inches at 25% Point

$$\text{Mean Length} = \frac{\frac{\Sigma WL}{\Sigma W}}{16} = \frac{15.019}{16} = 0.939 \text{ Inches}$$

$$\text{Variance} = \frac{\Sigma(WL^2)}{\Sigma W} - \left(\frac{\Sigma WL}{\Sigma W}\right)^2 = 243.59 - 225.57 = 18.02$$

$$\text{Standard Deviation} = \frac{\sqrt{\text{Variance}}}{16} = \frac{\sqrt{18.02}}{16} = \frac{4.24}{16} = 0.265$$

$$\text{Coefficient of Variation} = \frac{\text{Standard Deviation}}{\text{Mean}}(100) = \frac{0.265}{0.939}(100) = 28.22\%$$

Sorted By MS Calculated By MS Checked By [illegible] Date 10-2-55

Fig. 12.6. Fiber length calculations.

into one group and classified as 19/16 inch in length, and all fibers from 20/16 to 22/16 would be placed into one group and classified as 21/16 inch in length. Measuring and grouping should be done in descending order with respect to fiber length; that is, the longest fibers are measured and grouped first, and the procedure continued until all fibers are properly grouped. Fibers measuring from 0 to $\frac{1}{8}$ inch in length are grouped last and classified as being $\frac{1}{16}$ inch in length.

B. Weighing. All groups are then weighed on a suitable balance to the nearest 0.05 milligram and the weights recorded as shown in Fig. 12.6. The total weight of the groups should be within ± 2 milligrams of the original sample weight.

Calculations

A. Preliminary. After recording the weight of groups, complete the form by filling in the proper columns. From this accumulation of data, it is customary to calculate the mean or average length of fibers, length at 25 per cent point, standard deviation, and coefficient of variation. In some cases the fiber length distribution is also drawn. (Calculations are shown in Fig. 12.6.)

B. Mean Length

1. Multiply the weight (W) of each group by the midpoint (L) of the length class and get the sum of the resulting values. This value is represented by $\Sigma(WL)$. Thus, $5(21) = 105$, $10.55(19) = 200.45$, etc. $\Sigma(WL) = 1109.64$.
2. Divide the value $\Sigma(WL)$ by the total weight (ΣW) of the fibers. This is represented by $\Sigma(WL)/\Sigma W$ and is the average number of sixteenths of an inch length of the sample.

$$\Sigma(WL)/\Sigma W = \frac{\Sigma(WL)}{\Sigma W} = \frac{1109.64}{73.88} = 15.019$$

3. This value is then multiplied by the value of the interval, and the result is the average length of the fibers.

$$15.019(\tfrac{1}{16}) = 0.939 \text{ inches, or } \tfrac{15}{16} \text{ inches}$$

C. Length at 25 Per Cent Point

1. Divide the total weight (ΣW) of the fibers by 4.

$$\frac{73.88}{4} = 18.47$$

2. Beginning with the longest length group, add the weights of each group until the value just calculated is either equalled or included in one of the groups. In this case, the weights of groups 21 and 19 are used.

$$5.00 + 10.55 = 15.55$$

The value 18.47 is not included in the first two groups, so the third group (17/16 midpoint group in this example) is added.

$$5.00 + 10.55 + 23.80 = 39.35$$

3. Record the lower limit of the last group (17/16 midpoint) included in step 2. The lower limit for this group is 16, or (16/16) = 1.000 inch.
4. Determine the excess of the partial weight over the $\frac{1}{4}$ total weight. This is done by subtracting the value calculated in step 1 from the weight of the three groups calculated in step 2.

$$39.35 - 18.47 = 20.88$$

5. Since the range of the group is $\frac{1}{8}$, the next step is to find what portion of $\frac{1}{8}$ of an inch is to be the lower limit (1.000 inch) of the group.

$$\frac{20.88}{23.80}\left(\frac{1}{8}\right) = 0.110 \text{ inch}$$

6. Add the result of step 5 and step 3, and the result is the length at the 25 per cent point.

$$1.000 + 0.110 = 1.110 \text{ inches or } 1\tfrac{7}{64} \text{ inches}$$

D. Standard Deviation and Coefficient of Variation

1. Calculate the variance by use of the following equation:

$$\text{Variance} = \frac{\Sigma(WL^2)}{\Sigma W} - \left[\frac{\Sigma(WL)}{\Sigma W}\right]^2$$

$$\text{Variance} = \frac{17996.12}{73.88} - \left[\frac{1109.64}{73.88}\right]^2$$

$$= 243.59 - 225.57 = 18.02$$

2. To calculate standard deviation (σ), extract the square root of the variance and multiply by the midpoint value taken from the data sheet.

$$\sigma = \sqrt{\text{Variance}}\,(\tfrac{1}{16})$$
$$= \sqrt{18.02}\,(\tfrac{1}{16})$$
$$= 4.24\,(\tfrac{1}{16}) = 0.265 \text{ inches}$$

3. To calculate the coefficient of variation (V), express the standard deviation as a per cent of the mean length.

$$V = \frac{\sigma}{\text{Mean length}}(100)$$

$$= \frac{0.265}{0.939}(100) = 22.82\,\%$$

The coefficient of variation is a relative measure of the fiber length variation. The larger the value, the greater the variation in fiber lengths, and, conversely, the smaller the value, the more

TABLE 12.3

Rating of Cotton Samples with Respect to Fiber Length Variation [a]

Coefficient of variation %	Rating
Below 27	Low variability
27 to 34	Average variability
35 and above	High variability

[a] Source: United States Department of Agriculature.

uniform the fiber lengths. A large variation in fiber lengths will result in increased waste (especially at the card), greater processing difficulties, and lower quality yarn.

The values in Table 12.3 are suggested ratings for cotton with respect to the coefficient of fiber length variation.

Number and Frequency of Tests. The number of tests made per sample and the frequency of such tests depend entirely upon the use to be made of the results. If the results are to be used in connection with research work, the number and frequency of tests are generally much higher than when using the results for routine quality control.

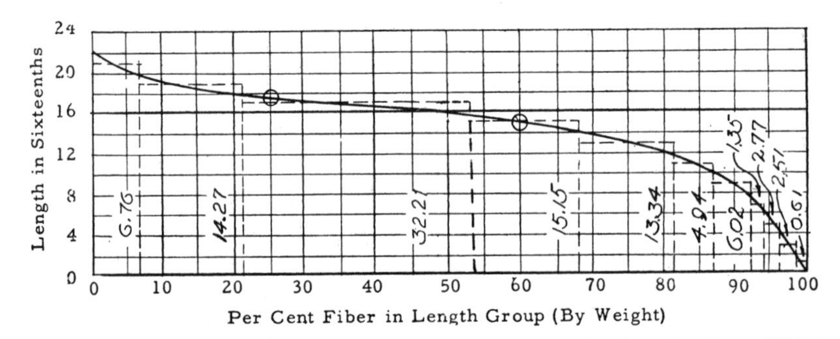

Fig. 12.7. Array diagram for the example calculated using the Suter-Webb sorter (see Fig. 12.6). Length at 25 per cent point is 1.110 inches.

Figure 12.7 is the array diagram for the example previously calculated using the Suter-Webb Sorter.

CHAPTER XIII

Fiber Strength

The importance of testing cotton fibers for strength and the best method to use in measuring the strength have been topics of discussion for years throughout the textile industry. The importance of strength testing has been agreed upon since the strength of the fibers has a direct effect upon the strength of the finished product, whether yarn or fabric. The strength of the fibers will also affect the hand, drape, and similar characteristics of a fabric. With other properties remaining fairly constant, the stronger the fiber, the stronger will be the yarn or fabric. The high strength of cotton fibers is associated with a high degree of crystallinity, and for this reason the high strength fibers are stiffer than low strength fibers. Therefore, for fabrics that are to have an exceptionally soft hand or feel, it would be advisable to use fibers of average or less strength. This does not necessarily mean that the fibers must be in the "very weak" class to make a good fabric.

Webb and Richardson (1) reported the following relationship between fiber properties and the strength of 22s carded yarn:

In net effect on yarn strength, the fiber properties that were important contributors in each staple length group were:

$\frac{7}{8}$ and $\frac{29}{32}$ inch	fiber strength only
$\frac{15}{16}$ and $\frac{31}{32}$ inch	fiber strength and fineness
1 inch	fiber strength, fineness, and uniformity ratio
$1\frac{1}{32}$ inch	fiber strength and grade index
$1\frac{1}{16}$ inch	fiber strength, uniformity ratio, and grade index
$1\frac{3}{32}$ inch	fiber strength, uniformity index, and per cent of mature fibers.

The work reported above is the result of a study on a few varieties of cotton and does not necessarily mean that the same relationship would be true for all cottons. However, it is very strong support for the point of view that fiber strength testing can be of value to the mill, especially in selecting cotton for a particular type of end product.

The strength of the fibers is determined by the flat bundle method using the Pressley Tester [1] or the more recently-developed Scott-Clemson and Stelometer instruments. In this test a sample of cotton is prepared by hand combing to parallelize the fibers to a small ribbon of $\frac{3}{16}$ to $\frac{1}{4}$-inch width. This ribbon of fibers is clamped in the miniature jaws, and the fibers protruding from the outside edges of the jaws sheared off so as to give a definite length of fibers (depending on gauge spacing, which is usually zero gauge or $\frac{1}{8}$-inch gauge). The jaws are then placed in the testing instrument — they fit either the Pressley or the Scott-Clemson — the fibers are broken by the instrument, the load at rupture is recorded, and the fibers weighed. The ratio of the load in pounds to the weight in milligrams gives a ratio which can be converted into either thousands of pounds per square inch or a fiber strength index. Until recently, practically all fiber strength tests have been made with zero gauge length, i.e., the two clamps have been themselves clamped together in a small vise so that there was a minimum clearance between them. Recently, however, the U.S.D.A. has adopted the $\frac{1}{8}$-inch gauge spacing between the jaws, as the standard test method for their laboratories. The basic reason for the adoption of the $\frac{1}{8}$-inch gauge test by the U.S.D.A. is their belief supported by their own tests that the fiber strength obtained is more highly correlated with yarn strength than results obtained with zero gauge. The basic method of calculations for zero gauge jaw spacing will be given under the Pressley Tester, together with the analysis of results for $\frac{1}{8}$-inch gauge as recently adopted by the U.S.D.A.

[1] As of February, 1955, the U.S.D.A. laboratories use the Pressley Tester for their fiber work.

Pressley Tester

The Pressley Tester is the instrument in widest use up to the present time throughout the textile industry for measuring the strength of cotton fibers. It is a relatively simple instrument to operate; however, it requires a considerable amount of operator technique. The results from the tester can be influenced by such things as the condition of the jaws, the tightening of the jaws,

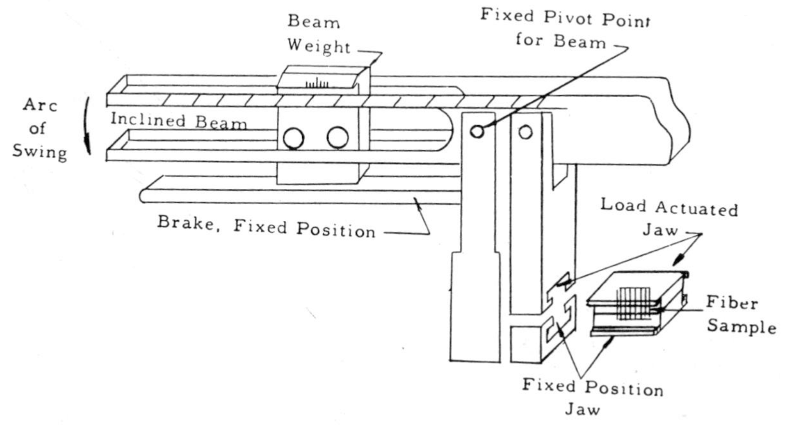

Fig. 13.1. Schematic diagram of Pressley Tester.

the ability of the opeator to select and prepare the proper size of test specimen so that it will break within the proper range on the instrument, and operator technique in preparing and mounting the sample.

The Pressley Tester works on the principle of applying a load to the fibers which are clamped between a pair of jaws, the load acting to separate the two jaws. The load is applied by the travel of a beam weight down an incline, the load being increased with movement of the weight away from the fulcrum until rupture occurs. At rupture of the fibers, release of the beam causes the beam weight to come to rest immediately. The position of the beam weight indicates the load sustained by the specimen at rupture. This is shown in the schematic sketch, Fig. 13.1. The bottom jaw is held in slots in a fixed position. The top jaw is held in slots by the lever arm attached to the right side of the fulcrum of the inclined beam. The sketch shows the approximate position of the beam weight just prior to the rupture of the fibers. When the fibers do break, the beam is free to swing in an arc, the left side of the beam dropping far enough to allow the beam weight to drop onto a bar that serves as a brake. Thus, the total travel of the weight to cause rupture of the fibers gives a means of getting the corresponding load in pounds. This load is indicated on the beam load scale opposite an index line on a small vernier on the carriage weight.

Procedure for Preparation and Breaking of Test Specimen

A. Preliminary

1. Level the machine.
2. The beam containing rolling weight tracks should be at an angle of 1.5 degrees below horizontal with the beam weight locked in zero position.
3. Place a pair of clamps in the vise and tighten the vise.
4. Open the clamps.

B. Preparation of Sample

1. Condition the fibers in the laboratory a minimum of four hours prior to testing.
2. Test samples may be taken from the Fibrograph or from samples prepared on the mechanical blender. If the test sample is to be taken from the original laboratory sample, several pinches should be taken at random from the large sample and then worked into a composite test sample repeating step A2 under Procedure for Preparation of Test Sample for Fibrograph.
3. Comb a small bundle, taken from the sample in B2 above, with the coarse comb until the fibers are fairly straight. Discard the short and tangled fibers which were combed out.
4. Comb two or three times through the fine comb attached to the vise. Discard the short and tangled fibers from the comb.
5. Place the remaining ribbon of long straight fibers across the

open jaws so that the ends of the fibers will extend an equal distance on each side of the jaws. The ribbon should be approximately $\frac{1}{4}$ inch wide. (The actual size of the ribbon depends upon the strength of the fibers being tested. The operator must develop the ability to approximate the size of the ribbon through experience with different cottons.)

6. Hold the left end of the ribbon against the side of the vise with the left thumb. With the left forefinger, lower the vise clamp. Hold the clamp in place with the left thumb.
7. Apply sufficient pressure to the right end of the ribbon with the right thumb and lower the top clamps with the left forefinger. Continue to apply pressure to top clamps with the left forefinger until the clamps are locked in place.
8. Lock the top clamps in place with the torque wrench. This is done with the right hand by tightening first one locking screw and then the others.
9. Use only enough pressure to prevent the fibers from slipping during the test. This can be done by starting with a low pressure and continuing to increase the pressure on subsequent samples until no slippage occurs.
10. Remove the jaws from the vise.
11. Cut off the ends of the fibers protruding from the jaws. Use a special knife provided for this purpose. Use the steel part of the holding surface as a cutting edge. *Do not* cut against the leather.

C. Test Procedure

1. Raise the left end of the beam and slide the clamps into the slots at the right end of the base.
2. Release the beam weight by gently raising the locking lever.
3. When the weight stops, read the breaking strength in pounds of the fibers to the nearest 0.1 pound. If the sample breaks less than 10 pounds or greater than 20 pounds, disregard the test and repeat until the break occurs between 10 and 20 pounds.
4. Remove clamps from the machine and place in the vise.
5. Unlock the top jaws.

6. Remove the broken fibers with tweezers and weigh to the nearest 0.1 mg. Caution: Extreme care should be taken not to lose any fibers.

D. Calculations. The Pressley index is calculated by dividing

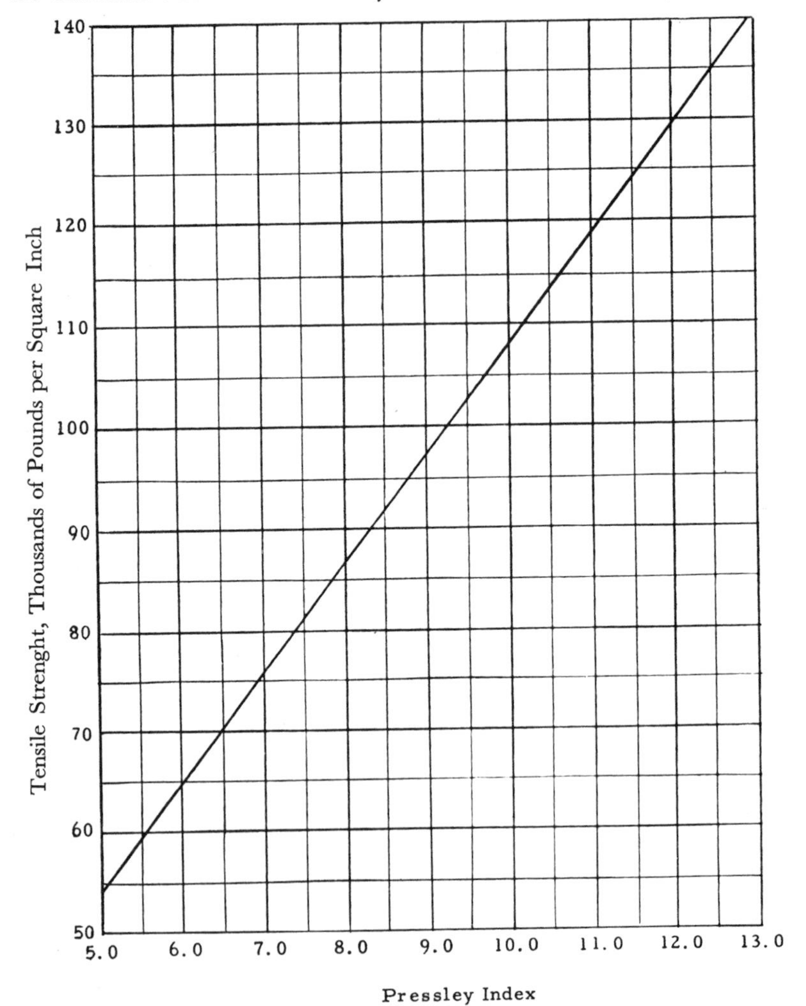

Fig. 13.2. Relationship of tensile strength to Pressley index for cotton fibers.

the breaking strength in pounds of the fibers, as read from the tester, by the weight of the broken fibers in milligrams. This value, the Pressley index, is the pounds required to break a milligram of cotton of a standard length of 0.464 inches. The index ranges from approximately 7.00 for the weak cottons to 11.00 or more for the very strong cottons.

The Pressley index can be converted into an estimation of strength in units of thousands of pounds per square inch by the following relationship which was established by the United States Department of Agriculture. (This is for zero gauge results.)

$$1000 \text{ pounds/square inch} = 10.8116 \text{ (Pressley index)} - 0.1200$$

This relationship is shown in Fig. 13.2.

The U.S.D.A. adopted following the 1954 crop new methods of calculation of fiber strength, based on the use of a $\frac{1}{8}$-inch gauge length.[1] The use of $\frac{1}{8}$-inch gauge length appears to have a good deal of merit. On the other hand, the new calculation methods have not been in use sufficiently long to allow thorough analysis.

In the new U.S.D.A. procedure, a Pressley ratio for $\frac{1}{8}$-inch gauge test is calculated by dividing the specimen strength by its weight, and the correcting for the standard level of the check-test cotton.

$$\text{Pressley ratio } (\tfrac{1}{8}\text{-inch gauge}) = \frac{\text{Strength, lbs.}}{\text{Weight, milligrams}} \text{ (Check level)}$$

This ratio when divided by 3.19 and multiplied by 100 gives a fiber strength index based on 100 as the average level. Thus, cottons with a value of over 100 are stronger than average, and below 100 are weaker than average.

$$\frac{\text{Pressley ratio}}{3.19}(100) = \text{index, where 100 is average}$$

Actually, the 1954 U. S. crop gave an average Pressley ratio of 3.19 with $\frac{1}{8}$-inch gauge spacing. For this reason, based on the 1954 commercial crop, a Pressley ratio of 3.19 for any sample

[1] The new calculation methods have not at the time of this writing been accepted by the A.S.T.M. or other groups, and so the methods used are given here only as a guide.

means that the sample is average, and thus has an index rating of 100.

$$\frac{3.19}{3.19}(100) = 100 \text{ index}$$

Note that with $\frac{1}{8}$-inch gauge, the ratio of pounds Pressley to weight is called the Pressley ratio; for zero gauge tests it is called the Pressley index. Fiber strength ratings have been prepared by the U.S.D.A. for interpretation of results. These are shown in Table 13.1. Concurrently with the $\frac{1}{8}$-inch gauge tests, zero gauge

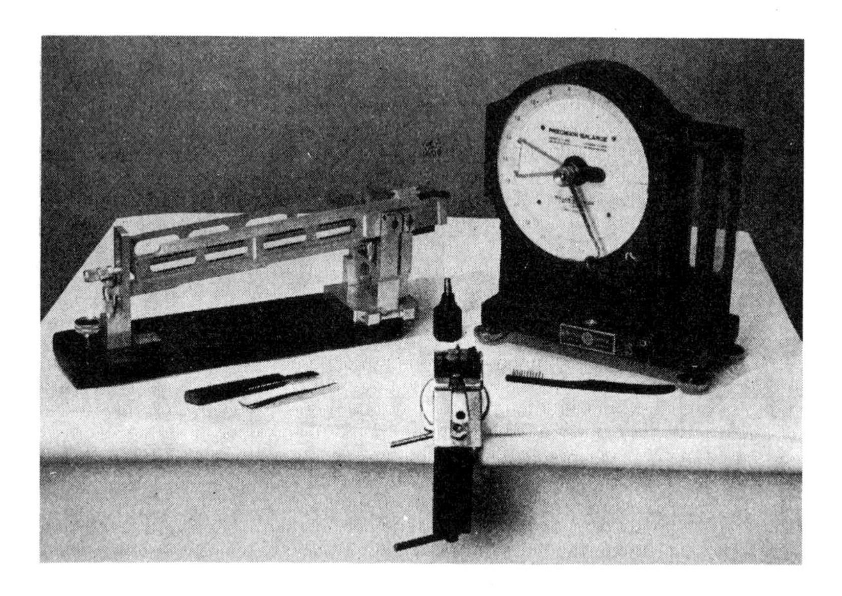

Fig. 13.3. Pressley Tester and accessories.

tests were made, and these resulted in an average of 84,000 pounds per square inch for the 323 samples tested. An estimated strength in thousands of pounds per square inch can be figured by multiplying the index by 84 ÷ 100. Thus, average cotton with an index of 100 has an estimated strength of 84,000 pounds per square inch.

It is doubtful if the use of an index is any more valuable than the actual strength level, or than the use of the original Pressley index as shown in Fig. 13.2. For this reason, there are grounds to believe that this method of using an index will not be accepted readily.

In Table 13.1, an additional listing of the strengths in thousands

TABLE 13.1

Fiber Strength Index

Index	Rating	Thousands lbs. per sq. in.
Above 115	Very Strong	Above 97
106 to 115	Strong	89–97
96 to 105	Average	81–88
86 to 95	Fair	72–80
85 or less	Weak	71 or less

of pounds per square inch has been made, based on the conversion factor of 84. This tabulation does not agree with previous ratings of fiber strength.

Check Tests. Check tests should be made daily before any routine testing is begun. These check tests should be made on a standard laboratory sample or a sample supplied by the Department of Agriculture which has been subjected to enough tests so as to establish within very close limits the actual breaking strength of the fibers. The object of the check tests is to aid the technician in maintaining a uniform technique. A series of three successive breaks should be obtained within the limits for the Pressley Index and strength before the operator begins testing the unknown samples.

The variation in results of the check test from standard, if such variation is found and is not eliminated by recheck, can be used to correct the level of results of the unknown sample. For example, if the standard laboratory check is found by an individual operator to average 3 per cent above its normal, then all tests by the same individual on the unknown should be raised by 3 per cent. The correction of 3 per cent should be used only over short periods of time and rechecks made frequently to verify the correction factor.

Number and Frequency of Tests. The accuracy and precision

required will determine to a great extent number and frequency of tests to be made. The American Society for Testing Materials recommends as follows: "Two breaks shall be made from each test specimen. It is recommended that six to ten breaks be made for each laboratory sample and that two technicians participate with each, making a strength test from each tuft."

Scott-Clemson Tester

The Scott-Clemson Tester, known also as the Clemson Flat Bundle Fiber Tester, is an instrument that has been developed to measure the strength of fibers. The design incorporates certain innovations over the widely used Pressley Tester. At the same time, the instrument uses the same jaws as are used on the Pressley, and in fact, they are interchangeable. Inasmuch as skilled operator technique is required in preparing the samples and placing them in the jaws properly, it can be said that results from the use of this tester are influenced to just as great an extent by operator technique as is the case with the Pressley Tester.

Principle of Machine. The greatest differences between the Scott-Clemson Tester and the Pressley Tester lie in the following innovations which have been applied to the Scott-Clemson Tester. A constant rate of load is applied to the sample; the sample gauge length may be varied from zero to 10 millimeters; the instrument is motor-driven; it is arranged for either a single tensile test or continuous hysteresis cycles; and the breaking load and elongation are recorded on a graph which provides a permanent record.

One of the main differences is that of the method of loading. On the Pressley Tester, the load is applied by the action of a weight which rolls down an inclined surface; obviously, the greater the strength of the sample, whether due to inherent fiber strength or to the number of fibers in the sample, the greater the distance that the weight rolls; consequently, the greater is its acceleration. It can be theorized that at low levels of load application on the Pressley, the weight, being at a relatively low velocity, stops quite promptly at the fiber rupture. However, on high loads with the resulting high velocity of the beam weight, there is a tendency for the momentum to cause the weight to move further along the beam at fiber rupture. In practice, the size of the fiber sample is regulated so that fiber breakage always occurs over a rather narrow range. However, with the Scott-Clemson Tester,

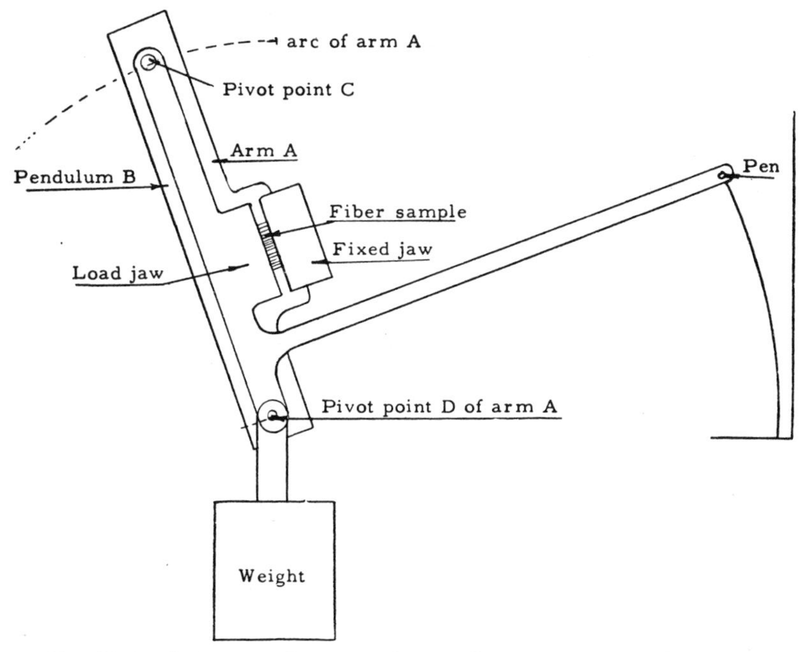

Fig. 13.4. Schematic diagram of Scott-Clemson Flat Bundle Tester.

the load is applied at a constant rate through an ingenious application of the pendulum principle. The loading system is a dead weight and a constant rate of increase gives a load rate of 1,000 grams per second. The action of the machine is shown in Fig. 13.4.

The arm A is motor driven, pivoting about point D to swing in the arc shown. The pendulum B swings about C. At its lower end, a weight is suspended; the suspension point is in line with point D, but is a separate bearing independent of the pivot point D for arm A. The fiber sample is held in the Pressley jaws as shown. As the arm A is moved from its initial position (a vertical position, not shown), there is an increasing load on the sample due to the horizontal component of the weight acting on the pendulum B and thus on the load jaw. The load and elongation are indicated by the curve drawn by the pen on a tensilgram chart. At rupture of the fibers, the load jaw moves away from the fixed jaw, with the result that the pen draws a pip on the chart. The movement of the weight pivot point is restricted to prevent damage to the instrument.

On the Clemson instrument, it will be noted that the sample clamps are positioned in a vertical plane as opposed to the horizontal plane used in the Pressley Tester. The instrument makes an autographed chart of the strength-elongation characteristics of the fibers being tested. The chart is designed so that five successive positions of a chart can be made in the holder so that five tests can be made on a single chart. As previously indicated, the sample clamps can be placed in the machine so that there is a zero-gauge sample length or any other gauge length up to 10 millimeters. There is a scale on the clamp holder which gives the amount of clamp separation; this simplifies the job for the operator. Another feature of the tester is that it can perform hysteresis tests. This is done by a very simple adjustment that can be made so that the load is alternately applied and removed.

Procedure for Preparation and Breaking of Test Specimen

A. Preparation of Sample.

Follow the same procedure as outlined under the Pressley Tester, using Pressley jaws. When using a greater than zero gauge length, measure carefully the spacing between the clamps and duplicate this on the stationary clamp holder.

B. Preliminary

1. Level the machine, as indicated by the centering of the bubble in the engraved circle on the glass. In leveling, back off the locknuts on leveling screws, adjust the screws until the instrument is level, and tighten the locknuts against the bottom of the base flange.
2. Slide a tensilgram (See Figs. 13.5 and 13.6) into place on the platen, so that the left side of the tensilgram rests against

the chart stop and the bottom against the base casting. For additional tests on the same tensilgram, move the tensilgram card to the right until the left edge coincides with the vertical index line on the platen. This can be repeated for a total of five tests.

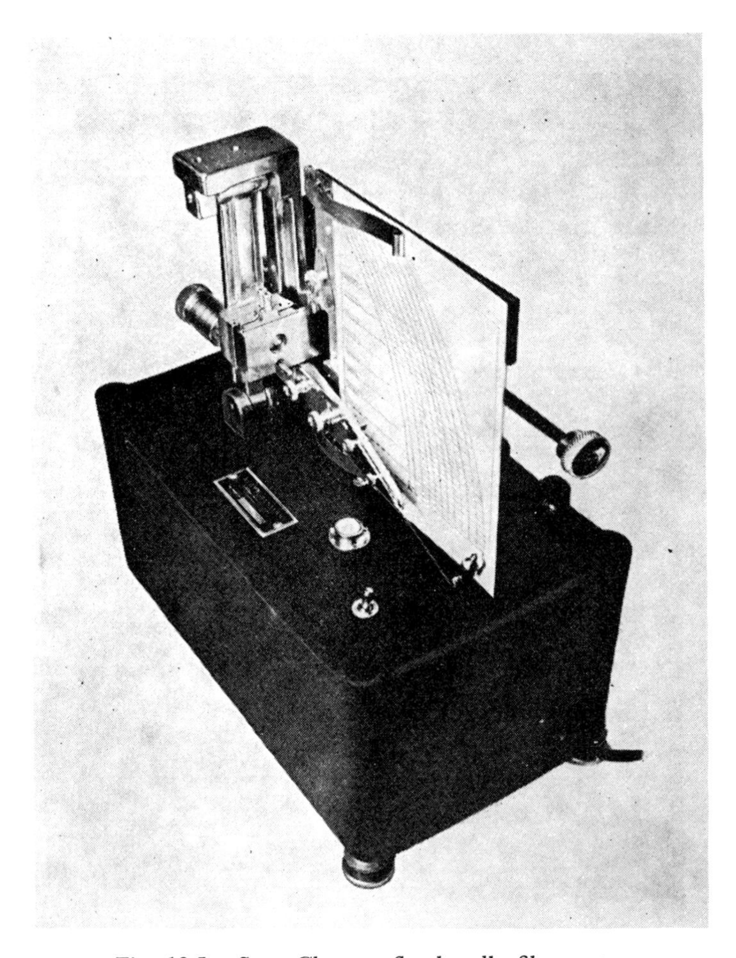

Fig. 13.5. Scott-Clemson flat bundle fiber tester.

3. Fill the pen cup and replace the pen cap.
4. Adjust the pen to zero. The pen is dropped against the tensilgram to indicate its position. In making the zero adjustment,

the two large knurled knobs on the pen holder bracket are temporarily loosened, and the pen holder arm moved to the correct position.

5. The gauge length is adjusted to the corrected spacing (0 to

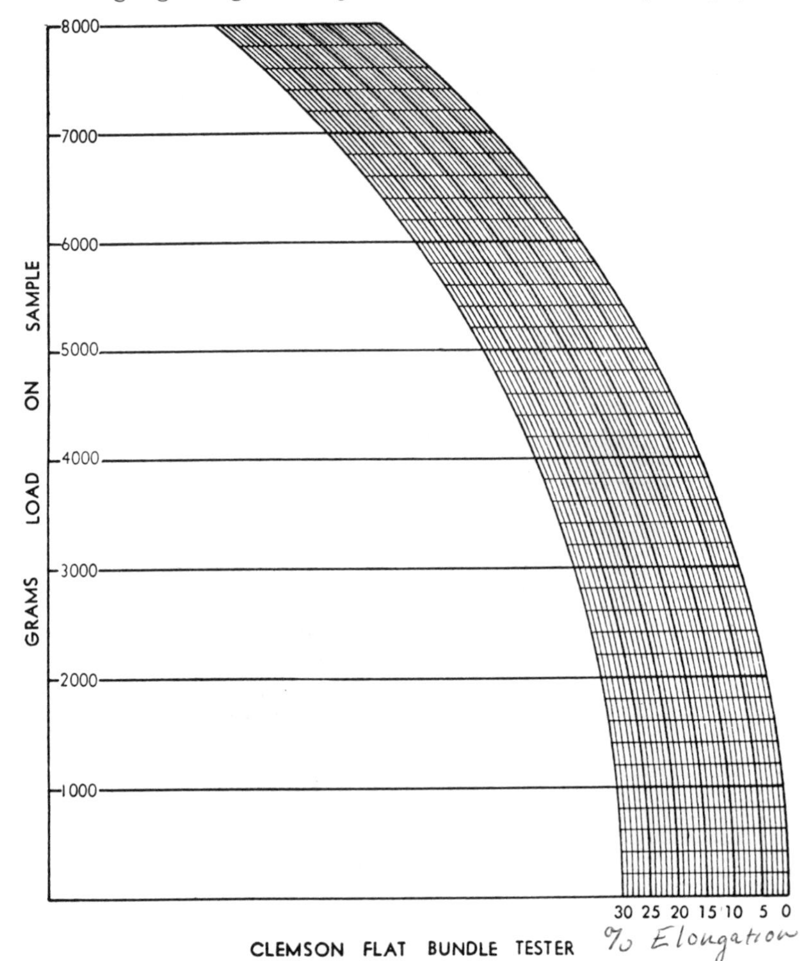

Fig. 13.6. Tensilgram for use on Scott-Clemson Fiat Bundle Tester.

10 mm. possible). The adjustment is made by loosening the two screws that hold the stationary clamps with the *T*-handle socket wrench provided, and then moving the clamp until the

desired length is exactly in back of the edge of the rocking lever. The screws should be re-tightened securely.

C Test Procedure

1. Throw the toggle switch to the left to *ON* position.
2. Insert the sample for test.
3. To start the tester, push the operating switch shaft to the left and release. (Check to see that the actuator for the starter switch located behind the platen is in the notch of the operating switch shaft.) The tester will reverse itself at the end of the loading cycle and return to zero.
4. The pen will draw a stress-strain curve. To measure the breaking strength, note the sharp reversal and hook in the test line. The top of the jog is the breaking load and is measured by the scale on the left of the tensilgram.
   Samples of greater than zero gauge length will elongate. The pen will drift to the left over one or more arcs. These arcs are spaced 1 mm. apart.
5. For cyclic or hysteresis tests, load the Pressley clamps with samples too large to be broken at maximum capacity of the tester, and rotate the operating switch one-quarter turn forward. The tester will continue the cycle until the switch shaft is rotated one-quarter turn backward. In order to record the cyclic or hysteresis curves, move the pen knock-off out of position by swinging its hooked lower end to the left.

D. Calculations

Clemson index (same value as Pressley index)

$$= \frac{\text{Breaking strength, gm.}}{\text{Sample weight, mgm. } 453.6}$$

Tensile strength (thousands lbs./sq. in.)

$$= (10.8116)(\text{Clemson index}) - 0.1200$$

$$\text{elongation} = \frac{\text{Stretch, no. of arcs}}{\text{Gauge length, mm.}}(100)$$

Stelometer

The Stelometer was developed by Hertel (2) for measuring the strength and elongation of fibers. The loading unit is of the pendulum type; however, the pendulum weight remains stationary and the pendulum axis moves through an arc. The loading speed is 500 grams per second based on a time limit to break of 20 seconds.

The Stelometer uses the Pressley jaws to hold the test specimen and the gauge length between the jaws is adjustable. The instrument does not draw a stress-train curve but the readings are indicated by force and extension indicators.

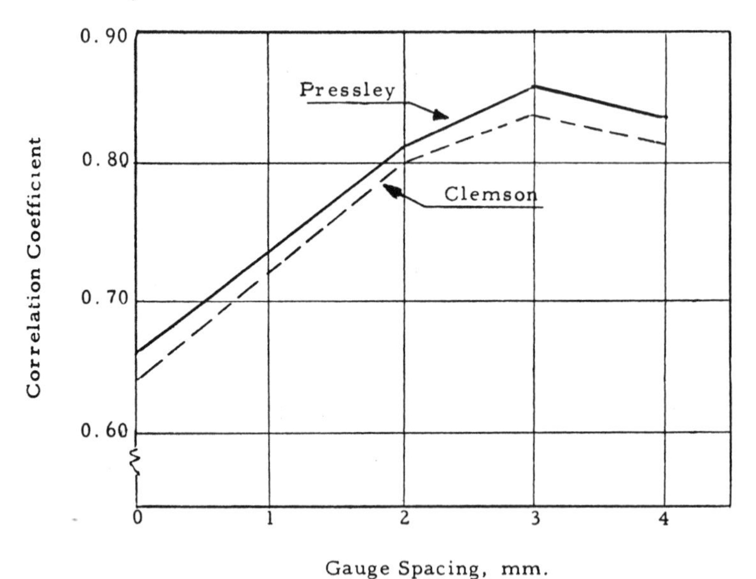

Fig. 13.7. Coefficient of correlation between 22s yarn strength and fiber tensile strength at various gauge spacings for Pressley and Clemson tester. (Source: U. S. Dept. of Agriculture.)

Analysis of Gauge Length

Until quite recently in using the Pressley Tester to determine the strength of cotton fibers, the distance between the jaws was zero. Phillips (3) advocates that the zero gauge length is undesirable when testing cotton fibers for strength. He points out that when a zero gauge length is used, the probability of including weak spots in the test specimen is very low. For this reason, he proposes the use of greater-than-zero gauge length. In comparing test length of zero and 4 millimeters, he shows a difference in strength when using the 4-millimeter length ranging from 42 to 55 per cent. In other words, testing the same cotton by two methods, the results when using a 4-millimeter gauge length were from 42 to 55 per cent less than for the zero length. Part of this difference can be explained as being due to the inclusion of more weak spots in the longer test specimen. It can also be partially explained by the fact that the Pressley Tester approaches an impact type of test with an accelerated rate of loading as compared with the inclined plane type of tester used for the 4-millimeter gauge tests.

A recent study of the three strength testers described above (Pressley, Clemson, and Stelometer) has been made by Burley and Carpenter (4). The following is quoted from their summary and conclusions.

> It was found that strength test results of maximum significance could be obtained at slightly more than 3 mm., hence ⅛″ (3.2 mm.) gauge spacing was adopted as standard by the Standards and Testing Branch of the Cotton Division, U. S. Department of Agriculture . . .
>
> In the preliminary work it was found that bundle widths from ⅛″ to ¼″ gave essentially the same results when the 3 mm. gauge spacing was used. Continued use of the narrow bundle widths, however, was found to damage the leathers of the clamps, hence the ¼″ bundle width has been adopted as standard procedure for testing by laboratories of the Cotton Division of the Agricultural Marketing Service.
>
> A torque within the range of 10 to 18 inch-pounds when the 3 mm. gauge spacer was used, gave comparable results.
>
> Fiber length seems to have little, if any, effect on optimum gauge spacing within the range of 0 to 4 mm.

Fig. 13.7 shows the correlation found by Burley and Carpenter between 22s yarn strength and fiber tensile strength.

References

1. Webb, and Richardson, *Textile World*, 100, Feb. (1950).
2. Hertel, K. L., Fiber Research Lab., Univ. of Tennessee.

3. Phillips, *Textile Research J.*, 18, Nov. (1948); *Textile Research J.*, 19, May (1949).
4. Burley, Jr., S. T., and Carpenter, F., "The Evaluation of Results Obtained on Available Types of Fiber Strength Testers using Various Gauge Spacings and their Relation to Yarn Strength," U. S. Dept. Agr. (1955).

CHAPTER XIV

Fiber Fineness and Maturity

Fiber fineness and maturity seem to be associated closely, not only from the physical properties themselves, but from some of the testing methods developed for their evaluation. For this reason, the two properties are discussed together.

The term fiber fineness is defined as the weight fineness of

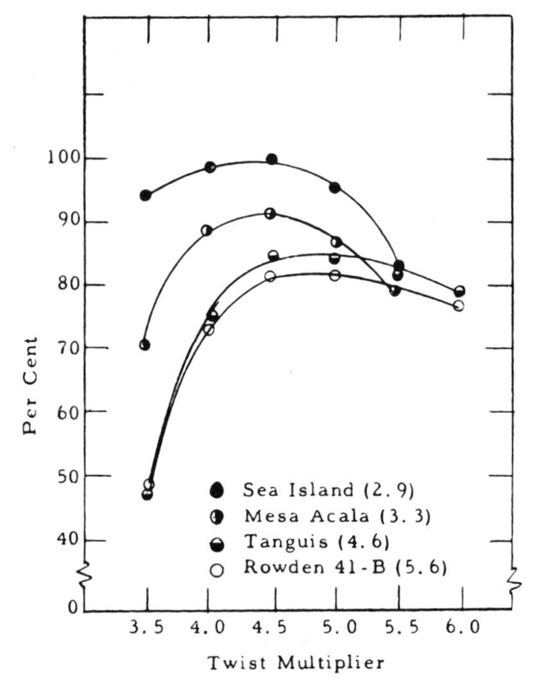

Fig. 14.1. Relative skein strength (per cent of Sea Island optimum) at different twist multipliers for 16/1 yarns. (Source: U. S. Dept. of Agriculture.)

cotton fibers and is usually expressed in units of micrograms per inch, that is, the average weight of 1 inch lengths of fiber expressed in micrograms (0.000001 gram). The comparable term for synthetic fibers is denier, which is the measure of grams per 9000-meter length; for wool, the fiber diameters of a known sample size are measured in microns (0.001 mm.) to get a group indication of fineness, which in turn determines wool quality.

The fineness of cotton fibers is governed by variety, and within varieties the fineness is affected by fiber diameter and the actual

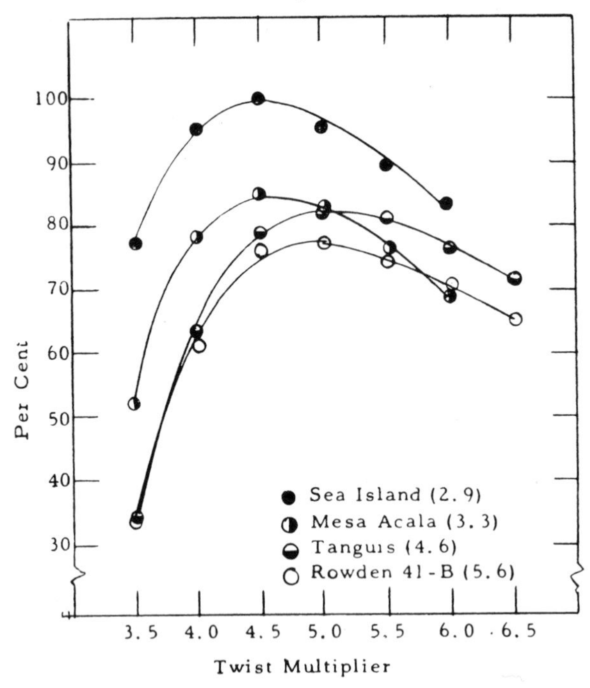

Fig. 14.2. Relative skein strength (per cent of Sea Island optimum) at different twist multipliers for 24/1 yarns. (Source: U. S. Dept. of Agriculture.)

amount of cellulose present. Generally speaking, short-staple cottons tend to be coarse, and long-staple cottons tend to be fine-fibered.

The measuring of cotton fineness is one of the most important fiber tests that can be made in a testing laboratory. The fineness of the fibers controls to a large extent the processing difficulties which are likely to be encountered, the strength of the yarn, and the appearance of the finished material. One of the biggest factors which fineness affects is formation of neps. With other variables remaining constant, cotton with a low microgram per inch weight will produce a substantially greater number of neps than will a cotton with a high fineness rating subjected to the same manufacturing standards. This factor is of great importance to mills producing high-quality yarns and fabrics where the presence of neps is detrimental to finished product quality.

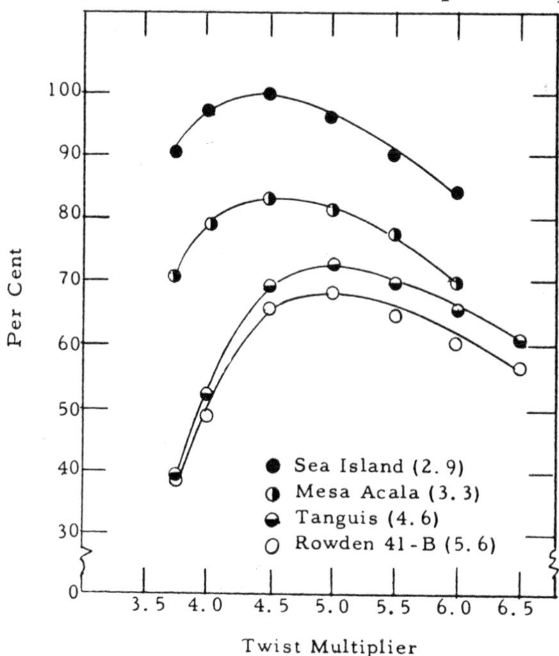

Fig. 14.3. Relative skein strength (per cent of Sea Island optimum) at different twist multipliers for 36/1 yarns. (Source: U. S. Dept. of Agriculture.)

In addition to affecting nep count, the fineness also has a marked effect upon the spinning performance and strength of the yarn. With other fiber properties remaining constant, fine fibers will produce a yarn with higher strength than will coarse fibers. Also, the finer fibers can be spun into higher yarn numbers than coarse fibers.

Fiori and Brown (1) conducted a research project in which all fiber properties were held as constant as possible with the exception of fiber fineness in order to get the effect of fineness on the strength of yarn. As shown in Figs. 14.1, 14.2, and 14.3, the finest cotton, Seaberry Sea Island (2.9 micrograms per inch), has a substantially greater strength than the other cottons which ranked in the following order with respect to fineness: Mesa Acala (3.3), Tanguis (4.6), and Rowden 41-B (5.6).

It was also confirmed in this study that yarns spun from fine

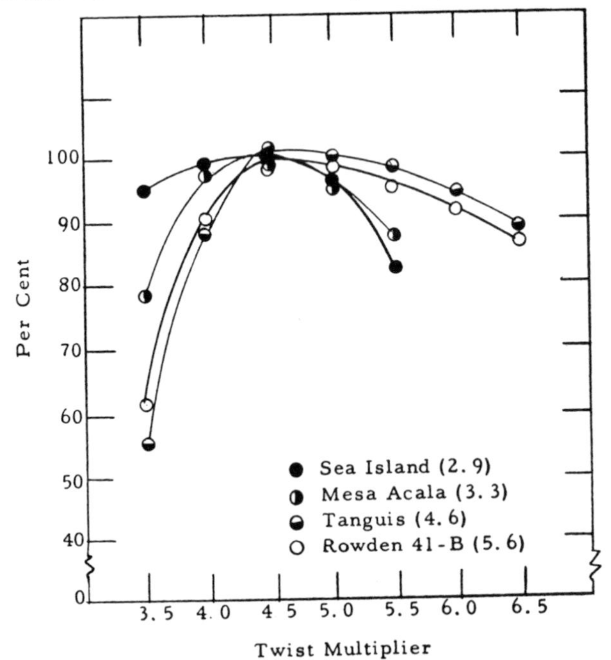

Fig. 14.4. Percentage of skein strength based on optimum strength of each cotton for 16/1. (Source: U. S. Dept. of Agriculture.)

fibers will reach maximum strength with less twist than is required for the coarser fibers, with the exception of the medium coarse yarn number (16/1). This is shown in Figs. 14.4, 14.5, and 14.6. Another interesting relationship shown in these charts is the rate of increase and decrease in yarn strength with respect to twist. The fine fibers appear to be less sensitive to twist up to the optimum strength point of the yarn and to be more sensitive beyond this point, whereas yarn spun from coarse fibers requires more twist to reach the maximum strength; however, the yarn loses strength at a slower rate once the maximum point has been exceeded.

Fineness is also important from a processing and finished product quality point of view. If an extremely fine cotton is used to increase yarn strength, the nep count will also increase; if a coarse fiber is used to reduce neps, the yarn strength will also be

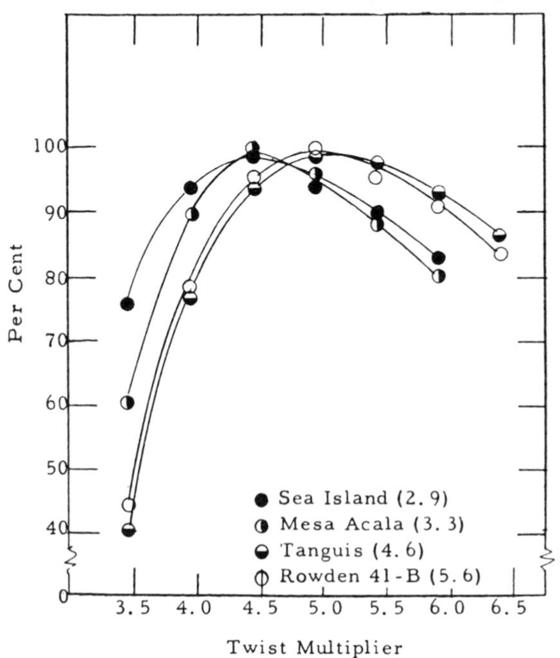

Fig. 14.5. Percentage of skein strength based on optimum strength of each cotton for 24/1. (Source: U. S. Dept. of Agriculture.)

reduced. Therefore, a balance between these extremes must be established and maintained by blending the fibers with different fineness ratings in the proper proportions in the original cotton mix at the opening department. It has been proven rather conclusively that these problems can be solved by a properly established testing program and intelligent use of test results. For example, some mills have found that by controlling the fineness of the mix, it was possible to control both the nep count and the ends down at spinning (2).

Fiber Maturity

Much is known about the planting, growth, and harvesting of cotton, for probably no other agricultural product has been studied by scientists any more carefully. Yet, with all the research and investigation, much remains to be learned, and especially

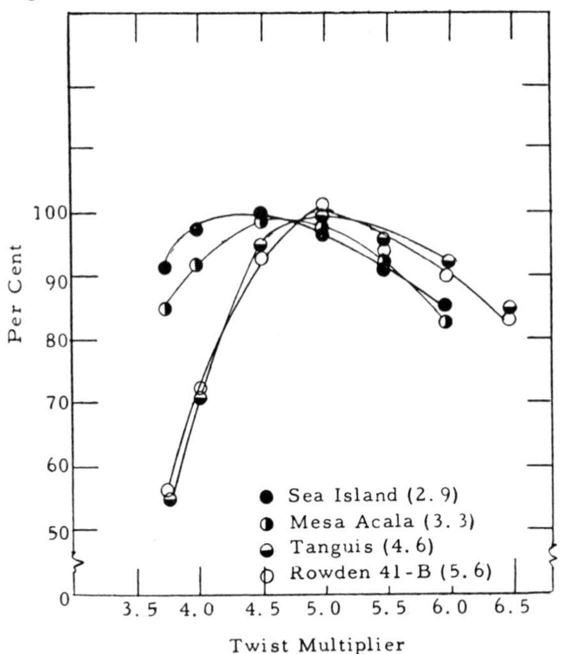

Fig. 14.6. Percentage of skein strength based on optimum strength of each cotton for 36/1. (Source: U. S. Dept. of Agriculture.)

much to be done in genetics to control the maturity characteristics of the fibers produced.

Maturity indicates ripeness or full development; immaturity indicates lack of full development. As applied to cotton, maturity refers to the development of individual fibers within a boll, rather than to the plant or boll itself. Thus a ripened, fully mature boll contains fibers both mature and immature.

After planting the cotton seed, the first sign of growth occurs within a period of two weeks when two heart-shaped leaves appear, with a stalk between; the leaves drop off shortly. In about five to six weeks fruiting buds appear, which are followed in two to three weeks by the first blossoms. The blossoms are white for two days, red for one day, and then fall off, leaving the small bolls. On the day the flower opens to full bloom, the seed hairs start to grow from the outer surface of the tiny seeds. The fibers grow in length for about sixteen to eighteen days; short staple cottons reach their length earlier, and long staple cottons may take up to twenty or more days. During this first period of growth, the fibers reach full length and diameter but contain only a thin primary wall. During the following twenty-two to fifty or more days, there is formed the secondary wall or filling out of the fiber, accomplished by the daily depositing of successive layers of cellulose internally. There is an undeveloped section within the fiber called the lumen, and a skin or cuticle forming a thin film of wax and pectin covering the primary wall. At maturity, the fibers die, the boll dries and cracks open. The individual fibers dry out and shrivel, collapsing to form the ribbon-like tube.

If any of the seeds in the boll are not fertilized, or if fertilized seeds are killed or die for any reason, then these partially developed seeds known as motes produce fibers that are immature. These immature fibers are thin-walled: that is, they have not developed thick secondary walls. Other thin-walled fibers are produced from the broad end of each seed; these fibers may not reach full growth before the boll opens. However, they generally do not give trouble in processing.

Mature fibers are characterized by fully developed secondary walls: the thicker the wall, the more mature the fiber, and the thinner the wall, the more immature the fiber. It is difficult to judge maturity from the lumen size, for the robust fibers crowd the lumen to a small slit-like size, and the immature fibers collapse to make an apparently small lumen.

What is known about the effects of the presence of immature fibers, for the average bale of cotton contains about 30 per cent of such fibers? Actually, knowing that immature fibers are present in such numbers, and knowing that millions of bales of cotton are processed each year, it can be assumed that the industry has had to tolerate an unwelcome guest for a long time. On the other hand, it is known that as the ratio of immature fibers is increased, the spinning of yarn deteriorates with accompanying lowering of efficiencies in processing; that dyeing and finishing problems multiply; and that neppiness in yarn and fabric increases. Neps are small tangled aggregates of fibers; it is only logical that

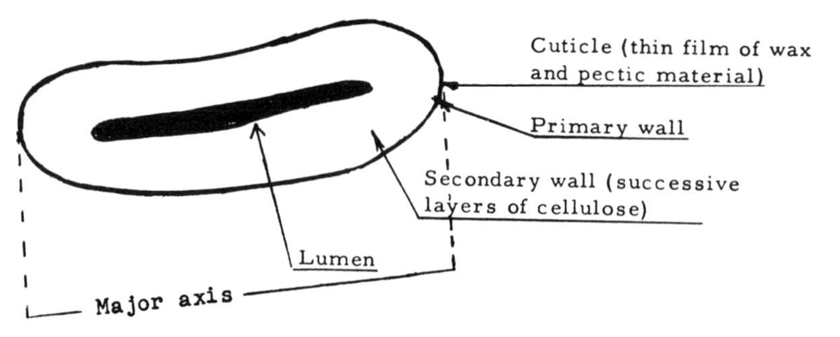

Fig. 14.7. Diagram of a normal cotton fiber cross section.

any very soft and flexible fiber should tangle more easily than a rigid fiber. Immature fibers are weak, so almost invariably one or more of these fibers can be found in a nep. (By the same token, very fine cottons like Sea Island and Egyptian, and very fine denier synthetics form neps more easily because of their fineness than do the coarse-bodied fibers.)

The actual importance and use of maturity test data are still points of discussion in the industry. This is partially due to the fact that it is difficult to obtain cotton of high maturity consistently. However, those mills studying the problem, and who are getting mature cotton, are benefiting substantially.

Measurement of Fiber Fineness by the Sheffield Micronaire

The most widely used method in quality control laboratories for measuring fiber fineness is the Micronaire method. This is due partially to the speed with which the measurements can be made and the simplicity of operation.

Theory of Machine. The Micronaire (See Figs. 14.8 and 14.9) operates on the principle of resistance to air flow. Air is forced through a sample of fibers and the resistance of the sample to air flow is indicated to give a measure of fiber fineness.

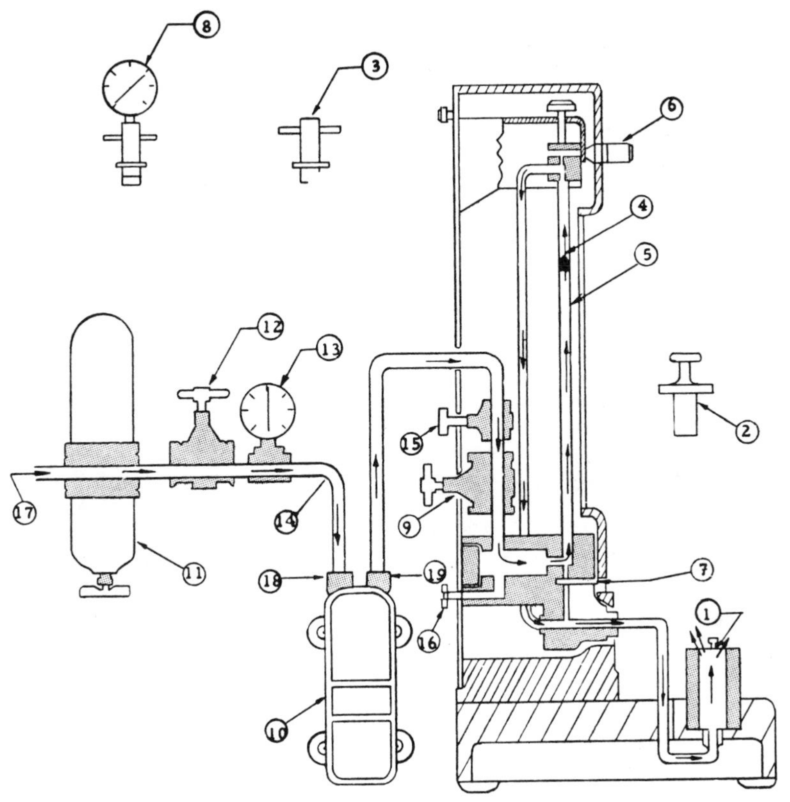

Fig. 14.8. Diagram of Sheffield Micronaire.

The rate of air flow through a given weight of a sample of fibers is dependent on the surface resistance of the fibers. This resistance to flow is greater for fine fibers than it is for coarse ones, just as the ratio of surface area to weight is greater for fine fibers than for coarse. For example, knowing that the diameter of a fiber is proportional to the square root of its fineness or weight, μ, in micrograms per inch of length, then the ratio of surface area to weight is equal to

$$\frac{kd}{\mu} = \frac{k\sqrt{\mu}}{\mu} = \frac{k}{\sqrt{\mu}}$$

(The value of k can be neglected and unity used in this example.) If a given total weight of fibers is assumed, it is possible to com-

Fig. 14.9. Micronaire for measuring fiber fineness. At right, the sample is being ejected from the instrument.

pute (see Table 14.1) the relative resistance to air flow for different finenesses; the values calculated are the ratios of surface

TABLE 14.1

Computation of Relative Resistance to Air Flow

μ Fineness	w Total wt.	n Fibers in sample	$\frac{1}{\sqrt{\mu}}$ Ratio, surface to wt./fiber	$\frac{1}{\sqrt{\mu}}(n)$ Ratio, total surface to wt.
3	24	8	0.58	4.64
4	24	6	0.50	3.0
6	24	4	0.41	1.64
8	24	3	0.35	1.05
12	24	2	0.29	0.58

area to weight multiplied by the number of fibers in the assumed sample weight.

Thus, fibers of 3 micrograms per inch have a relative resistance of 4.64 as compared to that of 3.0 for cotton of 4 micrograms per inch, and so on. If the values in the last column, i.e., the ratios of total surface area to weight, are plotted on log-log paper against the fineness in each case, a straight line results. This shows a curvilinear relationship between fineness and rate of air flow, with the fine-fibered cottons offering the greatest resistance, and the coarse-fibered cottons the least resistance to air flow.

Procedure for Preparation of Test Sample

A. Preliminary

Select at least three pinches or small tufts of fibers at random from laboratory or classer's sample. Remove all large pieces of foreign matter such as leaf, stems, etc., and randomize or mix the composite sample. The mixing is done by hand unless a mechanical fiber blender is available.

B. Weigh sample

From the composite sample, weigh a 50-grain test sample. Randomize the test sample by hand and work into a fluffy form by opening up any knotty or compressed sections of sample. With this completed, the sample is ready for testing. The Shadowgraph scale designed primarily for weighing 50-grain samples will speed up the test and insure consistent accuracy.

Procedure for Calibration and Use of the Micronaire

A. Calibration (See Fig. 14.8)

1. Open primary air regulator (12) until pressure reads 25 pounds on air pressure gauge (13).
2. Insert air pressure indicator (8) into fiber compression chamber (1).
3. Press foot pedal (10) so air will flow through the instrument, and adjust pressure regulator (9) until the needle on the air

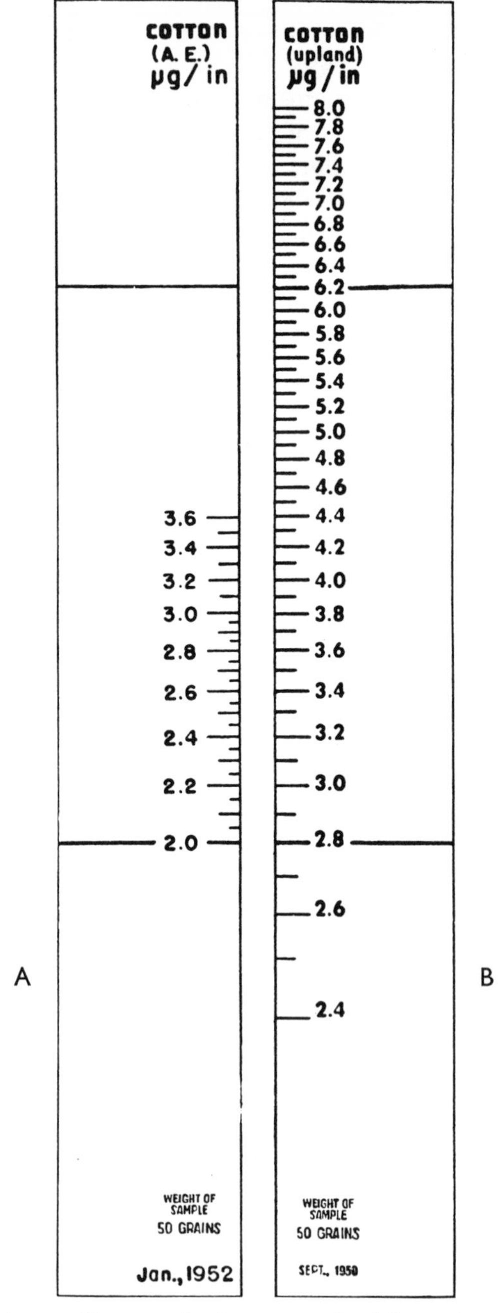

Fig. 14.10. A: curvilinear scale for use on the Micronaire with American Egyptian cotton; B: curvilinear scale for use on the Micronaire with American Upland cotton. The distance between horizontal reference lines should be four inches on scales to be used on the instrument. (Source: U. S. Dept. of Agriculture.)

pressure indicator coincides with the line marked "cotton." If the Micronaire is equipped with the mercury "U" type manometer, adjust the pressure regulator (9) until a difference of 6.0 pounds per inch is obtained between the right and left-hand columns. The right-hand column will fall, and the left-hand column will rise when pressure is applied. Be sure to tighten the locknut when the desired pressure is reached. *Caution:* Excessive pressure will blow the mercury out of the manometer. The first time the manometer is used, adjust the pressure regulator (9) until no air is flowing through the Micronaire and then adjust to the desired pressure.

4. Release the air pressure and remove the air pressure indicator.
5. Insert master with 6.0 stamped on the top. Apply air pressure and read float position. The top of the float should coincide with 6.0 if the instrument is equipped with a linear scale, and 6.2 if it is equipped with a curvilinear scale (See 14.10). If the top of the float does not coincide with the proper position on the scale, adjust with the float positioning knob (6).
6. Remove 6.0 master and insert 2.0 master.
7. If the float position does not correspond with 2.0 on the linear scale or 2.8 on the curvilinear scale, use the amplification adjustment screw (7). If the spread of the float travel limits is not sufficiently wide and the float is too high, turn the amplification adjustment screw clockwise until the float rises about $1\frac{1}{3}$ times the distance it was above the desired position on the scale. Re-position the float by float positioning knob (6), and check the spread by inserting 6.0 master. Repeat until properly calibrated.

   If the spread of the float travel limits is too wide and the float is too low, turn the amplification adjustment screw counterclockwise until the float falls about $1\frac{1}{3}$ times the distance it was below the desired position on the scale. Remove 2.0 master and insert 6.0 master. Observe the float position and make the limits of the float travel coincide with the scale limits by repeating the foregoing adjustments.

B. Use

1. Place the sample in the fiber compression chamber.
2. Insert the fiber compression plunger and lock in place by twisting. This compresses the cotton into a cylinder 1 inch in diameter and 1 inch long.
3. Step on the foot pedal valve.
4. Take the reading directly from the scale, which is calibrated in units of micrograms per inch. During steps 3 and 4 the float will usually act as follows: (a) When air is turned on (step 3), the float will immediately rise to the top of the tube. (b) It will then fall back and drift downward slightly, then come to rest, fluctuating slightly. Reading should be taken at this time. (c) If the sample is left in the compression chamber for any length of time with the air turned on, the float will fall due to loss of moisture by the sample.
5. Remove the fiber compression plunger before turning off the air so that the sample will be blown from the chamber.
6. Turn off the air with the foot pedal before inserting the next test specimen.

C. Calculations. As outlined previously, the reading taken directly from the scale on the Micronaire is in units of micrograms per inch.[1] For this reason, no calculations are necessary when using this instrument.

TABLE 14.2

Rating of Cotton Fiber with Respect to Fiber Fineness

Microgms./in. of fiber	Fineness
Below 3	Very fine
3.0 to 3.9	Fine
4.0 to 4.9	Average
5.0 to 5.9	Coarse
6.0 and above	Very coarse

The standards which are used throughout the textile industry for rating fibers with respect to fineness were developed and

[1] The scale has been developed empirically and occasional readings may reflect some influence of fiber maturity as well as fiber fineness.

issued by the United States Department of Agriculture and are shown in Table 14.2.

Micronaire Scales. During the first years that the Micronaire instruments were used, they were equipped with linear scales. However, after a considerable amount of research and investigation, the United States Department of Agriculture installed curvilinear scales on all of the instruments owned by the Department. As a result of this action by the United States Department of Agriculture, most instruments now in use are equipped with the curvilinear scales. The curvilinear scales are more accurate than the linear scales due to the fact that the relationship between air flow and pressure drop is not linear but is curvilinear. Since this is the principle used to measure fineness with the Micronaire, it then is evident that a curvilinear scale is more accurate than a linear scale.

Based upon the research work done with the Micronaire, the United States Department of Agriculture has established the following relationships between the linear and curvilinear scales:

$$Y_1 = 1.514 + 0.0413X + 0.101X^2$$
$$Y_2 = 2.314 + 0.0413X + 0.101X^2$$
$$Y_1 = Y_2 - 0.8$$

where Y_1 = American Egyptian fiber fineness
Y_2 = American Upland fiber fineness
X = Micronaire linear scale (inches)

Number and Frequency of Tests. The number and frequency of tests depend entirely upon the use of the test results. In mills where the test results are used as a means of determining and laying out mixes in the opening department, it is usually customary to test each bale as it is received. In this way the bale can be labeled with respect to its fineness property and stored in the warehouse according to some definite plan.

In mills where the testing is done merely to gain background material or to get some general estimate of the type of stock being processed, the frequency will be as low as 10 per cent of the bales received.

The number of tests to make per bale or shipment again depends upon the use of the test results. Mills testing 100 per cent of the bales received have found it satisfactory to take one sample per bale. The reason for this is that a sufficient number of samples are taken to get a reasonably reliable estimate, even though only one sample per bale was taken. For mills testing much less than 100 per cent, it is recommended that a composite sample be taken and well blended before testing.

The Causticaire Method

The Causticaire method for obtaining cotton-fiber maturity and fineness was developed and adopted by the U. S. Department of Agriculture in evaluating these properties for their own reports. The method involves the use of the Micronaire instrument. However, in addition to the usual test for fineness previously described, there is an additional test of the sample after it has been treated with caustic soda and properly re-prepared for test in the Micronaire. Thus, the Causticaire method is a combination of the Micronaire test on raw cotton and on cotton treated with caustic soda (sodium hydroxide, 40 Tw) to give fineness as well as maturity.

Theory of Test. When cotton fibers are treated with caustic soda, there results a swelling of the fibers with the mature fibers reacting more readily and to a greater extent than the immature fibers. Thus, the action causes an exaggeration of the original dimensions of the fibers. Mature fibers swell quite visibly, approaching in many cases a more nearly round section, with obliteration of the lumen. Immature fibers swell much less, retaining their characteristic thinwalled flat section and thus can be distinguished from the mature. Immaturity is determined by comparison of the wall thickness with lumen diameter: If wall thickness is one-half or less than the diameter of the lumen, as measured along its major axis, the fiber is immature.

Procedure for Testing by the Causticaire Method

For the procedures for preparation of the test sample, for calibration, and for the use of the Micronaire instrument, refer to the outline under the Micronaire test. The changes in procedure for the Causticaire test are summarized below.

1. The original laboratory sample should weigh from 7 to 10 grams. The sample is conditioned and run through a mechanical blender so as to give uniformity to the sample.
2. From the master sample, the usual 50-grain specimen is weighed out; as previously indicated, a Shadowgraph balance simplifies the weighing. For the Causticaire, it is recommended that two 50-grain specimens be weighed out for the test.
3. Run a Micronaire test on each sample. The instrument should have the Causticaire scale, or else readings in micrograms per inch should be converted to the equivalent Causticaire value by use of a suitable table or chart.
4. Treat the two 50-grain specimens plus any residue from the original blended sample in a solution of sodium hydroxide (40 Tw) with a $1\frac{1}{2}$ per cent wetting agent added. The sample can be kept in an Orlon marquisette bag during all wet treatments. The time of immersion of the sample in the caustic soda should be sufficient to permit thorough saturation.
5. Wash the sample thoroughly in tap water.
6. Dry out the sample in an oven with a temperature not in excess of 220°F.
7. Condition and reblend the sample on a mechanical fiber blender so that the treated cotton has the same degree of fluffiness as in the original portion of the test.
8. Weigh out two 50-grain specimens.
9. Make another pair of tests on the Micronaire, again using the Causticaire scale, or converting to Causticaire values.

Calculations. The Causticaire maturity index is calculated from the formula:

$$\frac{\text{Ave. Causticaire reading, untreated}}{\text{Ave. Causticaire reading, treated}}(100) = \text{maturity index.}$$

The Causticaire fineness in micrograms per inch is calculated from the formula:

$$+1.185 + 0.00075(T^2) - 0.020\ (\text{MI})$$

where T is the average reading on the treated specimen; MI is the maturity index.

The descriptive terms that can be applied to the Causticaire maturity are shown in Table 14.3.

TABLE 14.3

Fiber Maturity Index, Causticaire [a]

Maturity index	Rating
82 and above	Mature
76 to 81	Average
70 to 75	Immature
Below 70	Very immature

[a] Source: U. S. Department of Agriculture

Discussion. It is claimed that Causticaire fineness and maturity values represent all gradations in fiber weight per inch and cell wall development from fiber to fiber throughout a sample, and from base to tip of individual fibers. It is believed that this method of obtaining maturity is fast, simple, and gives good results and therefore should be of real value to the industry. However, the method of calculating fineness based on maturity is open to question and has not been accepted as yet by technical groups.

In using the Causticaire test, a special Causticaire scale is used with the Micronaire. It is suggested that the original curvilinear scale can yet be used for Upland cottons, for example, for mill use in getting fineness. An additional Causticaire scale can be mounted on the instrument if desired so that in those cases where the maturity is wanted, the additional test on the sample treated with caustic soda can be made. It is also possible to convert readings on the usual curvilinear scales calibrated in micrograms per inch to an equivalent Causticaire reading by the simple expedient of preparing a dual scale for desk use, or by preparing a table which lists Causticaire values for equivalent conventional Micronaire readings. The testing for fineness by textile mills has grown in use. One reason for this is the speed with which fineness values can be obtained to allow the laying out of proper mixes. It is believed that in many cases the additional time delay in getting maturity would be prohibitive. Thus, it would seem logical to retain the usual accepted method for determining fineness, with the provision that maturity could be determined as and when desired.

Frequency of Test. The frequency of test for fineness has been discussed under the Micronaire test.

It is recommended that maturity tests be made on a spot-check basis of all bales received by the spinning mill. The number of bales to test might be set at 5 to 10 per cent of each shipment. Also, it is possible to test composite samples representing 10 to 20 bales to give an indication of the average level of maturity.

Arealometer

The Arealometer is a null air-flow instrument developed by Hertel and Craven, University of Tennessee, to measure both the fineness and maturity of cotton fibers. Hertel and Craven developed the instrument so that changes in air flow due to shape of the fiber could be measured. Most instruments using the air-flow principle of measuring fineness of fibrous materials are capable of measuring only the amount of air passage, using a constant pressure through the test specimen. This flow of air through the specimen is affected not only by the size or fineness of the material, but also by the shape of the fibers. The flat, thin-walled, ribbon-like fibers will offer a different air resistance than the mature thick-walled fibers. This is explained by Hertel and Craven (3) as follows:

> . . . the Arealometer is responsive to two properties: specific area, defined as the ratio of the external surface of the fibers to the volume of fibrous material and immaturity ratio, defined as the ratio of the area of a circle having the same perimeter as an average fiber to the actual cross-sectional area of the fiber. A primary cause of resistance to the flow of air is the amount of external fiber surface exposed to the air flow, and this resistance is used as a measure of specific area. The resistance, however, is modified by the distribution and orientation of these surfaces; hence, on suitably compressing the sample, that change in resistance which is essentially due to reorientation of flattened fibers becomes a measure of the immaturity ratio.

Principle of Machine. The principle used in measuring the fineness and maturity with the Arealometer is described by Hertel and Craven as follows (see Fig. 14.11):

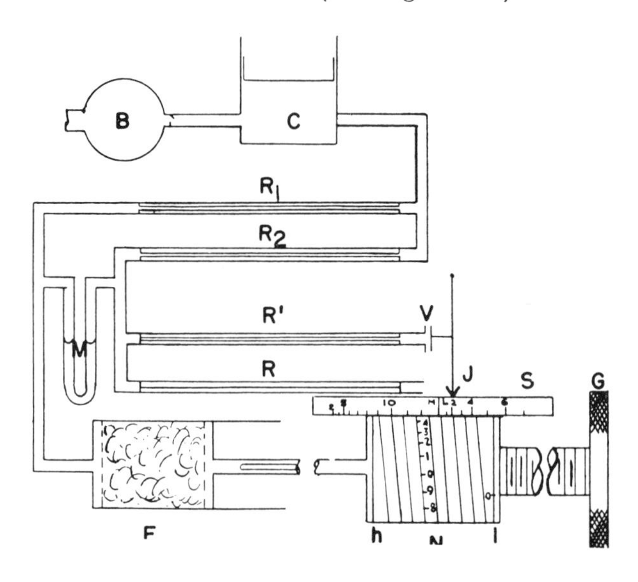

Fig. 14.11. Arealometer.

The method adopted for the Arealometer is to fix the resistance to air flow, R, and find by experiment the length, L, to which a plug of fiber must be compressed in order that it shall have this resistance. In electrial measurements, the Wheatstone bridge has provided a very reliable and accurate method for comparing resistance. The Arealometer is similar to a Wheatstone bridge in which three of the resistances consist of copper capillaries and the fourth is the compressible porous plug of fiber being tested. Fig. 14.11 shows a diagram of the instrument. An aspirator bulb, B, forces air into a cylinder, C, containing a loose-fitting weighted piston. This cylinder supplies air to the bridge at a steady pressure of about one inch of water. It is clear from the diagram that the circuit consists of two parallel branches with each branch containing two resistances in series. All of the air entering R_1 passes on through the fibers in F to the atmosphere, while that entering R_2 reaches the atmosphere through R or R and R' depending upon the position of valve V. Since R_1 is essentially equal to R_2, the inclined manometer, M, will indicate zero pressure difference when the fiber sample in F has a resistance equal to that of R (or R and R' combined). The necessary degree of compression is achieved by driving a piston into the sample chamber by means of a screw which carries a calibrated drum. The end of the piston is perforated, as is also the bottom of the chamber. Two spiral scales engraved on the drum give direct readings of specific area in square millimeters per cubic millimeter. One scale is read at low compression when the valve, V, is open, and the other scale is read at high compression when V is closed. A movable screen (not shown in the diagram) whose motion is synchronized with the opening or closing of valve V uncovers the appropriate scale.

Calculations. Since the specific area, A, is read directly, it is reported as an average value for the specified number of test specimens. The units are mm^{-1}.

The immaturity ratio, I, is calculated by the formula

$$I = \sqrt{0.07D + 1}$$

where D is the average difference between the readings at low compression and high compression. The immaturity ratio has no units.

The immaturity ratio will usually range from approximately 1.5 to 1.75 for very mature cottons to 2.5 or 3 for very immature cottons. The relationship of per cent maturity and immaturity ratios is shown in Fig. 14.12.

The fiber fineness, W, expressed in micrograms per inch, may be calculated by the formula

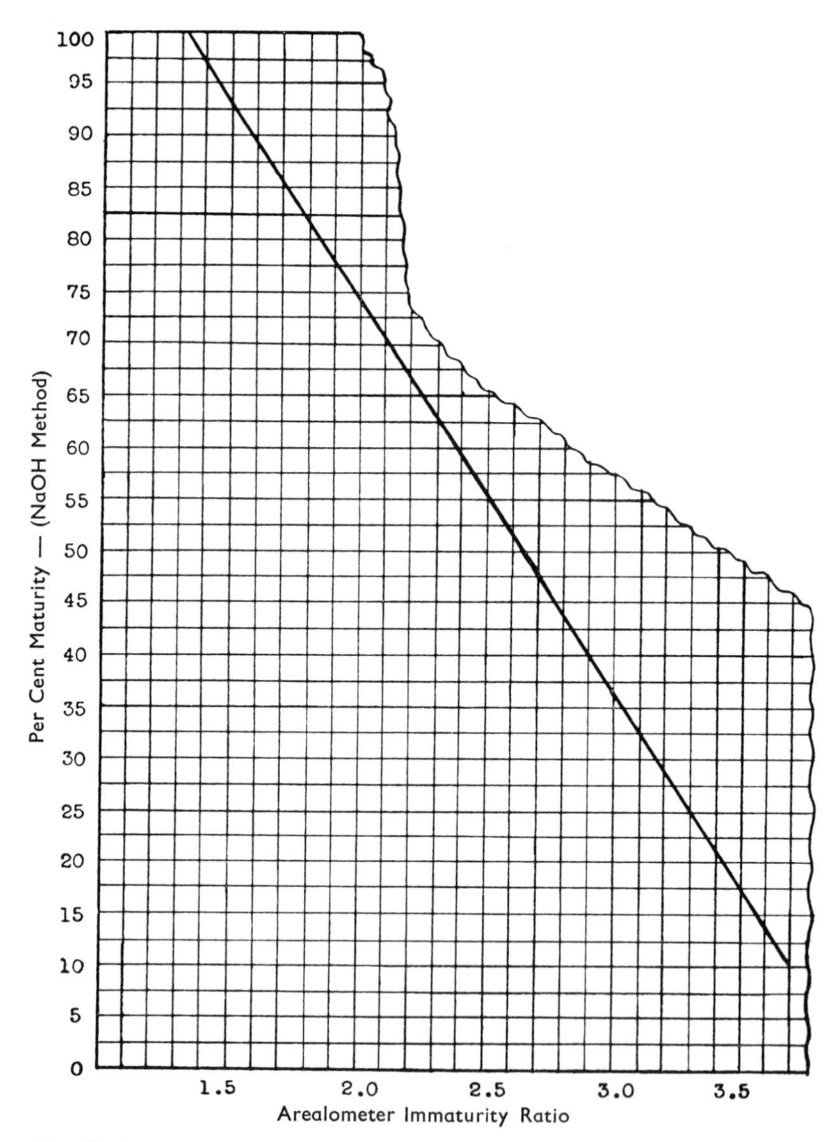

Fig. 14.12. Relationship of Arealometer immaturity ratio to array per cent maturity. Line is represented by the formula:

$$\% \text{ maturity} = 150.5 - (38.1)I$$

(This formula is good for approximations only.)

$$W = 485{,}000\left(\frac{I}{A^2}\right)$$

In addition to the above, such values as the average perimeter of the fiber expressed in microns may also be calculated. The average perimeter, p, is calculated by the formula

$$p = 12{,}566\left(\frac{I}{A}\right)$$

The average wall thickness, t, is calculated by the formula

$$t = 1{,}000\left(\frac{T}{A}\right)$$

where

$$T = \frac{2}{1 + \sqrt{1 - \frac{1}{I}}}$$

Substituting,

$$t = \frac{2000}{\left(1 + \sqrt{1 - \frac{1}{I}}\right)A} \text{ in microns}$$

Fiber Fineness and Maturity by Weight of Known Lengths

Array Method with Sodium Hydroxide Swelling

The Array method is not used extensively by quality control laboratories because of the tedious and time consuming nature of the test, and the prerequisite of having fiber length arrays already prepared. Research laboratories, however, use this test method because of the accuracy made possible by consideration of individual fibers. Fiber weight per unit lengths in micrograms per inch (abbreviated as μg/in) is determined from the total length and weight of the fibers tested. Per cent maturity is arrived at by inspection and classification of individual fibers as either mature or immature, and calculation of the per cent mature fibers of the total tested. Treatment with caustic soda solution resulting in varying degrees of swelling of the fibers enables the technician to differentiate more easily between mature and immature fibers. The array method may be carried out for both fineness and maturity, or for one of the two properties alone.

Procedure for Preparation of Test Specimen and for Test

A. Preparation of Test Specimen

1. The test specimen is selected from the length groups of two arrays formed in connection with the Suter Webb Sorter test. The fibers in matching length groups should be combined and mixed to provide a single array. Groups of $\frac{3}{16}$ inch length and shorter, and any length group in which the total weight of fibers is less than 2 mg., should be excluded from this test. Each of the remaining groups can then be divided in half, and two determinations can be carried out, preferably by different technicians.
2. For each test at least 100 fibers should be taken from each of the length in one sub-sample. The total number of fibers selected from all the length groups in one sub-sample should approximate 1,000.
3. Weigh each specimen group selected to the nearest 0.01 mg.

B. Test Procedure for Fineness Only

1. Place a specimen group upon a black background and under a fixed magnifying glass.
2. Accurately count the number of fibers in the group.
3. Repeat for all specimen groups.

C. Test Procedure for Maturity and Fineness. If maturity and fineness are both to be determined, the following procedure is used in place of B above.

1. The fibers in one specimen group at a time should be gently pulled apart by using a needle and mounted reasonably parallel and in a single layer upon a microscope glass slide; the fibers should then be covered with a cover glass.
2. After mounting, flood the fibers with an 18 per cent solution of NaOH (caustic soda). This is done with a dropper or pipette by introducing the solution at the edge of the cover glass.

The cover glass can be tapped to remove air bubbles and excess solution; the excess solution should be wiped off.

3. The microscope should be equipped with the fiber micrometer and euscope. The micrometer is used to facilitate making accurate measurements of fiber dimensions and the euscope to make the counting less tiresome.
4. Adjust the microscope to a magnification of approximately 400×; place the glass slide on the stage of the microscope.
5. Examine and classify the fibers by bringing them successively into view with the use of the mechanical stage of the microscope. Fibers having a wall thickness less than one-half the diameter of the lumen are considered immature, and those with a wall thickness greater than one-half the lumen diameter are considered mature. In longitudinal microscopic examination, the dimensions at maximum ribbon width are considered, that is, where the major cross sectional axis appears flattened to the observer. Fibers which appear immature in any section of their length within the field of vision are rated as immature. All fibers must be classified in either category according to the A.S.T.M. method.
6. Examine all fibers in the test specimen and record the number of mature and immature fibers.

D. Calculations for Fineness. To facilitate the recording and calculations involved in this test, it is recommended that a form similar to that shown in the A.S.T.M. manual be used. The calculations involved are demonstrated in Table 14.4.

TABLE 14.4
Basic Fineness Calculations

Length groups $\frac{1}{16}$ in.	Weight of fibers, mg.	No. of fibers	Fineness, μg per in.
19	0.50	100	4.2
17	0.48	125	3.6
15	0.39	94	4.4
13	etc.		

For 100 fibers in the $\frac{19}{16}$-inch group length, the total length of fibers would be:

$$\tfrac{19}{16}\,(100) = (1.1875)(100) = 118.75 \text{ inches}$$

With 1,000 mg. per microgram, the fineness in micrograms per inch would be:

$$\text{Fineness} = \frac{(0.50)(1{,}000)}{118.75} = 4.2\ \mu\text{g per inch.}$$

To simplify this calculation, A.S.T.M. recommends the use of the midpoint of the length group in recipocal inches. For the $\tfrac{17}{16}$-length group, this value would be 0.941, and the calculation is:

$$\text{Fineness} = \frac{(0.941)(0.48)(1.000)}{125} = 3.6\mu\text{g per inch.}$$

For the $\tfrac{15}{16}$ inch group:

$$\text{Fineness} = \frac{(1.067)(0.39)(1.000)}{94} = 4.4\ \mu\text{g per inch.}$$

To be technically correct, the weighted average for fineness must be calculated. One method of determining the weighted average is shown in this example:

Fibers	μg/in.	Product
100	4.2	420
125	3.6	450
94	4.4	414
$\Sigma = 319$		1284

$$\text{Average fineness} = \frac{1284}{319} = 4.0\mu\text{g per inch.}$$

Another method for calculating the average fineness as shown in the A.S.T.M. manual is as follows:

$$\text{Average fineness} = \frac{\Sigma(WF)}{\Sigma W}$$

where W is the weight of the group; F is the fineness of the group.

The calculation using the previous data is:

W	F	WF
0.50	4.2	2.10
0.48	3.6	1.73
0.39	4.4	1.72
$\Sigma = 1.37$		5.55

$$\text{Average fineness} = \frac{5.55}{1.37} = 4.0\ \mu\text{g per inch.}$$

Unless the fineness of the sample has a high variation, it may not be necessary to calculate a weighted average. The number of fibers can be obtained from the count in either B or C test procedure.

E. Calculations for Maturity. The number of mature fibers found in the test specimen expressed as per cent of the total number of fibers in the test specimen represents the per cent maturity. This is shown in the following relationship:

$$\text{Per cent maturity} = \frac{F_t - F_i}{F_t}(100)$$

where F_t = total number of fibers examined; F_i = total number of immature fibers.

EXAMPLE: If 780 mature fibers and 220 immature fibers were found in a test specimen, the per cent maturity would be calculated as follows:

$$780 + 220 = 1{,}000 \text{ total number of fibers inspected}$$

$$\%\ \text{maturity} = \frac{1{,}000 - 220}{1{,}000}(100)$$

$$= \frac{780}{1{,}000}(100) = 78.$$

The United States Department of Agriculture has issued

ratings with regard to maturity as determined by the Array method, and these are given in Table 14.5.

TABLE 14.5
Array Fiber Maturity

% maturity	Rating
Above 85	Mature
76 to 85	Average
66 to 75	Immature
65 and below	Very immature

Number and Frequency of Tests. The frequency of maturity testing is influenced by several factors. If the results are to be used in an effort to correlate this property with certain processing difficulties or finished product defects, it may be found necessary to test on a rather frequent schedule, such as for each cotton mix. If the results are used simply as background material or as general information as to the over-all quality of the cotton being used, the tests can be made less frequently.

If during fineness testing some of the samples are found to have an abnormally low weight per unit length, it is advisable to make maturity tests on the samples, for the larger variation in fineness may be due to immaturity.

It is usually satisfactory to make one test for maturity from each master laboratory sample. The number of fibers to be examined will depend upon the individual laboratory. However, the fibers selected for classification should be representative of the master sample. See also the discussion of this subject under the Causticaire method.

Random Sample Method for Maturity

This test is similar to the array method already described, but does not provide information on fiber fineness, nor does it require specimen preparation from a fiber length array. It may be applied to raw cotton, or cotton already processed but not subjected to chemical treatment.

A. Preparation of Test Specimen

1. The test specimen shall be obtained from a laboratory blended sample prepared according to the Fiber Blending Method (see page 251) or a similarly blended sample.
2. In drawing fibers from the sample, a large tuft is first extracted, and from this approximately two-hundred randomly selected fibers are coaxed from the unclasped fringe of the tuft. The fibers are directly deposited evenly onto a glass microscope slide, and covered with a cover glass. This procedure is repeated six times for a total specimen size of roughly twelve hundred fibers (six slides).

B. Test Procedure

The procedure for determining maturity is identical to that described for the Array method, Section C.

B.S.I. Cotton Fiber Maturity Test

The British Standards Institution *Cotton Fibre Maturity Test* distinguishes three classes of fibers, with the two extreme classes of "normal" and "dead" fibers being identified, while the middle group, termed "thin-walled" fibers, are accounted for by subtracting the fibers in the other groups from the total number of fibers tested. The maturity determination in this case may be subject to fewer errors of judgment, since only the easily recognized highly immature and very mature fibers must be distinguished. After treatment in the 18 per cent caustic soda solution, dead fibers are defined as having a wall thickness which is one-fifth or less of the maximum ribbon width; normal fibers are those which appear rod-like, have no continuous lumen, and lack well defined convolutions. In Fig. 14.13, *a* would be considered as a dead fiber, while *b* would be classified as normal. Thin-walled fibers are those which do not fit either of these definitions. The results of this test are expressed as the per cent normal and per cent dead fibers, as 60 per cent normal and 18 per cent dead; the per cent thin-walled can be found by subtracting these totals from

100 per cent (thin-walled equal 22 per cent in this example).

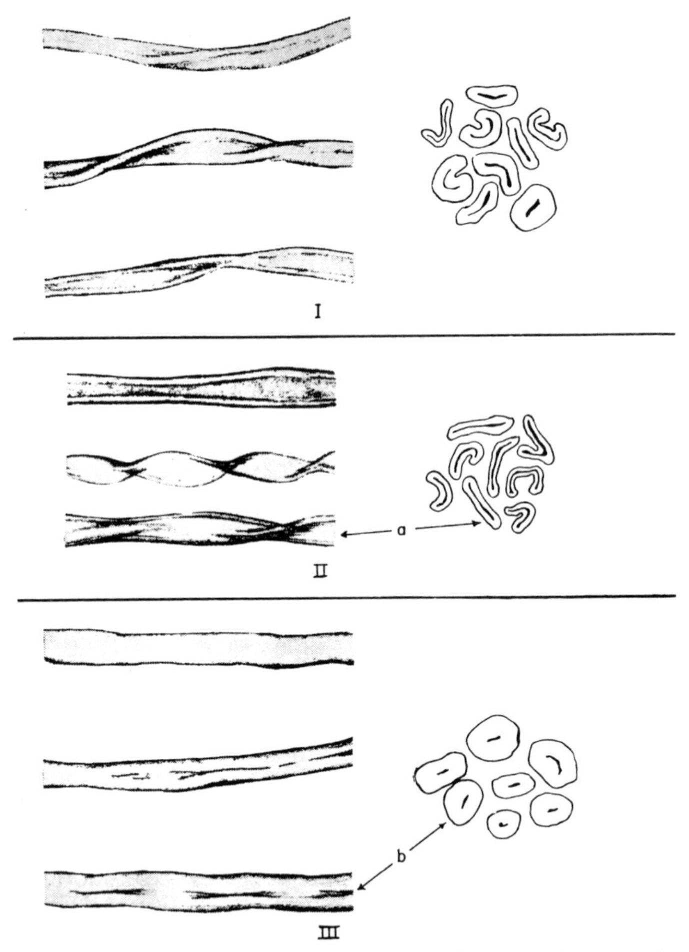

Fig. 14.13. Longitudinal and cross-sectional views of cotton fibers (Lefkowitz). I. Randomly selected untreated fibers. II. Immature fibers treated with caustic soda. III. Mature fibers treated with caustic soda.

The maturity ratio can be calculated from this equation:

$$M = \frac{(N - D)}{200} + 0.70$$

where M = maturity ratio; N = normal fiber per cent; D = per cent dead fibers.
For the example,

$$M = \frac{(60 - 22)}{200} + 0.70 = 0.89.$$

Polarized Light Method for Maturity

Theory of Test. In using the Polarized Light method to determine maturity, the ability of the fibers to transmit light is measured. The immature fibers will transmit the polarized light, and the lumen in this case will appear to be of the same color as the background. The mature or thick-walled fibers do not transmit light as do the immature fibers and will appear as a complementary color to the background.

Procedure for Preparation and Counting of Test Sample

A. Preparation of Test Sample

1. Follow steps A and C1 under Procedure for Preparation of Test Sample for the Array method.
2. Flood the fibers with water using a dropper.

B. Procedure for Counting

1. Place the slide on the mechanical stage of a polarizing microscope.
2. Place a dark blue selenite plate in the proper position between the fibers and prisms. The fibers on the mount should make an angle of 45° with the prisms.
3. Adjust the microscope to a magnification of 100× and 400× and bring the fibers successively into view by use of the mechanical stage.
4. Classify the fibers with respect to color. The immature fibers will appear blue, purple, purplish red, or approximately the color of the background if some other background is used rather than the blue selenite. Mature fibers will appear yellow, green, or the complementary color to the background if some other background is used rather than the blue selenite.

5. Record the number of mature and immature fibers observed.

C. Calculations. See Calculations under the Array method p. 231).

Number and Frequency of Tests. See Number and Frequency of Tests under the Array method (p. 232).

Differential Dyeing Method for Maturity

Theory of Test. The theory of the differential dyeing technique as explained by Armfield and Boulton (4) is: "The net result of the differential dyeing rates of the two dyes, the differential optical appearance of the two dyed fibers, and the influence of fiber surface on both the rate of dyeing and the ease of stripping is to give, under the conditions of test, a fine filament (thin-walled cotton) which looks green and a coarse (thick-walled cotton) one which appears flat red. . . ."

The test consists of dyeing three-gram samples of the cottons to be tested in a bath of distilled water, diphenyl fast red 5BL Supia 1, and Chlorantine fast red BLL, then redyeing twice in the same bath plus sodium chloride, and washing. The detailed procedure for preparing and dyeing the samples is outlined in *Textile World* (5).

The color of the test samples after dyeing is the index for determining per cent maturity, and using this method there are no calculations involved. When using this method to determine maturity, a set of master samples of known maturity should be prepared. The test samples of unknown maturity are compared with respect to color with the masters, and in that way an accurate estimate of the maturity of the unknown is obtained.

References

1. Fiori, L. A., and Brown, J. J., "Effects of Cotton Fiber Fineness on the Physical Properties of Single Yarns," *Textile Research J.*, **21**, Oct. (1951).
2. "Joanna Blends Cotton by Fineness," *Textile Ind.*, **116**, Jan. (1952); Denmark, J. B., "How Pepperell Blends Cotton by Fiber Fineness," *Textile World*, **101**, Dec. (1951).
3. Hertel, K. L. and Craven, C. J., "Cotton Fineness and Immaturity as Measured by the Arealometer," *Textile Research J.*, **21**, Nov. (1951).

4. Armfield, W. and Boulton, J., "Application of Viscose Rayon of the Differential Dyeing Test for Cotton Maturity," *Textile Research J.*, **19**, Apr. (1949).
5. Goldthwait, C. F., Smith, H. O., and Barnett, M. P., "New Dye Technique shows Maturity of Cotton," *Textile World*, **97**, July (1947).

CHAPTER XV

Non-Lint Content and Nep Potential

The determination of the non-lint content of cotton taken from the bale or at any stage in processing up through carding or drawing can be of value to mill management. The measurement on the raw stock gives an accurate estimation of the amount of trash in the bale, and the non-lint content of the cotton at different steps in the opening and carding areas is a good measure of the efficiency of the cleaning equipment.

Just as valuable is the analysis of the waste removed at the different cleaning points in opening and picking: What proportion of the waste is lint? As a matter of fact, the analysis of waste is frequently of more interest in quality and cost control than is the analysis of the cotton itself.

Shirley Analyzer

To measure the amount of trash in a sample of cotton fibers, the Shirley Institute developed a machine known as the Shirley Analyzer, which separates the fibers from the trash.

Theory of Machine. The Shirley Analyzer uses the air flotation principle for separating lint from trash. The sample is fed into the machine with a licker-in and feed plate arrangement. When the fibers leave the licker-in, they are carried by an air stream and deposited on a cage similar to a condensing screen. The air stream is adjusted so that it will carry only the cotton fibers and dust; it is not strong enough to carry the trash, so as the cotton fibers move from the licker-in to the cage, the trash falls to the lower section of the machine. When the fibers and dust reach the cage, the dust passes through and out the exhaust.

Location of Shirley Analyzer and Recommended Settings

The following suggestions and recommendations are quoted from the instructions issued by the Shirley Institute:

It should not be situated for use in any position in the testing room where undesirable draughts from open windows or doors or strong convection currents from a nearby heat supply can interfere appreciably with the air entering or leaving the machine. For the same reason, the user of the machine, while the feed roller is operating, should not approach within six inches of the feed plate and two feet from the open outlet of the fan.

The fan is provided with an outlet shaped to disperse the dust-laden air into the testing room. No guarantee is offered that the machine will function efficiently if the outlet is extended by additional ducts leading directly to the outside atmosphere or to a mill dust chamber, whereby the pneumatic system in the machine would be subject to the effects of variable back draughts. Usually, the period of actual operation of the machine is so discontinuous, and the quantities of dust liberated so small as to make additional exhaust ducts unnecessary.

To give the best results, the temperature of the testing room should not be less than about 65° F and the relative humidity not more than about 60 per cent, that is, normal conditions for an office. Both temperature and humidity should be maintained as constant as possible. Excessive coldness and dampness will cause licking of lint to the licker-in teeth and other such working parts and will interfere with the performance of the machine. The machine should be immediately stopped and the licker-in teeth cleaned if such licking occurs.

Auxiliary apparatus required for a test consists of a balance capable of weighing 200 grams. To give the most accurate results, the sensitivity of the balance should be $^{1}/_{100}$ gram.

The weighing, conditioning of the sample, and its actual testing by the machine should all be done in the same testing room.

Beware of excessive greasing and oiling in such regions where working surface might be contaminated. For example, (a) grease or oil from the feed roller bearings should not be allowed to creep to the fluted surface of the roller which is in contact with the test sample during feeding, and (b) the cage surface collecting the clean lint should similarly be free from grease or oil creeping from the cage bearings.

The driving motor supplied with the machine is of ½-hour rating and should not be run continuously for longer than that period.

All the working parts of the machine should be kept clean to give the correct machine performance. In particular, the working face of the streamer plate must be maintained brightly polished and free from burrs. Similarly, the outside surface of the cage must not be allowed to become dirty and lose its bright polish. The inside of the cage can be cleaned through the portholes. The striking face of the dish feed plate should be cleaned occasionally and freed from all accumulations of dust and waxy matter. The choke valve on the fan will need occasional cleaning. The whole outlet unit can be removed from the fan housing to do this. It should always be sufficiently clean to allow the level operating valve to be moved freely to either extreme end of the scale.

No guarantee is offered that the machine will function efficiently if the speeds and settings are altered from the standards determined by the Shirley Institute and outlined below.

(a) Speeds:

Licker-in cylinder	900 r.p.m.
Feed roller	0.9 r.p.m.
Cage	80 r.p.m.
Fan	1500 r.p.m.
Motor	1400 r.p.m. approximately

(b) Settings:

1. Feed plate to licker-in	$\frac{4}{1000}''$
2. Streamer plate (lead-in edge) to licker-in	$\frac{5}{1000}''$
Streamer plate (lead-off edge) to licker-in	$\frac{7}{1000}''$
3. Stripping knife (bottom edge) to licker-in	$\frac{4}{1000}''$
4. Stripping knife (bottom edge) to cage	$\frac{5}{16}''$
5. Licker-in to cage	$\frac{7}{32}''$
6. Separation sheet (top edge) to cage	$\frac{1}{4}''$
7. Separation sheet (top edge) to licker-in	$\frac{9}{16}''$
8. Delivery plate to cage.	$\frac{1}{16}''$

The machine will not function efficiently if certain working parts are damaged to the slightest extent. Notable amongst such parts which must retain their smoothness and freedom from burrs are:

1. Striking face of the feed plate.
2. Lead-in and lead-off edges and outer working face of the streamer plate.
3. Lower edge and working face (facing the cage) of the stripping knife.
4. Outer surface of the cage.
5. Inner surface of the celastoid cage cover.
6. Lead-in edge and outside surface of the delivery plate.

The licker-in should be examined occasionally and all bent or damaged teeth repaired.

Procedure for Preparation of Test Sample

A. Preliminary

1. Collect representative sample of the material to be tested. Handle the sample as little as possible to avoid losing any trash and to prevent the trash from settling to the bottom of the sample.

2. Select a 100-gram sample by taking small tufts from all parts of the master sample. Some laboratories have found that a 200-gram sample is more satisfactory when testing cotton of extra heavy trash.

B. Preparation

1. No additional preparation of the test sample is necessary.
2. Spread the 100-gram sample as uniformly as possible over the entire area between the guides on the feed plate.

Procedure for Use of Machine

1. Make sure that the machine is clean of all fibers and trash.
2. Start the motor with clutch out (feed roll not operating) and exhaust valve open to the fullest extent. Run the machine for two or three minutes with the clutch out.

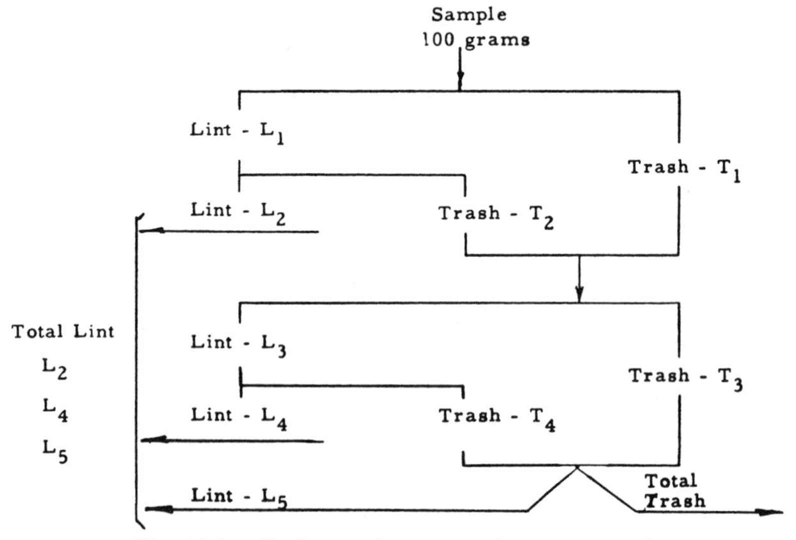

Fig. 15.1. Shirley analyzer procedure — normal.

3. Engage the clutch and observe the trash as it begins to fall into the tray. Only small amounts of unopened lint should be falling with the trash in the first passage through the machine. For hard cotton it may be necessary to tighten the loading springs on the feed rolls.

4. When all the sample has passed through the machine, close the valve and stop the motor.
5. Carefully collect all of the lint in the delivery box.
6. For raw cotton, laps, strips, and so forth, the procedure outlined in steps 1 through 5 are followed and the material is rerun through the machine in the following manner as recommended by the Shirley Institute (see Fig. 15.1):

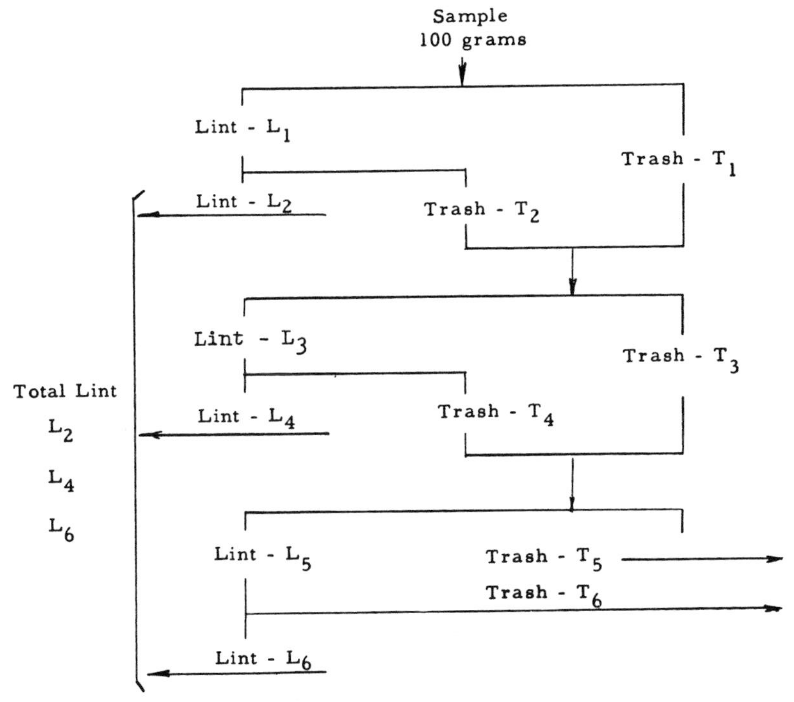

Fig. 15.2. Shirley analyzer procedure — high trash content sample.

The 100-gram sample S is passed through the Analyzer giving L_1 gms. of lint which is then given a second passage yielding L_2 gms. of lint. The lint L_2 is placed on one side and the trash T_1 and T_2 from the two passages are collected together from the settling chamber and fed through the machine yielding L_3 gms. of lint which is again passed through to give L_4 gms. of lint. The L_2 and L_4 are weighed together. Trash T_3 and T_4 is also weighed and then given a final passage yielding a further small amount of lint L_5 which is weighed. This weight added to that of L_2 and L_4 gives the total lint content of the 100-gram sample as a percentage, and the same weight L_5 subtracted from that of T (or T_3 plus T_4) is the trash extracted from the 100-gram sample, also expressed as a percentage. Finally the percentage cage loss is the difference between the original 100 grams and the sum of the trash and lint content.

7. In testing slivers they should be fed longitudinally side by side on the feed plate.
8. For cotton with exceptionally high trash content, the procedure shown in Fig. 15.2 may be found necessary.

Calculations. The calculations for the procedure outlined in step 6 above and shown in Fig. 15.1 are as follows:

$$T = T_3 + T_4$$

(this is weighed and passed through the Analyzer to give L_5, which is weighed).

$$\text{Trash Content} = T - L_5$$

$$\text{Lint} = L_2 + L_4 + L_5$$

$$\text{Cage Loss} = S - (T + L_2 + L_4)$$

(the difference between the original weight of the sample and the sum of the trash and lint contents).

The scale in Table 15.1 has been established to represent percentages of non-lint removed by the Shirley Analyzer from different grades of Upland cotton.

TABLE 15.1

Shirley Analyzer Ratings [a]

Grade	Non-lint content %
Strict Good Middling	2.0
Good Middling	2.4
Strict Middling	2.9
Middling	3.7
Strict Low Middling	5.1
Low Middling	7.6
Strict Good Ordinary	11.0
Good Ordinary	17.0

[a] Source: U. S. Department of Agriculture.

Number and Frequency of Test. The Shirley Institute recommends that four 100-gram samples be analyzed and the average of the four reported as a complete test. The frequency of this test is usually less than for the fiber tests. If the cleaning efficiency of opening and cleaning equipment is being checked, this can usually be done on a rather long testing cycle; however, if raw cotton is being tested, it is usually done on a more frequent schedule.

Fig. 15.3. Nepotometer.

Nep Potential by the Nepotometer (1)

The ability to evaluate rapidly and accurately from a small sample the tendency of a cotton to form neps in processing is one which has been long desired by those concerned with the various aspects of cotton: its breeding, growing, distribution, and manufacture into yarn and fabric. Each segment of the industry is anxious to do the best job possible. But desire has no accomplishment. Only after a thorough understanding of the qualities and failings of presently available materials can one suggest areas for concentration on improvement. A cotton bred for improved qualities of fiber length, strength, maturity and fineness may still fail miserably in manufacturing if it also possesses a high "Flab Factor," that undesirable property which causes one cotton to curl up and die when subjected to the rigors of current manufacturing methods, at the same time that another, with identical fiber characteristics as measured by current tests, will resist the tangling forces.

. . . .

Neps are aggregates of tangled fibers and caused by the effects of mechanical manipulation. The actions necessary for their formation exist in many forms. The actions of the revolving spindles of the mechanical picker, the saws of the gin, the beaters of a modern opening and picking line are but a few of the more obvious ones. It is rather remarkable that cotton fibers, so minute in cross-section, are capable of resisting these forces as effectively as they do.

Simple mechanics can explain why a coarse fiber is more resistant to nepping than a fine fiber; why a fiber, with a rounder cross-section, is more resistant to nepping than an immature fiber with its flat, ribbon-like cross-section. But why do fibers with identical degrees of fineness and maturity differ in their resistance? The answer must lie in structural differences which are not usually measured. One may consider identical metals in two forms which have the same weight per unit length; one in the form of a rod and one in the form of a chain. The difference in the ease with which they can be bent and entangled is apparent.

Anyone who has observed the effects of reprocessing on the quality of the stock has noted the increase of tangled fibers with the increase in the amount of mechanical treatment. Like the stonecutter's rock, a fiber may resist forty-nine blows, only to be shattered into a nep at the fiftieth.

It is on this principle of repetitive action on the same fibers that the final model of the instrument for predicting the nepping potential of cotton is based. It also incorporates the actions of carding and stripping found in a full-scale cotton card.

Basically the instrument consists of a feeding arrangement, three rolls covered with metallic card clothing to manipulate the cotton, a presser plate to prevent fiber plucking at one of the fiber transfer points, a pre-set timer to insure a standard degree of manipulation, a means of transferring a sample of the web for inspection, and the necessary means of actuation in the working direction, and in reverse for the purpose of clearing the instrument after the test is completed.

The test method consists of first obtaining an accurately weighed 25-grain sample which is representative of the bale of cotton to be evaluated. This is carded between a pair of hand cards into a pad 5″ × 5″, and the sample is placed on the feed table and tucked under the feed roll. The rotation of the mechanism is begun in the working direction with the timer set for four minutes. At the end of this time the machine stops automatically. The presser plate is removed and a piece of cardboard 6″ × 10″ covered with black velveteen is placed in contact with the web at the back of the top roll. By

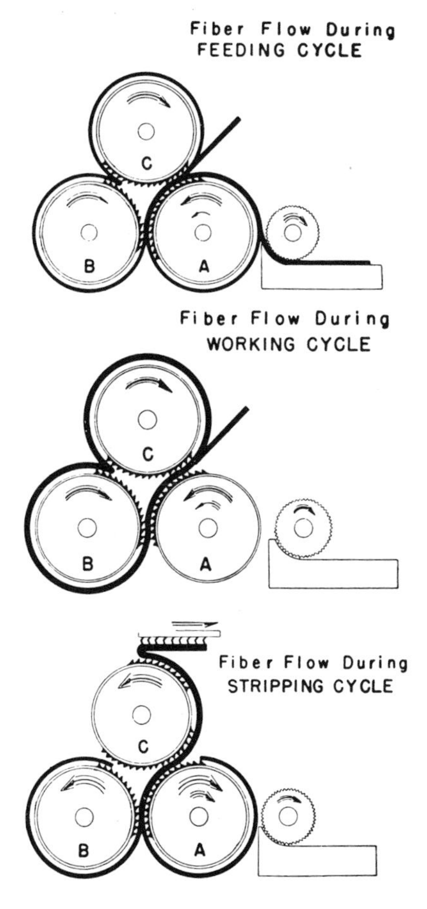

Fig. 15.4. Nepotometer action.

maintaining contact with the web as the board is swung forward, the sample is made to adhere to the pile of the velveteen. The sample web is broken at the edge by swinging the board away from the roll. The sample is set aside and first bottom cylinder is stripped of excess material by rotating the cylinder by hand in the reverse direction while a hand stripper is held in contact with the cylinder. A reverse switch permits the three cylinders to rotate in the reverse from that during the working cycle, and all of the cotton remaining on the bottom cylinder is transferred by the action of the wires to the top cylinder from where it is stripped readily with a hand stripper. After the presser plate is replaced, the instrument is ready to accommodate the next sample which was prepared during the time the first test was running.

The appearance of the sample web is compared for neppiness against a set of five standard photographs of webs which show a range of neppiness from very low to very high. The average weight per unit area is approximately the same as that for a card web weighing 50 grains per linear yard.

Direction of fiber flow during the feeding cycle is shown in Fig. 15.4. The fringe of cotton is torn away from the nip of the feed roll and feed plate by the rotation of cylinder *A* which has a relatively high surface speed. Cylinder *B* and *C* have the same surface speed, lower than *A*. As the fibers, held on the wire teeth of *A*, pass roll *B*, some fibers are transferred to *B* in a carding action, while others remain on *A*. Those fibers which are transferred are removed by cylinder *C* through a stripping action and, in turn, from *C* to *A* by a stripping action.

After all the material has been fed into the system, the fiber flow is illustrated in the second part of Fig. 15.4. The fibers are worked from *C* to *A*, back to *C* as long as desired. The presser shown near the nip of cylinders *A* and *C* tends to retard the very easy transfers of fibers from cylinder *C* and prevents plucking of fibers, thus permitting the production of a more uniform web.

At the end of the test when it is desired to clear the instrument, the rolls are reversed and the fiber flow is described by the diagram. Cylinder *B* strips all fibers from the under side of *C*. Cylinder *A*, because of its higher surface speed, brushes fibers from *B* and impales them on the points of cylinder *C*, from which they are removed by a hand stripper.

References

1. Bogdan, J. F., "The Nepotometer," *Textile Forum*, **11**, Feb. (1954).

CHAPTER XVI

Other Fiber Tests and Methods

Moisture Content

All fiber tests should be performed under standard conditions of temperature and humidity. In addition, it is necessary in some cases to determine the moisture content of the cotton. To do this, the usual procedures discussed in Chapter X (Humidity and Moisture, page 141) are followed. Selected samples are dried in ovens, and the moisture content or regain is calculated from the formulas:

$$\text{Moisture content, \%} = \frac{\text{Original weight} - \text{dry weight}}{\text{Original weight}} (100)$$

$$\text{Moisture regain, \%} = \frac{\text{Original weight} - \text{dry weight}}{\text{Dry weight}} (100)$$

Sugar Content (Honeydew) Analysis

Honeydew is a term used to describe a condition in cotton resulting from the secretion of plant lice on the cotton fibers, bolls, and plant. The presence of honeydew will cause the fibers to stick to the rolls during processing, frequently causing roller laps. A key to the presence of honeydew is that the rolls on the different drafting elements will have dark brown or black spots in the form of a soft sticky substance.

The United States Department of Agriculture made a series of tests on the 1950 and 1951 crops of cotton and found that less than 3 per cent of the 1950 crop and approximately 1 per cent of the 1951 crop had sufficient honeydew to cause trouble during processing. Even though this indicates that a very low per cent of the cotton produced is affected, it is nevertheless advisable to recognize that such a condition exists since the fibers that are contaminated can cause serious trouble during processing. The presence of honeydew can be determined in raw cotton by testing for the amount of sugar present. This may be done by any of several methods. Clinitest tablets may be used and can be purchased at most drug stores. The material used by the U.S.D.A. for their investigation was Benedict's copper solution. This solution is prepared by dissolving 173 grams of sodium citrate in 100 grams of anhydrous sodium carbonate in about 800 milliliters of water. The two solutions are mixed and diluted to one liter.

To test the cotton, place approximately 1 grain of cotton in a 100 cc. beaker and wet out with 20 cc. of the Benedict's solution and immerse the beaker in boiling water. The beaker should remain in the water bath until the reaction is complete, which usually requires three to five minutes. The solution is then compared with a standard color chart to get an indication of the sugar present.

Cavitomic Cotton

It has been shown that some cottons can be subject to deterioration of their fiber properties by the action of microorganisms or their metabolic products. This alteration of fiber properties has been called "cavitoma," and the cotton exhibiting this condition, "cavitomic cotton." (1) The changes in properties most evident have been the increased swelling of fibers in caustic soda and fiber breakage resulting in shorter staple lengths. In many cases apparently good spinnable cotton can be cavitomic.

A method of determining the extent of cavitomic cotton is to measure the pH and reducing material: a high pH together with a small amount of reducing material is associated with the presence of cavitoma. A simple color test described by Hall and Elting in *Textile Industries*, July, 1953, consists of passing small samples of about $\frac{1}{2}$ gram of cotton through 3 milliliters of a prepared solution in a test tube and noting the color changes during the passage and after the cotton is at the bottom of the tube. The solution used is made of 10 milliliters of Gramercy Universal Indicator from Fisher Scientific Company and 1 milliliter of Santomerse S from Monsanto Chemical Company dissolved in a liter of distilled water; this solution is light green in color. Cavitomic cotton turns the solution blue; highly degraded cotton turns the solution violet. Noncavitomic cotton causes either no change in color, or the original green to turn yellow. Samples of cotton with both good and bad cotton show blue streaks as the cotton is pushed to the bottom of the test tube, but the over-all color change may be so slight to be indecisive.

Fiber Blending

It has become very evident that the increase of fiber testing by the cotton industry points to the need of specimens properly pre-

Fig. 16.1. Fiber blender.

pared and representative of the entire sample. The preparation of a proper blended sample should mean that individual specimens could be used for different tests. The U.S. Department of Agriculture constructed and tested several different mechanical blenders, and emerged with a relatively simple unit that produced a satisfactory blend with the minimum time required for preparation of the sample. Blueprints of this mechanical blender shown in Fig. 16.1 are available from the U.S.D.A., or machines of the same design can be purchased commercially. Blended samples are used for tests of moisture, fiber array or Fibrograph test, strength, fineness, maturity, and non-lint.

Operation of Unit. The heart of the blender is the rotary blending cylinder which is covered with card stripper-wire clothing. The cotton is fed at a predetermined rate to the wire-clothed cylinder by the combination of a fluted feed roll and an adjustable feed plate. The ratio of surface speeds of the feed roll to the blending cylinder is about 1 to 135. The adjustment in the feed plate is made so as to accommodate short (up to 1 inch) or long (over 1 inch) staple; this adjustment consists of a hinged extension plate on the feed plate which alters the distance between the roll bite and the cylinder. Maximum blending without fiber damage is attained through three passages through the blender. The blended fibers are removed by hand as described below.

Procedure for Test

A. Preparation of Sample

1. A total of 32 pinches of cotton are taken from different parts of the master sample to give a total weight of about 150 grains (10 grams).
2. The pinches are manipulated so as to form a rough bat about 15 inches long and a little less in width than that of the feed roll length.

B. Operation of Machine

1. Set the cam (i.e., the adjustable hinged extension) according to the staple of the cotton. (a) For short staple setting of cottons less than 1 inch in length, the feed plate extension is swung

away from the face or outside surface of the feed plate. (b) For long staple setting of cottons more than 1 inch in length, the feed plate extension is brought in contact with the face of the feed plate.

Caution. The blending cylinder may be damaged if the cam and feed plate extensions are not carried out simultaneously. To change the setting of the blender from short staple to long staple operation, grasp the feed plate extension with the hand and swing it forward under the feed plate. Change from long staple setting to short staple setting by reversing the above procedure.

2. Start the blender and feed in the rough bat. Care must be taken so that large tufts of fibers will not be taken by the blending cylinder from under the feed roll.
3. Remove the bat from the blending cylinder. To do this, rotate the cylinder backward with the handwheel until the traverse slot appears in the wire clothing. Insert a rod in the slot and lift it until the cotton bat breaks. Continue to rotate the cylinder backwards with the right hand while rolling the bat from the wheel with the left hand.
4. Blend a second and third time for optimum blending of the sample. To do this, unroll and fold the bat over lengthwise once and tuck in the sides to conform with the length of the feed roll. Then repeat the second and third steps.
5. After the sample has been removed following the third blending, clean the blending cylinder. To do this, press a hand card, covered with stripper clothing, against the cylinder while it is rotating in the reverse direction either by hand or motor.

References

1. Hall, L. T. and Elting, J. P., "Cavitomic Cotton," *Textile Research J.*, **21**, Aug. (1951).

CHAPTER XVII

Staple Synthetic Fiber Tests

The manufacturers of the synthetic fibers operate under very rigorous quality control programs that cover not only the manufacturing processes but also the physical and chemical properties of the fibers produced. The manufacturer in his fiber quality control program tests for denier, fiber strength, crimp, amount of finish, staple length, dye affinity, moisture regain, and other individual properties. The textile manufacturer buying this product usually limits his quality control program insofar as the staple is concerned to such tests as moisture regain (which is the most frequent test made, for synthetics are bought on a commercial regain basis), amount of finish, staple length, and occasionally dye affinity, actual denier, and fiber strength. Probably because the variables are so well controlled by the staple producer, the mill testing of staple synthetics has never reached the proportions of the programs devoted to the testing of cotton staple. For example, the length of staple synthetics is very uniform because it is machine cut to definite specifications. Thus, length as such is of little concern to the spinner once the lot or shipment has been checked, and for these reasons, the program for testing staple length can be quite simple and economical. It should be noted, however, that occasionally lots of staple display fibers that have not been cut, thus resulting in discrete groups of fiber of two or three times the normal length. This type of length variation does cause concern to the spinner.

Another reason for the reduced need for testing the synthetics in comparison to the natural fibers is the absence of trash and foreign matter. Thus, grade analysis and waste content analysis tests such as are performed on the Shirley Analyzer are not necessary with the synthetics.

While these examples should not be interpreted as giving a clean bill of health to all staple synthetics, they do indicate the trend of thinking. There are some tests, however, which should be made on staple synthetic fibers which are not necessary on natural fibers. One of these is the amount of finish applied to the fiber. The finish is applied by the fiber manufacturer to control such processing characteristics as static electricity and drafting performance. The type and amount of finish applied is a problem continually investigated and studied by the fiber manufacturers and especially the companies who make the hydrophobic fibers.

Fiber crimp is another property that is not generally of any concern to the natural fiber spinners but is important in man-made fibers. No man-made fiber has natural convolutions or crimp to the same proportion as found in cotton. For that reason the crimp must be formed either chemically or mechanically after the fiber has been spun.

In order to identify individual units of production having the same physical and chemical properties, lot or merge numbers are assigned by the producer. These individual batches must be processed by themselves or merged in accordance with the producer's instructions in order to avoid defective quality. The troubles that result from mixing different lot numbers of material are amplified when the yarn or fabric is finished or dyed because of such reactions as differences in dye affinity or shrinkage characteristics.

Amount of Finish and Moisture

The testing for per cent of finish or moisture regain is not a matter of quality control by the spinning mill, but rather one of financial control. Since the user has no direct responsibility over the quality or quantity of finish applied, or over the regain in the staple, the results of the tests on these quantities are the means for checking the commercial weight of the shipment. The commercial weight of any shipment made by the producer is based on the dry weight plus an amount of moisture (and finish) based on the commercial regain for the particular fiber. This commercial regain is that amount agreed upon by the industry through such organizations as the American Society for Testing Materials. The weight that is invoiced by the producer to the buyer is the commercial weight based on the dry weight plus the accepted commercial regain. The actual weight of the shipment, neglecting package tare weight, would differ from the commercial weight if the actual regain of the fiber differed from the commercial regain. Therefore, the purpose of the moisture and finish quantitative analysis is to give the data to determine the reliability of the invoiced weight. If the invoiced weight does not match the commercial weight calculated from dry weight plus the accepted regain, then the matter becomes one of arbitration between the producer and the customer mill.

For fibers such as glass and the acrylic type which have a moisture regain to the order of 1 per cent, the commercial weight is usually accepted as the actual net weight of the shipment. Therefore, this test is of importance with hydrophilic fibers such as viscose and acetate.

Procedure for Preparing Test Sample. The sample to be used for this test should be a composite sample from at least 20 or 25 per cent of the cartons or bales. Duplicate samples should be prepared and each sample should weigh approximately 300 grams. As the samples are collected, they should be placed in an airtight container immediately to prevent moisture change.

Procedure for Drying and Extracting. After the composite samples are collected and placed in airtight containers, the containers are weighed. The tare weight of each container is subtracted from the gross weight, and the remainder is the net weight of the test specimen. After this net weight is determined, the samples can then be removed from the containers for testing.

For determining only the per cent of regain, the samples are placed in an oven and dried at 105 to 110° C. until three consecutive readings taken at ten-minute intervals do not differ by more than 0.1 per cent.

After drying and weighing, the samples are then removed from the oven and the finish extracted. The samples may be placed in a soxhlet extractor with such solvents as carbon tetrachloride, xylene, ether, or alcohol. The fiber manufacturer will usually recommend the solvent for any particular type finish. After extracting for four or five hours, the samples are removed and placed in the oven. The drying procedure is the same as outlined above and also discussed in the chapter on moisture. The amount of finish is calculated as a per cent based on the dry weight.

This method of determining the amount of finish has several disadvantages. One is the time necessary for the test. Another is the number of extractors necessary to handle the two samples. It usually will take several extractors to accommodate the large samples necessary for this type of test. Third and most important is the possibility of error. Since the amount of finish applied to staple fibers is usually 1 per cent or less, any fiber loss will influence the results to a large extent.

To overcome these disadvantages, a boil-off method is generally used. With this method, which is recommended by A.S.T.M., the two samples are placed in separate bags which have been cleaned, desized, and bleached. For viscose, the bags are placed in a bath and boiled for thirty minutes. For acetate, the temperature of the bath is 71 to 77° C., and the time is thirty minutes. The recommended weight of the bath is twenty-five times the combined weight of the two samples, and the bath should contain 0.5 grams of neutral soap per liter.

After the boiling period, the bath is removed by overflowing with soft water until the surface of the bath is free of scum. The bath is then drained and the specimens given three ten-minute rinses in soft water baths at 87° C., 71° C., and room temperature respectively. The bags are passed through a wringer or extracted after each rinse. After a final rinsing, the samples are dried at 105 to 110° C. until three consecutive readings do not differ by more than 0.1 per cent in weight.

Calculations. A = net weight of test specimen; B = oven-dry weight of boiled-off specimen; C = oven-dry weight of original test specimen (before boil-off).

$$\% \text{ of finish} = \frac{C - B}{C}(100)$$

$$\% \text{ of moisture and finish} = \frac{A - B}{A}(100)$$

$$\text{Commercial weight} = \frac{(W - t)(100 + R)B}{100A}$$

where W = gross weight of shipment or bale; t = tare weight of shipment or bale; R = commercial regain of fiber.

EXAMPLE

W	(gross weight of shipment)	=	20,000 pounds
t	(tare weight of shipment)	=	1000 pounds
A	(net weight of test specimens)	=	500 grams
B	(oven-dry weight of boiled-off specimens)	=	430 grams
C	(oven-dry weight of original specimens)	=	436 grams
R	(using viscose commercial regain)	=	11 %

$$\% \text{ of finish} = \frac{C - B}{C}(100)$$

$$= \frac{436 - 430}{436}(100) = 1.37\ \%$$

$$\% \text{ of moisture and finish} = \frac{A - B}{A}(100)$$

$$= \frac{500 - 430}{500}(100) = 14\ \%$$

$$\text{Commercial weight} = \frac{(20{,}000 - 1000)(100 + 11)(430)}{(100)(500)}$$

$$= 18{,}137 \text{ pounds}$$

The above formula is a simple means for calculating the oven-dry clean weight of the net shipment. With a net weight of 19,000 pounds and 14 per cent moisture and finish, the weight of clean dry fiber would be 19,000 − .14(19,000), or 16,340 pounds. With a commercial regain of 11 per cent, the commercial weight would be 16,340 + .11(16,340), or 18,137 pounds.

Staple Length

The measure of staple length at the mill is usually confined to pulling a sample in a fashion similar to stapling cotton and measuring the length with a ruler. This technique is used primarily as a check test and is sufficient unless processing difficulties are encountered and a complete fiber array is needed. Conditions such as this are unusual since the fiber manufacturer continually tests the sample for length as well as other properties.

Most manufacturers use the array method for measuring the length of the fibers. The technique for man-made fibers is similar to that used for cotton and wool. The fibers are placed in a Suter-Webb Sorter and arrayed into group length cells in a fashion similar to cotton fiber testing. This technique is described in detail in the A.S.T.M. Manual.

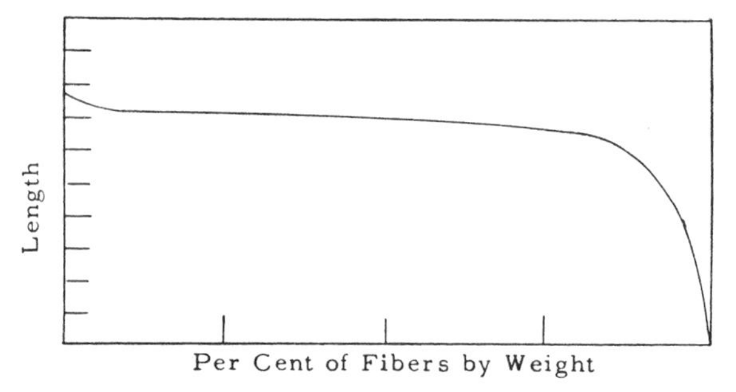

Fig. 17.1. Typical fiber array for staple synthetics.

The length of staple synthetics will normally be considerably more uniform than for the natural fibers. A typical fiber array is shown in Fig. 17.1.

Fiber Strength and Elongation

The measurement of fiber strength is also a test which is performed by the manufacturer on a routine basis but is seldom necessary for the spinner. The strength naturally varies with the type of fiber but within a given type from any one manufacturer, the strength generally will be within close limits.

Flat Bundle Method

There are two accepted methods of measuring the fiber strength. The preferred method suggested by A.S.T.M. is the Flat Bundle Method. To use this method, a small sample of fiber is selected and combed until the fibers are parallel. The bundle is then reduced in size by removing fibers from the side of the bundle until the weight is 15 to 20 mg. when cut to the specified length of 22.5 mm. After the bundle is cut to 22.5 mm., the bundle is weighed to the nearest 0.1 mg. A piece of masking tape approximately 1 by 3 inches is placed over the bundle in such a way as to cover one-half of the exposed surface from one end. The longer edge of the tape is placed perpendicular to the length of the fibers. A second piece of tape is placed over the other half of the fibers in the same manner as described above. The two pieces of tape must not overlap, but should be approximately 1 mm. apart. The bundle should be turned over and two more pieces of tape placed on the other side as described above.

The bundle is then placed in a suitable type of testing instrument equipped with 1 by 2-inch jaws. The jaws are mounted with the 2-inch dimension perpendicular to the force, and the distance between the jaws is adjusted so that the top and bottom jaws just fail to touch. To place the sample in the tester, move the bottom jaw away from the top jaw about 2 or 3 inches and fix the sample in the bottom jaw. The sample is clamped by the jaw in such a way that the portion of the fibers not covered by the tape is barely above the bite of the jaw. The bottom jaw is returned to the zero position, and as it returns, the other end of the sample is guided into the top jaw. When the bottom jaw stops, the top jaw should be tightened and the load applied to the specimen. The breaking load is recorded in grams.

Calculations. The strength is calculated in terms of tenacity with units of grams per denier. Knowing the length and weight of the specimen, the gross denier is calculated. From this, the tenacity can be determined.

EXAMPLE. A 22.5 mm. length weighing 18 mg. breaks at 8,000 grams load. Determine the fiber tenacity.

The denier is calculated from the equation

$$\text{Denier} = 9{,}000\left(\frac{W}{L}\right)$$

where W = the weight in milligrams; L = the length in millimeters. (In this equation milligrams and millimeters are substituted for meters and grams in the usual equation.)

$$\text{Denier} = 9{,}000\left(\frac{18}{22.5}\right) = 7200.$$

The grams per denier are calculated using the total denier and strength.

$$\text{Grams per denier} = 8000 \div 7200 = 1.11 \text{ g.p.d.}$$

Note that the total denier would be equal to the product of the individual fiber denier times the total fibers in the bundle. Using the total denier eliminates the need of counting the number of fibers.

The entire calculation can be simplified by using the equation

$$\text{g.p.d.} = \frac{ML}{9000\,W}$$

where M = the average breaking load in grams; W = the weight in milligrams; L = the bundle length in millimeters.

In the foregoing example, the substitution becomes

$$\text{g.p.d.} = \frac{8000(22.5)}{18(9000)} = 1.11$$

Single Fiber Method

The alternate method for determining the fiber strength is to break single fibers. The distance between the jaws varies with the length of the fibers being tested and ranges from 15 mm. for fibers shorter than 35 mm. to 100 mm. for fibers of 125 mm. or longer.

The strength of the fibers is expressed as the grams per denier strength, calculated by dividing the average strength in grams by the average denier of the fibers. This gives a measure of the strength usually designated as tenacity or breaking length. The value can be compared with the strength of cotton fibers by using the appropriate conversion factor.

More recently the Instron tensile testing instrument has proven to be excellent in measuring single fiber strength and elongation characteristics. This machine is fully described in Chapter 21, p. 383. It is equipped with a high-speed graphic recorder, enabling stress-strain curves of the fiber under load to be drawn and analyzed for different fiber characteristics such as ultimate strength, elongation, elastic stress and strain, stiffness, and toughness.

Fiber Cohesion in Drafting

Draftometer

An important phase in determining the processability of staple synthetic fibers is the evaluation of fiber crimp and finish. These two factors greatly affect the interfiber friction or drag and, therefore, the force and smoothness of the drafting operation.

In drafting, the fibers are caused to slide past each other fairly easily, but the real difficulty consists in sliding the fibers forward uniformly so that they will not advance before their proper turn and yet will do so easily when required. This sliding process is not continuous but usually proceeds with a jerky motion in which the sliding fiber surfaces alternately stick and slip due to fiber friction. This phenomena is usually termed the "stick-slip" effect and depends mainly on the coefficient of friction between fiber surfaces, the mechanical interlocking of fiber surfaces and the effective surface area in contact. Other fiber characteristics which might be expected to influence interfiber friction are fiber finish, crimp, length, fineness, luster and roving twist.

Description of Draftometer. A combination mechanical and electronic instrument for measuring the forces of drafting has been developed by the Department of Textile Research, School of Textiles, N. C. State College. This instrument, tentatively called the Draftometer, was designed by C. M. Asbill, Jr., head of the Department of Textile Machine Development, and is a modification of a similar instrument developed by Tennessee Eastman Corporation. (1)

The instrument, which is shown in Fig. 17.2, consists of two lines of drafting rolls of which the lower pair is mounted in a pivoted cradle in such a way that the cradle is displaced when a force is applied. These lower rolls are flexibly supported on two thin spring-steel strips and counterbalanced by an off-center weight on an extension rod in the rear of the instrument. The force to pull the fibers past each other and the variations in this force result in mechanical displacements of the lower rolls, and these displacements are converted into an electrical signal by a strain gauge.

Fig. 17.2. Draftometer.

This strain gauge is connected to an amplifier and recorder which produces a continuous graphical record of the drafting force. The purpose of the dash pot is to damp out the tendency of the pivoted cradle unit to oscillate at its natural frequency when any shock or sudden change of load occurs.

The upper rolls are adjustable in the vertical position to accommodate different staple lengths, and in general the gauge length is

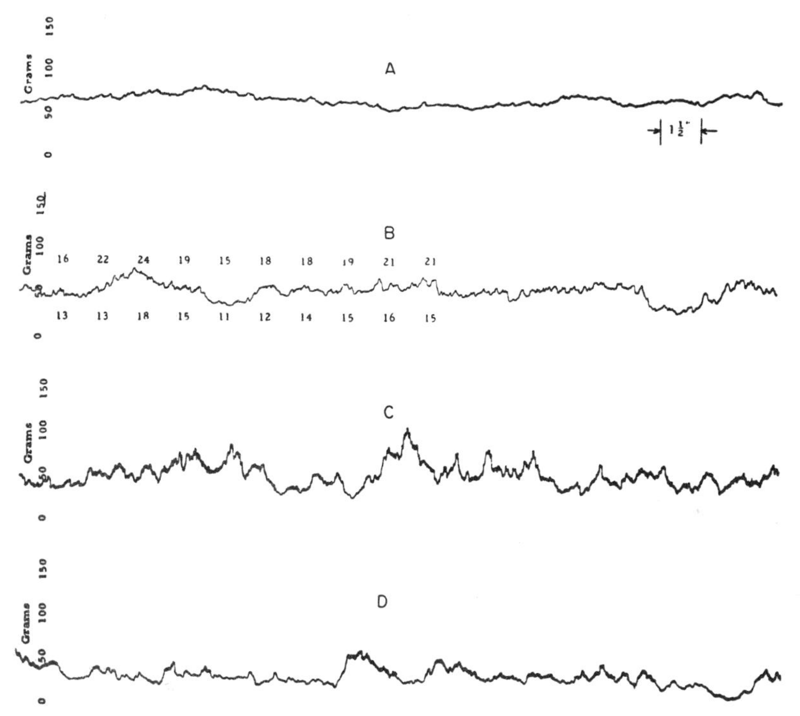

Fig. 17.3. Draftometer charts. A, 50 grain card sliver, bright viscose, $2\frac{1}{4}''$, 3 denier. B, 50 grain first drawing sliver, bright viscose, $2\frac{1}{4}''$, 3 denier. C, 3.00 H. R. Dacron, $1\frac{1}{2}''$, 3 denier. D, 3.00 H. R. Dacron (tinted), $1\frac{1}{2}''$, 3 denier.

set to the staple length plus $\frac{1}{8}$-$\frac{1}{4}$ inch. The speed of the lower rolls is 2 r.p.m. and that of the top rolls, 1.5 r.p.m., giving a draft of 1.33, which simulates the break draft of conventional drafting systems. These low speeds are necessary to allow the pivoted cradle system to accurately follow the variations in drafting force. The surface speed of the lower rolls is 6 inches per minute and the chart speed is 60 mm. per minute, resulting in a 2.5 to 1 material-to-chart ratio.

Sample charts obtained from the Draftometer are shown in Fig. 17.3. Charts A and B, illustrate the increase in variations of drafting force with subsequent processing. Charts C and D show the change in drafting force obtained when Dacron stock is sprayed with an oil base tint. A sample calculation is shown below for B in which the chart was analyzed for average drafting force and smoothness ratio. A $1\frac{1}{2}$-inch sensitive length (length in which the high and low peaks are chosen) is used and high and low peaks are recorded on the chart in mm. above the zero or bottom of the chart.

$$\text{Ave. high} = 19.3$$
$$\text{Ave. low} = 14.2$$
$$\text{Total} = 33.5$$
$$\text{Ave. drafting force} = \tfrac{1}{2}(33.5) = 16.7 \text{ mm.}$$
$$\text{Smoothness factor} = \frac{14.2}{19.3}(100) = 73.4\ \%$$
$$\text{Smoothness ratio} = \frac{19.3}{14.2} = 1.36$$

The smoothness ratio indicates the smoothness of drafting and is the ratio of the maximum and minimum forces encountered during drafting. A smoothness ratio of 1 would represent no variation of drafting force while larger values represent more uneven drafting. The smoothness ratio can also be considered a ratio of static friction to kinetic friction since it is believed that the peaks represent static or starting friction and the lows represent kinetic friction.

Evaluation of Draftometer. A series of tests were run to investigate some of the variables that were expected to influence the precision of the results obtained on the Draftometer. The following results were found:

1. The actual sensitive length employed in calculating the charts does not affect the average drafting force, but has considerable effect on the smoothness ratio; an increase in sensitive length results in an increase in smoothness ratio. It was found that a sensitive length of $1\frac{1}{2}$ inches is sufficient to include most short-term variations which represent the "stick-slip" effect.

2. A sample length of 40 inches was found adequate to give a representative measurement of the average drafting force and smoothness ratio of a roving bobbin or can of sliver.
3 An increase in gauge length resulted in a decrease in both average drafting force and smoothness ratio. In comparing samples of different staple lengths, a gauge setting of staple length plus $\frac{1}{4}$ inch was chosen as the standard.
4. Good correlation was found between average drafting force as obtained on the Draftometer and cohesion as obtained on the Instron Tensile Tester.
5. An increase in twist resulted in an increase in average drafting force.
6. Variations in size or weight per unit length affect the average drafting force. For comparison purposes, the average drafting force should be divided by average weight per unit length to correct for variations in weight.

Additional Tests

In addition to the tests previously discussed, fiber crimp and dye affinity are also measured by the manufacturer. The evaluation of crimp can require a rather complex technique. The crimp in a fiber is three-dimensional, and for that reason does not lend itself to any common optical test. One technique which has been developed and reported by Roder (2) uses the principle of applying a load with a chain-o-matic weighting system and plotting a curve from which the force necessary to remove the crimp can be found. Such measurements as crimp and fiber deniers are not conducted in a mill laboratory because of the expensive equipment and time involved and also because the results would be of little value once the tests were made.

References

1. The information on the Draftometer is from papers and standards by William T. Waters, Dept. of Textile Research, School of Textiles, N. C. State College.
2. Roder, H. L., "The Properties of Fibres which Govern Spinning and How These can be Measured," *Enka and Breda Rayon Rev.*, **4** (1950).

CHAPTER XVIII

Control in Opening and Picking

The successful production of good yarns or fabrics is not determined entirely at the spinning frame or the loom. These processes must be operated efficiently and wisely, but the answer to success here lies in the opening room, in the picker room, and in the card room. To produce a good yarn, produce first a good lap, good card sliver, good drawing sliver, and then good roving. The degree of success of these primary operations can be measured by the quality of the yarn spun, but the control and measure should be performed at each step in the processing.

There are several tests that can be made in the opening and picking area which have been designed to measure quality and to provide methods of controlling this quality. It should be emphasized that not as individual tests, but rather as a coordinated group of tests does the control of quality come into its own. Although not all mills have laboratory facilities, the unification of a quality control plan under the responsibility of a single individual is most desirable, and moreover is usually justifiable in even very small plants.

Application of Oil to Cotton

Cotton is frequently oiled during the opening or picking process to improve running conditions in subsequent operations. In general, this application of oil is claimed to reduce static, provide lubrication to facilitate and improve drafting, and to reduce dust and fly. The amount of oil applied is of importance, inasmuch as an excess quantity might cause more trouble than good. For this reason, periodic tests using simple and accurate methods of checking on content are necessary.

There are a number of methods that can be used for oiling the cotton. The two methods in widest use are spraying of the oil onto the cotton at the hopper feeder and application of the oil by a "wiper" bar so located in the beater chamber of the picker that the cotton contacts the distributing bar immediately after leaving the feed rolls.

Oil Application at the Beater

In this system there is located near or adjacent to the beater a metering gauge provided with a needle valve to control the rate of flow of oil and a glass tube for permitting count of the flow by drops. The following description outlines one procedure that can be used to determine the oil content applied by this method.

1. Ascertain the weight of oil per drop by weighing about 300 or 400 drops of oil. This can be easily done by disconnecting the exit line at the drip valve, counting drops and weighing.
2. Ascertain picker lap weight (net standard weight of full lap).
3. Ascertain running time to produce one lap.
4. Calculate the number of drops per minute to give the desired percentage of oil content.

Inasmuch as percentages are in proportion to the rate of flow (drops per minute), a simple chart can be prepared to show percentage against rate of flow.

A sample calculation is shown herewith:

Weight of 100 drops of oil = 35 grains = 0.005 lbs.
Lap weight 42 lbs.
Lap running time 8 mins.

To obtain 0.33 per cent of oil in the cotton,

0.33 % of 42 lbs. = 0.1386 lbs. of oil to add.
0.1386 ÷ 0.005 = 27.72 hundreds of drops = 2772 drops for entire lap.

For one minute running time,

2772 ÷ 8 = 346 drops of oil per min. to obtain 0.33 % oil.

Oil percentage for any number of drops per minute would be in proportion. For example, if 200 drops per minute, then

$$\frac{200}{346}(0.33) = 0.19\%.$$

Oil Applications at the Hopper

In those installations in which oil is sprayed at the hopper feeders (or other opening points), a metering pump is employed at each machine to feed the oil, while compressed air at the spray nozzle gives the necessary means for atomizing the oil. The rate of flow of oil through the pump can be controlled; if more than one pump is used, as for example in installations where the oil is sprayed into each of several hoppers of blending feeders, then it is important to calibrate the rate of flow of each of these units accurately to prevent over-oiling in some hoppers, and under-oiling in others. The method of calibration is quite similar to that for the system described above. The average production of each feeder per operating hour must be determined. Secondly, the rate of flow of the pump at a given speed is measured. This can be done by collecting and weighing the oil pumped during any measured length of time. Then the two sets of values obtained can be compared to get a percentage value. The rate of flow can be regulated to give the desired oil content.

With this type of oil application, there is the danger that accumulations of lint will be formed around the oil spray heads. This can be caused by fly, or by the catching of fibers due to overfilled hoppers, in which the cotton in its tumbling action is brushed against the spray head. These accumulations become saturated with oil and tend to be dislodged as discrete groups of fibers. Such oil-soaked clumps can de definitely harmful to processing and equipment.

Amount of Oil Application. The usual limits of oil injection are from 0.20 per cent to a maximum of 0.60 per cent by weight.

Frequency of Test. Weekly spot check tests of the rate of application are recommended.

Picker Lap Weight Control

There are two definite measures of lap uniformity: the long-period variation as indicated by full lap weight, and the short-period variation as indicated by the weight of lengths of 1 yard or less.

It has long been mill practice to record the weight of each lap produced on finisher pickers. The weight, in pounds and ounces, usually includes the tare weight of standardized picker lap rods. This weight represents an average of the total yardage comprising the lap, and therefore represents only the long-period variation in weight or variation from lap to lap. It does not indicate any irregularity existing within the lap itself. Accurate scales designed for weighing laps are available. A tolerance in the weight of the lap can be established from experience, preferably using actual data to establish control limits by statistical analysis. For instance, a standard lap weight of 40 ± 1 pounds means that laps weighing less than 39 or more than 41 pounds are considered unsatisfactory for regular production and would be either reprocessed or processed as special lots.

Inasmuch as a comparatively close tolerance is usually set for these lap weights, consideration must be paid to the effect of atmospheric humidity on lap weight. In a picker room equipped with a humidification system, only little adjustment in the standard lap weight would be required to compensate for humidity changes, but in rooms with no such equipment, adjustments of standard weight must be made to compensate for actual regain variations. The regain adjustment procedure requires an instrument to measure the regain. Regain indicators permanently mounted in the picker room fulfill this requirement, as they contain an exposed batch of cotton arranged so that changes in regain of this small mass are indicated by the dial reading, expressed in terms of regain percentage. The standard weight of the lap is then adjusted to compensate for its actual regain, which can be assumed to approximate that of the regain indicator. As an example, if the standard weight of the lap is 40 pounds based on a regain of 7 per cent, and if the regain indicator showed 3 per cent regain, the standard lap weight must be corrected to $40(103 \div 107)$, or 38.5 pounds.

In order to simplify the job of the picker tender in weighing the laps, a regain target can be attached to the dial of the scales. The target location can be calibrated for regain, so that all that is necessary is to adjust the location of the target at least once a shift (and preferably more often). The picker tender then can accept or reject laps by noting whether or not the scale pointer comes within the target opening when he weighs his laps. See Fig. 18.1.

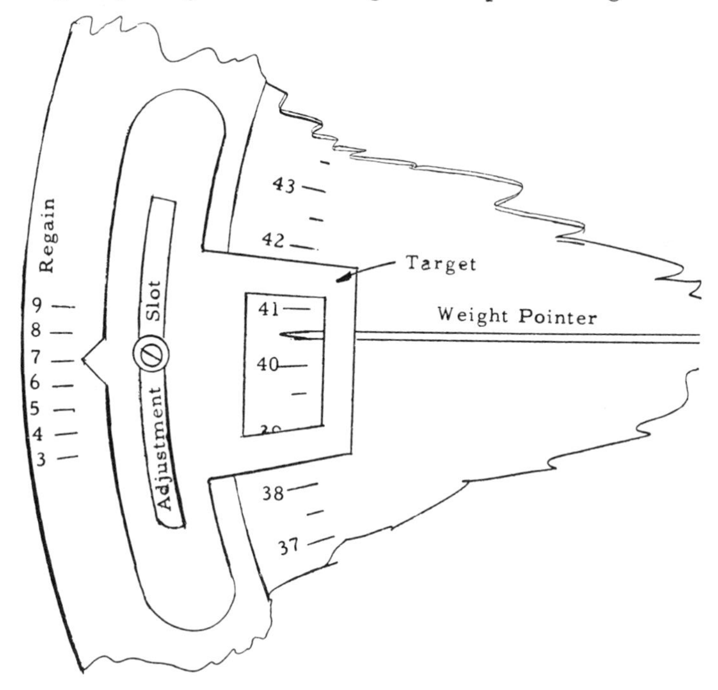

Fig. 18.1. Target for regain adjustment on lap scales.

Another method for adjusting the scales to the correct regain is to have a series of supplemental small weights made up that can be hung on the pan of the lap scales. Assume that the standard lap weight is 44 pounds, and that this includes a lap rod weighing 2 pounds. Then the net weight of the lap affected by moisture is 42 pounds, and this is standard at 7 per cent regain. The dry weight would be $42 \div 1.07$, or 39.25 pounds.

If the regain indicator showed a 7 per cent regain, then the scale dial should read normally at 42 plus 2, or 44 pounds. If the regain were down to 3 per cent, then the lap would record light; and if the regain were up to 12 per cent, then the scales would record heavy. In order to control for high humidity, the lap standard must be based on the highest likely humidity to be encountered. Assume this is 12 per cent. Then the corrected lap standard weight will be 39.25(1.12), or 43.96, nearly 44 pounds (plus 2 pounds for the lap stick). For any other regain, extra weight is added to a supplementary hook on the scales. Thus, for 7 per cent regain, a 2-pound weight must be added, so that the scale dial would read with a 7 per cent regain weight on the pan plus the lap and lap rod a total of $42 + 2 + 2$, or 46 pounds.

The regain instrument in widest use is the Aldrich Regain Indicator. This instrument is designed to be used in any operating section, so located away from abnormal drafts as to reflect the normal atmospheric conditions in the area. The instrument is in reality a sensitive balance, with the dial calibrated in units of per cent regain. The exposed element on one side of the balance is a batch of cotton contained in a fine net mesh bag, which in turn is protected by fine wire mesh to retard contamination by dust and fly. This is counterbalanced by a weighted lever; the entire system is damped by an oil dashpot. Changes in relative humidity affect the exposed cotton element; as its regain changes, these changes are reflected in weight changes, and so indicated in terms of regain by the dial reading. Two calibration weights are provided. It is wise to make up an additional calibration weight for the range most frequently encountered. For example, if the picker room varies normally in summer from 7 to 9 per cent regain, then an 8 per cent calibration weight should be made up and used. (The bone dry weight can be calculated from the weight of either of the other calibration weights supplied.) Accumulations of fly and dust on the exposed sample of cotton throw the indicator out of calibration. For this reason, the control element of cotton in the fine net holder should be replaced periodically, preferably at least once a month.

Other instruments that indicate either regain or relative humidity can be used. If a recording relative humidity instrument is used, a table showing regain at any level of humidity is required. For example, Minneapolis-Honeywell manufacture a portable unit that records the dry bulb temperature and relative humidity.

A more recent development that appears to have a good deal of merit is the Moisture Monitor (made by Strandberg Engineering

Laboratories, Greensboro, N.C.), which is an outgrowth of the automatic moisture control systems by the same company for use on slashers and other types of dryers. The Moisture Monitor is actually a picker tare weight indicator, so designed that the total tare weight indicated includes both the lap pin tare and moisture content tare. The method used is to contact the cotton itself intimately as it is being processed in the picker. The point of moisture measure is selected near the finish of the process, usually making use of the conventional rake in a blending reserve or some other suitable contact assembly where electrical contact can be made with the cotton. The location is important, for the high flow of room air through the cotton at the different screen sections tends to bring a condition of moisture equilibrium to the cotton at each location.

The instrument is designed for calibration based on moisture content and weight. For example, Fig. 18.2 shows a sample dial face for the Moisture Monitor. In this example, the scale calibration is based on a 44.7-pound lap with 7 per cent moisture content (approximately $7\frac{1}{2}$ per cent regain), but without a $2\frac{3}{4}$-pound lap pin. With the lap pin, the weight is 44.7 plus $2\frac{3}{4}$ or $47\frac{1}{2}$ pounds total.

In Fig. 18.2, a 7 per cent regain is indicated by the dial pointer. This gives a required lap scale reading of $47\frac{1}{4}$ pounds, which is made up of

41.60 lbs. net dry cotton per lap
2.90 lbs. moisture at 7 % regain
2.75 lbs. lap pin

47.25 lbs. total

The Monitor scale is designed so that (1) the required lap scale reading can be used, or (2) an extra weight can be added to the pan of the scale. In the first case, the picker tender uses the required lap reading, plus or minus the tolerance allowed for control (possibly plus or minus $\frac{1}{2}$ pound). In the second case, the tender hangs on the lap scale the extra weight indicated on the dial of the Moisture Monitor. This results in a total lap weight of $48\frac{1}{2}$ pounds in this example, for maximum moisture content is assumed to be

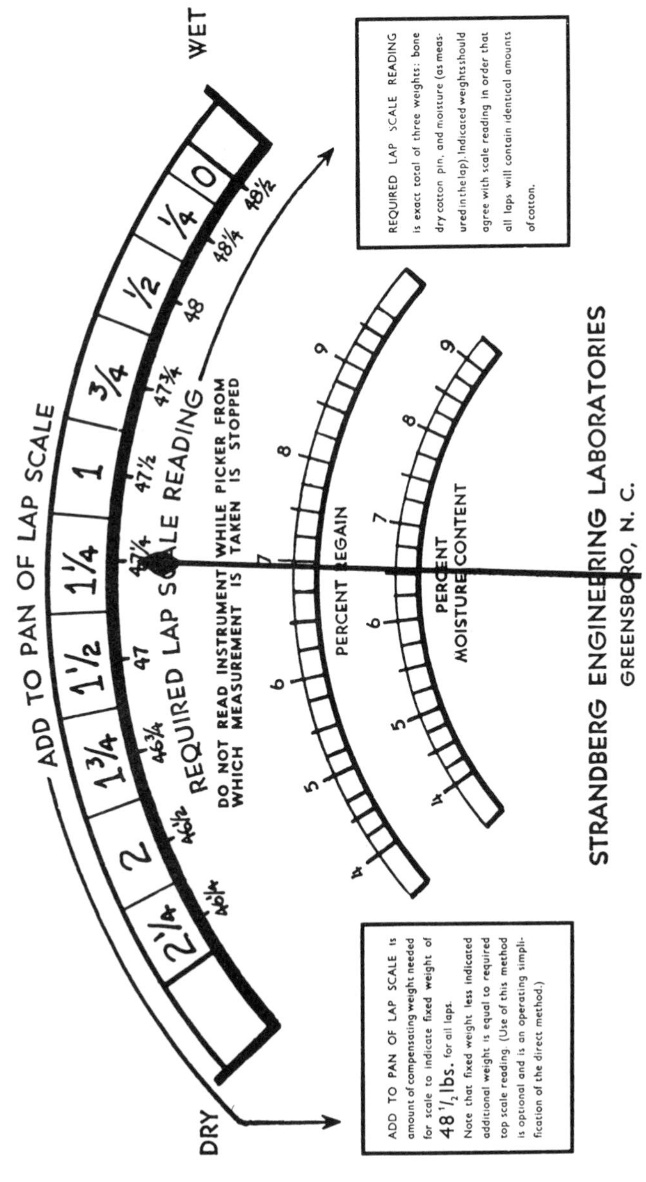

Fig. 18.2. Moisture monitor.

9 per cent (dry weight of cotton plus 9 per cent moisture content plus lap pin gives $48\frac{1}{2}$ pounds).

Fig. 18.3, Control Form, is suited for mill use in recording actual lap weights as they are taken from the picker.

Lap Weight Record
Gross Weight

To: ____________________ Date: ______________

Picker No.	1		2		3		4		
Stock									Regain Per Cent
Standard Weight / Lap No.	lbs.	oz.	lbs.	oz.	lbs.	oz.	lbs.	oz.	

Fig. 18.3. Control form.

Lap Uniformity

It is recognized that failure to control long-range and short-range evenness on picker laps means failure in the production of uniform yarn. In other words, if the first processing step is faulty, it is impossible completely to overcome the handicap in subsequent operations, especially with the trend in processing being toward fewer and fewer doublings. In addition, textile technicians are beginning to recognize the fact that widthwise variation in a lap is just as important as lengthwise variation, and efforts are being made to devise means of determining the extent of widthwise variation. Instruments for the measure of short-range variation in sliver, roving, and yarn are in wide use, and recently developments have been made to accommodate the electronic instruments for use with picker laps.

Mills equipped with semi-obsolete or out-of-date opening and picking machinery find it difficult to maintain as close a level of control on their laps as do mills with modern equipment. If the level of control that is established by any mill is set too close, i.e., with too tight a tolerance, then the obvious result will be an excessive number of rejected laps. Rejected laps in themselves mean possible trouble and a multiplying of the difficulties encountered in controlling the evenness in the picking process. It is for this reason that rejected laps should be processed in the most careful manner possible. One method of doing this is to utilize a single finisher picker with a lap apron. Overweight and underweight laps can be then intermingled or blended on the apron, say two light and two heavy laps, and these laps reprocessed through the one-beater process; the product of the machine should be within the tolerances required by the mill. A method of avoiding reprocessing of rejected laps is to funnel light laps to certain designated cards on which draft gearing is changed to accommodate the light laps, and in a similar way the heavy laps can be funneled to another set of cards where the opposite procedure is taken. In any event, it will be found that if the laps are carried back to the opening room and unrolled and refed through the entire opening and picking process, not only is the cotton overworked with a resulting increase in neps, but also it is much more difficult to control lap weight and yard-to-yard weight.

Yard-to-Yard Weight. The control of the weights of the individual laps is of prime importance in maintaining yarn quality, particularly in the area of holding average yarn numbers constant and at their respective standards. Just as important is the control of the ranges of the individual yarn numbers between spindles, between frames, and over periods of time. Excessive variations in the range can be traced to variations within picker laps, and thus the short-range control of lap weight is essential.

The analysis of yard-to-yard weight of picker laps is an accepted procedure. In some cases shorter lengths are analyzed, and with electronic instruments it is possible to detect variation within very short lengths.

One method of actual weight of short lengths is based on using the Saco-Lowell Lap Meter. By its use, 1-foot or 1-yard lengths can be weighed easily. The machine measures the required length and drops the sample onto the pan of a scale. The scale is designed to show variation in ounces and fractions over or under the standard weight of the lap. The individual readings for an entire lap can be plotted on a control chart, and limits of variation can be established statistically. The chart for analysis with typical entries and calculations is shown in Fig. 4.6, p. 57. This includes the method for calculating the coefficient of variation.

Lacking other means of analyzing picker lap uniformity, it is possible to weigh short lengths and analyze the data for the pertinent information. The lap to be tested is selected at random from the run-of-the-mill production so that it will be a true sample of the population. The lap is rolled out on the floor to the limit of space available, care being taken not to stretch or wrinkle the lap. A 1-yard template is placed over the lap, and a 1-yard section is separated from the lap by breaking away the lap to leave a close fringe. (Some cut the lap; this probably gives more accurate results.) The one-yard sections are weighed, and results plotted against standard weight. This gives an indication of the yard-to-yard evenness of the lap.

Frequency of Test. It has been well established that each lap produced should be weighed, and this procedure is recommended.

For short-term variation analysis, it is recommended that one lap from each picker be tested each week. If it is found that the laps tested do not meet the standards established by the mill, check tests should be made. A higher frequency of test is desirable, but inasmuch as the laps so tested must be reprocessed with the attendant lowering of quality, it is felt that the single test weekly should serve most mills' needs.

Standards. For average conditions, picker lap variation by weight control can be established on the following standards with the reservation that a statistical analysis might reveal that closer tolerances could be maintained. For total lap weight, a control of plus or minus 2 per cent is the outside limit, which means a total of 4 per cent variation in the lap weight. Thus, for a 50-pound lap, variation should be maintained within plus or minus 1 pound. A more acceptable level is a plus or minus 1 per cent, which means a plus or minus $\frac{1}{2}$ pound tolerance for a 50-pound lap.

There has been an increasing interest in the use of the coefficient of variation, V, as the method of expressing yard-to-yard evenness. This has the advantage of being a measure of relative variation and as such can be used as a method of direct comparison of laps of different weights. The average coefficient of variation for a picker lap should range from approximately 1.25 to 2.0 per cent. Some mills have been successful in averaging 1.25 per cent or slightly less with some laps having a coefficient of variation of less than 1 per cent. Quality control charts are frequently used in connection with this test, and the limits are established by the following formula:

$$\text{Control limits for } V = 3\left(\frac{V}{\sqrt{2N}}\right)$$

where N = the number of yards in the laps.

It should be understood that these standards for variation by weight are the average taken from experience, and as previously indicated, well-equipped, well-organized mills should be able to maintain slightly closer tolerances than given above, and that on the other hand, mills with inferior equipment might not be able to meet these standards without excessive rejection of laps.

Uster Vari-Meter

The previous methods of measuring the uniformity of picker laps have dealt with the net corrected weight of the lap and the yard-to-yard weight. The yard-to-yard analysis is considered by many as giving a measure of short-range variation; actually, when it is considered that the lap is drafted usually in excess of 100, a shorter sensitive length of measure of uniformity is needed. This need is satisfied by the Uster Vari-Meter, which is designed to give an inch by-inch measure of irregularity.

Operation. A condenser field which covers the full 40-inch width of the picker is mounted in the picker lap head at a spot just in front of the two bottom calender rolls and behind the lap winding unit. This is shown in Fig. 18.5 in place on a picker, but with no stock. Guide plates provide the means for guiding the lap between the two electrodes which form the measuring field. Once the lap has been fed through the measuring field, no other re-

Fig. 18.4. Uster Varisignal and centralized power supply.

threading is necessary. The condenser field, by means of a high frequency capacitance measurement, determines at any moment the weight of the approximate 1 inch portion of the lap which lies between the two electrodes. The measuring range is 10 to 20 ounces per yard.

An adapter unit which is mounted on or adjacent to the picker lap head is connected to the condenser measuring field by a short heavy-duty cable. This adapter converts capacitance variation resulting from variation in the lap weight into low frequency voltage variation. The dial on the adapter unit is calibrated in units of ounces per yard, so that in operation the needle pointer indicates

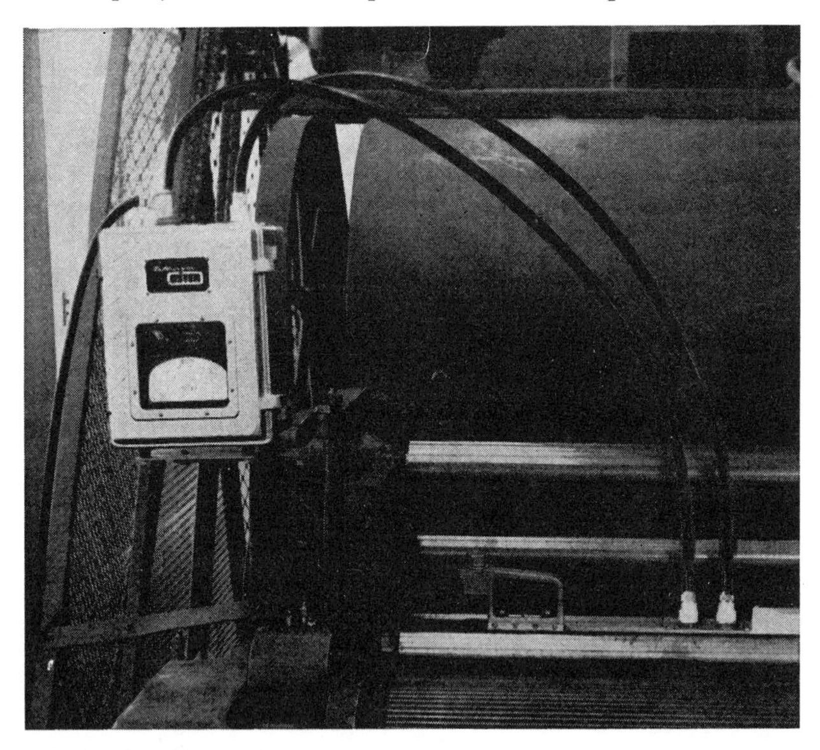

Fig. 18.5. Uster Vari-Meter and adapter unit.

the lap weight in ounces per yard. An adapter unit is shown in Fig. 18.5 mounted on the picker frame.

A standard type of Uster Evenness Tester is connected to the adapter by a multiwire cable. This permits the use of the auxiliary recorder so that a chart of the variation in lap weight can be made. A Power Supply Unit can be used, but this does not allow for the use of the Recorder. It is preferable, if the Evenness Tester is not available, to use the Uster Centralized Power Supply, for with this, up to eight Vari-Meters can be used to replace the Uster Evenness Tester. The Uster Tester and Recorder can be located in an area remote to the picker room, for example, in the testing laboratory.

Another unit in the Uster arrangement is the Varisignal. This device indicates by flashing lights variations in lap weight larger or smaller than the tolerance set by the mill technician. On the dial face of the instrument, which is calibrated in ounces per yard, are two indicator pointers: one is for the tolerance below average weight per yard, and the other for the tolerance over the average.

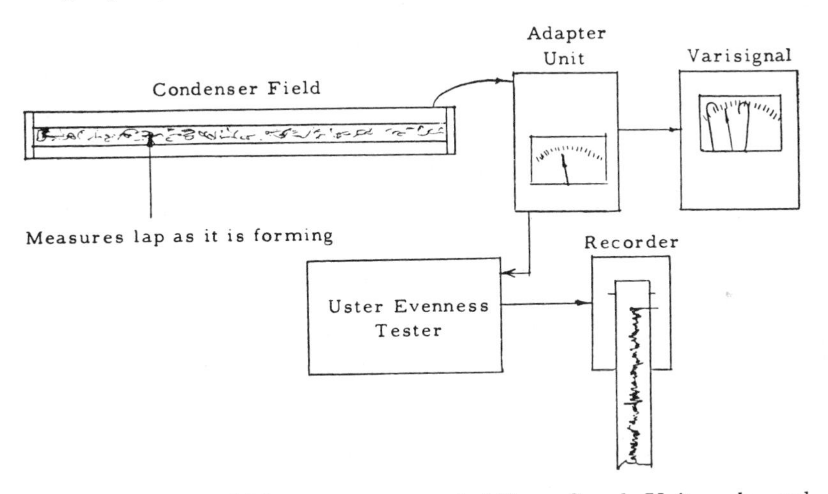

Fig. 18.6. Uster Vari-Meter arrangement. 1. A Power Supply Unit can be used in place of the Uster Evenness Tester, but without recording. The Varisignal can be used with this arrangement. 2. As a replacement for the Uster Evenness Tester, a centralized power supply can be used for up to eight Vari-Meters. The Recorder can be used for direct registering, and any picker in the group can be checked at any time. The Varisignal also can be used with this arrangement.

Both can be manually set or reset with no trouble. When in operation, each variation above or below the set tolerance field or spread is indicated by its own light, and in addition, the number of times the weight is above or below these limits is registered separately on counters. The Varisignal and a Centralized Power

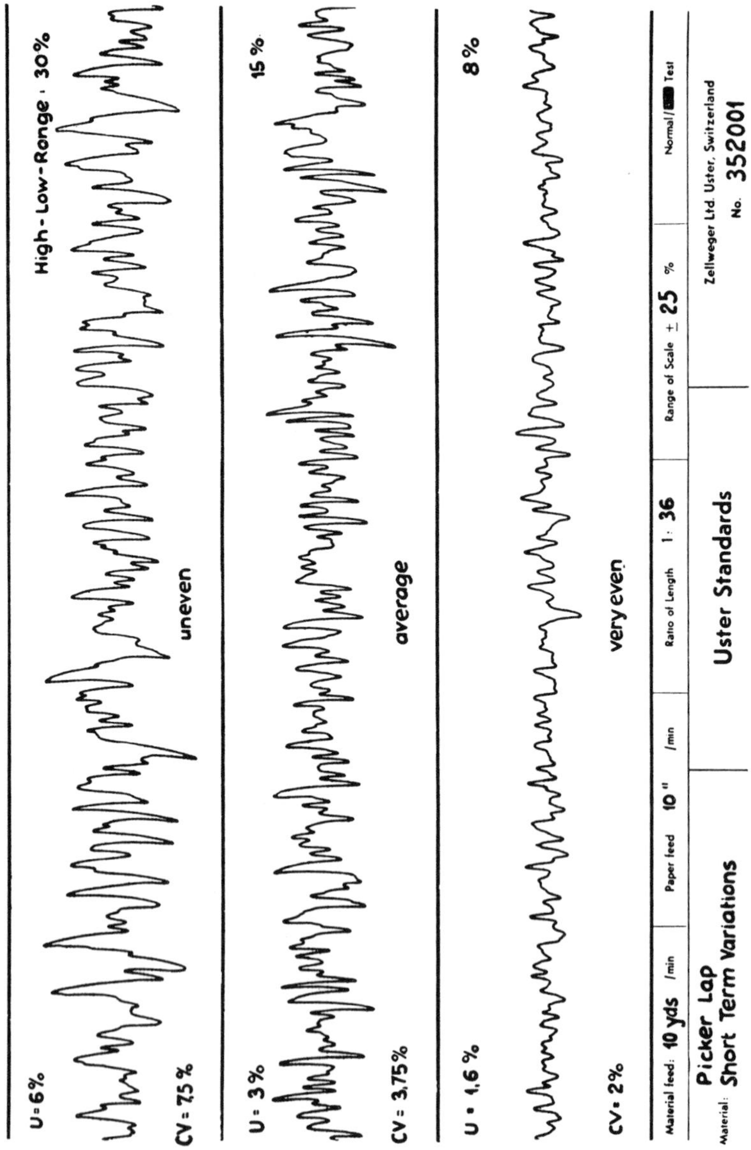

Fig. 18.7. Uster Vari-Meter standards.

Supply unit are shown in Fig. 18.4. A diagrammatic sketch of the setup is shown in Fig. 18.6.

Test Methods and Results. The Uster Vari-Meter can be a permanent installation on a picker. Group installations with a centralized power supply and recorder are desirable. The method of test if a graph is to be made is similar to the procedure used with the Uster tester, which is described later in this book. The Vari-signal unit may be left in operation continually, for a supplementary switch on the picker full-lap knock-off cuts the circuit when laps are not being formed. Because of the newness of these units, it is preferable to follow the manufacturer's procedures for calibration and test.

If desired, the Vari-Meter can be adapted to a Saco-Lowell Lap Tester. In this case, the lap meter is used primarily as a frame for the Vari-Meter, and a means for unrolling the lap. The usual breaking of the lap into 1-yard lengths is omitted. By using the Vari-Meter in this manner, the laps from all pickers can be tested and adjustments on pickers made to give optimum results. The same Vari-Meter can then be transferred to each picker, one at a time, and further adjustments made.

Fig. 18.7 is a reproduction of the graphic recording supplied by Uster to illustrate three levels of lap uniformity.

CHAPTER XIX

Control in the Card Room

Card Control

Weight Per Yard. In the everyday test for card sliver weight, usually a program is set up to check the size of the sliver from each card in operation. The frequency of the test depends upon the individual mill's requirements and facilities. A suggested program of control is to test the sliver from one-fifth of all cards daily, with each test consisting of three to five 1-yard sizings. These tests are performed, first, to show the average actual weight of the product, and second, to show excessive variation in size from the standard of any individual card, where such variation may be indicative of mechanical trouble.

Card Sliver Weight Record

Grains Per Yard

To: ______________________ Date: ______________

Card No.	Stock	Standard Weight	Actual Weight					Ave.	Range	Ave. Each Stock
			1	2	3	4	5			

Fig. 19.1. Control form.

One procedure is as follows: A laboratory representative goes to each card to be sized, usually accompanied by the head grinder or some card-room worker. From the can of sliver being produced on the card, he measures off with the 1-yard board three to five

1-yard lengths of sliver. These samples are identified and taken in a wire basket to the laboratory, where after conditioning for a few hours, they are weighed. Figure 19.1 shows the type of control form used.

It should be noted, also, that it is preferable to obtain the samples at a definite period; this period for sampling is usually not less than one-half hour after stripping, as stripping affects weight of the product.

Evenness of Card Sliver

It is sometimes desirable to make tests for uniformity of card sliver. An evenness test in this case would involve sizing each and every yard of sliver produced by any given card for the entire period between card strippings. It is more convenient to perform these tests in the laboratory, substituting a roving reel for the 1-yard template for measuring the lengths of sliver. The results of the test can be plotted on a control chart for analysis.

After a card has been stripped, there is a short period during which there is a progressive increase in weight of the card sliver. This is due to the fact that the card clothing becomes loaded with cotton following removal of previous accumulations by stripping. Tests for evenness performed on the sliver produced directly after stripping will show the length of time necessary for the sliver to return to approximately normal weight. The sliver produced during this interval is out of control and should not be used where high quality is desired.

Saco-Lowell Graphic Sliver Tester. One of the instruments in widest use for determining the evenness of card sliver is the Saco-Lowell Graphic Sliver Tester, shown in Fig. 19.2. This instrument, designed to test sliver and roving up to 1.50 hank, makes a continuous chart of sliver evenness by measuring the thickness of the material being tested. The instrument is valuable in measuring the uniformity of all types of sliver, inasmuch as it gives a record of short-period as well as the long-period variations in sliver.

It has been established that sliver from the card is uneven over long periods, i.e., the grains per yard for 1 yard or longer lengths vary quite substantially. When compared to the sliver after a single process of drawing, the card sliver is more uneven over these yard or longer lengths, and the sliver from the second drawing is more uniform for this long-period variation. Such improvement in average sliver weight is to be expected, due to the averaging out of differences by the many doublings. At the same time, the determination of what happens to short-range evenness remains obscure. By

Fig. 19.2. Saco-Lowell Graphic Sliver Tester.

short range, the inch-to-inch variation is meant. It has been shown that this variation is the least with card sliver and progressively deteriorates at each of the drawings and becomes worse with successive drafting processes. Thus, it has become apparent that the shortcoming of the usual mill test involving the weighing of yard lengths to control quality is that it has failed to show trends in short-length variation, which has a very definite effect on yarn evenness. The Saco-Lowell instrument, and more particularly other instruments developed subsequent to this one, have opened the doors to new concepts in quality control.

The Saco-Lowell sliver tester operates on the principle of mechanically measuring the bulk thickness of sliver or roving. The sliver to be tested is fed in between a pair of rotating rolls designed so that the bottom roll has a square groove into which fits the "splined" top roll. The top roll, being attached to a lever arm, is free to move in a vertical plane so that it can react to variations in the thickness of the sliver. Movements of this top roll are amplified mechanically. A compressive pressure of approximately 1600 p.s.i. for sliver or 2500 p.s.i. for roving is applied to the material between these rolls. This force is sufficient to press air out of the sliver and thus permits measurements of changes in bulk of the material itself. Movement of the upper roll is translated to a pen movement; the amplification is such that the pen moves $\frac{1}{10}$th of an inch in response to each $\frac{1}{1000}$th inch change in bulk thickness. Thus, the pen plots on a chart an autographic record of the uniformity. The chart speed is adjustable to two settings: first a speed equal to that of the sliver through the machine, and second, at a speed of 3 inches for each yard of sliver. The latter is used most frequently. In making a test, the pen is set by an adjustment so that it is centered on the chart, and with the pen centered, the thousandths of an inch separation of the two wheels is known so that sliver thickness is known. The variation in the sliver is measured by counting the tenths of an inch above or below the center. To aid in this, horizontal lines spaced at each $\frac{1}{10}$th inch are printed on the chart. For example, if in a given length the average thickness is set at 50, the actual thickness is $\frac{50}{1000}$ inch; and if the sliver varies from $\frac{5}{10}$ths above to $\frac{5}{10}$ths below the center line, then the total variation would be $\frac{10}{10}$ths or $\frac{10}{1000}$ inch of the cotton sliver. The per cent variation would be (10/50)(100), or 20 per cent.

In making a test of card sliver, 10 or 20 yards are tested, with a chart speed of 3 inches per yard. For convenience, the charts are ruled vertically every 3 inches. In each yard (within each 3-inch chart section), the thickest or highest point and the thinnest or lowest point are checked and tabulated in tenths of an inch. These values are added, then the totals for each yard of sliver added together and averaged to get the average variation from the center, as shown in Table 19.1.

TABLE 19.1

Variation in Thickness of a Card Sliver (in tenths of an inch)

Length, yards	Variation below	Variation above	Total variation
1	4.0	2.5	6.5
2	3.5	2.0	5.5
3	3.0	3.0	6.0
4	3.5	2.0	5.5
5	4.5	3.5	8.0
6	3.0	4.0	7.0
7	2.0	3.0	5.0
8	4.5	2.0	6.5
9	5.0	3.0	8.0
10	4.0	3.0	7.0
	37.0	28.0	65.0

Average thickness by setting before test = 50

$$\text{Variation} = \frac{65}{10}\left(\frac{100}{50}\right) = 13\,\%$$

Evenness Testers. There are three instruments in wide use in the U.S.A. for measuring sliver evenness. These are the Uster Evenness Tester, the Brush Uniformity Analyzer, and the Pacific Evenness Tester. In testing with the Uster or Brush instruments, the sliver, roving, or yarn is passed through one of the appropriate slots, each of which houses a set of condenser plates. Textile materials, being dielectric, cause variations on the capacitance of the condenser proportionate to the cross section and density of the material being tested. Variations in the cross section of the material are measured and amplified electronically, and results are indicated on a meter and simultaneously recorded in the form of a graph. Results of the tests are analyzed in much the same way as described for the Saco-Lowell sliver tester, although an integrator can be attached to the instrument to give average variation for longer lengths.

The Pacific Evenness Tester measures the bulk of the sliver in much the same manner as does the Saco-Lowell tester. However, the amplification of the signal by electromagnetic means with photoelectric recording is much more sensitive than the mechanical amplification of the Saco-Lowell tester.

These testers and their application for the measure of uniformity of sliver, roving, and yarn, are discussed in detail in a later chapter.

Nep Count at the Card Web

"Neps are small knot-like aggregates of tightly entangled cotton fibers, usually not larger than a common pin head, which are difficult to separate out of cotton and from which the individual fibers are generally difficult to remove." It takes some practice to identify a nep inasmuch as other characteristic defects in cotton are similar. One of these is *naps*; these are usually larger than neps, may show seed fragments, and are not as closely entangled, so that they can be pulled apart easily, and are more readily removed from cotton. Neps are usually associated with immature fibers or fine fibers; they are caused by tangling of these fibers in processing. Excessive working of cotton or of the finer staple synthetics, especially in opening, picking, and carding, is a source of neps. They are particularly objectionable in fabrics finished in solid colors, as the immature fibers in the neps take on a different shade, and the nep itself reflects light differently to cause a color imperfection. In spinning, the nep is a cause of end breakage because of interference with drafting.

Due to the fact that the classification of neps is difficult, any measurement of the number of neps in a sample of fibers [1] depends on the experience and judgment of the technician. However, where one technician makes all the determinations of nep count, results are comparable.

In grading yarn for appearance, neppiness is one of two important characteristics, the other factor being evenness. Thus, freedom from neps is desirable in the yarn and increases the uses to which the yarn can be put. The neppiness in a yarn can be predetermined

[1] The nep count of raw cotton can be made by following the specifications outlined in the A.S.T.M. Designation: D414–40 T.

at the card web, as there is a close relation between the neps in the card web and in the final yarn. It will be found that neps in yarn are greater in number than in the card web, except in combed work. This is due to the fact that drafting itself can be cited as a cause of neps; in other words, any manipulation of fibers is likely to cause this tangling that becomes a nep. In the case of combed work, many neps are removed in the combing process.

As indicated, neppiness is measured at the card web because of the fact that this is the one step in processing at which it is relatively simple to get counts that are representative. This nep count at the card web is one of the properties reported by the U. S. Department of Agriculture in its reports on selected varieties of cotton. The basis of evaluation given by the U. S. Department of Agriculture is given in Table 19.2.

TABLE 19.2

Number of Neps Per Hundred Square Inches of Card Web [a]

Number	Rating
1 to 15	Low
16 to 30	Average
31 to 45	High
Above 46	Very high

[a] Card production rate of $9\frac{1}{2}$ pounds per hour.

The procedure for sampling the card web and for making the count is as follows: The card to be sampled is selected. The doffer of this card is stopped, and the coiler gearing disconnected in the usual manner. A black board 4 × 9 inches is passed beneath the web close to the doffer; by turning the doffer forward by hand, a smooth sample can be lifted from the web. By using a magnifying glass, the neps showing on the surface of the board can be counted. The fringe of cotton web extending beyond the edge of the board should be trimmed off with scissors. Five samples are minimum requirement in order to obtain a reasonable estimate of the average.

A second method of sampling of the card web involves a similar technique with these differences. A velvet-covered black board is passed by hand under the web with the card in operation. By a quick movement, the board can be passed up through the traveling web without unduly hurting the web or breaking it down; in fact, if this is done with care and dexterity, the damage to quality is less than that resulting from breaking down the end and subsequently piecing it up. The sample board is then mounted in a special template (developed at the School of Textiles, N. C. State College) containing twenty holes, each one square inch in area. The template is designed so that the board with the web is locked in place on the back face of the template by sliding a simple locking device. By use of this template, the technician can more easily count the neps, and do this with less fatigue and error. Five such samples are taken, and they are selected so that the total width of the web is sampled and represented in the final analysis. Base results on neps per hundred square inches.

The neppiness of the web can be expressed in these different ways: (*1*) The number of neps for a given area, usually either 36 or 100 square inches. (*2*) The number of neps per unit weight, usually per grain or per gram. (*3*) The number of neps per 100 yards of any given yarn count.

It is recommended that all cards be checked for neps once a month of single-shift operation or approximately once a week of three-shift operation. Cards in the lower 10 per cent bracket should be rechecked and card adjustments made if they continue to show below standard. Also, a control chart can be used, and standards set to show limits poorer than standard.

Neps: Practice in Industry. A study of the replies to a questionnaire submitted to a group of cotton mills indicates that all are concerned with the problem of neps and that routine methods are employed by them to check the prevalence of this type of defect in their products. (The only mill reporting no concern with the problem employs cotton yarns for rug backings only.) There appears to be a lack of standardization of nep sampling techniques and procedures in industry.

The three quality factors reported to be affected most by neps are yarn appearance, cloth appearance, and dyeing. The effect on cloth appearance is judged as most serious. Neps are reported by about one-third of the mills to influence spinning and weaving end breakage.

The majority of mills check for neps at the card web. Many report checking at the yarn and cloth stages. The size of the sample inspected varies considerably between mills, but the most popular is an area of a hundred square inches of card web. Most mills express their nep counts in units of neps per hundred square inches, with others indicating a preference for a neps-per-grain value. Standards vary, with an average value of thirty neps per hundred square inches indicated as being acceptable.

Five mills report that their card web samples are taken halfway through the stripping cycle. Others show little agreement between the periods of the carding cycle at which samples are inspected. This makes it difficult to evaluate the standards reported. Ten mills report making their nep counts once every week.

The majority of the mills take corrective measures at the carding stage by grinding and/or checking settings. Only two report checking their raw stock for neps.

The responsibility for checking the neppiness of the cotton stock is delegated in the vast majority of mills to the quality control laboratory or its equivalent.

Neps: Synthetics. The procedures outlined in regard to neps in cotton apply equally to spun staple. The neps in staple rayon are somewhat similar in outward physical appearance to those found in cotton. However, manufacturers of staple rayon have gone further in their classification of defects. For instance, they call the smallest defect a "point," which in cotton might be classified as a very small nep. In all, they classify six different types of defects in the staple. In making a measurement of these defects, one method is based on the analysis of a 14-gram sample of top, card sliver, or picker lap prepared by taking pinches of fibers at random from at least twelve different parts of the submitted sample, Thin webs of fibers are drawn from this specimen, and any defects found are graded according to the classes shown in a standard photograph. The count of each type is kept on a six-position counter which shows totals for each class as well as the grand total. The results are reported as the number of defects of each class as well as the total number of defects, all reported as defects per one ounce or per 28 grams of sample. However, for spinning mill use, the methods and standards used for cotton can be used for comparisons.

Card Waste Control

One of the quality control tests that is perhaps neglected and disregarded more than other tests is that of determining the amount of flat waste at the card. The test itself involves the evaluation of the flat waste weight for individual cards, and more specifically, the weight of comparatively few individual flat strips. Usually ten strips, the strips from ten individual flats taken in sequence, are evaluated for two characteristics: first, the weight per strip, and second, the determination of variation from left to right or from one side to the other of the strip.

One of the values resulting from this test is the accumulation of information relating to the amount of waste to be expected from any given type of cotton and for any given set of machine specifications. Under the same operating conditions, the amount of waste per flat is a good indication of the quality of cotton and of the total amount of waste that is being made at the card, for the flat strip is correlated to the amount of waste removed at other points on the card. Therefore, with this correlation, it is a very simple matter to get an indication of overall waste content and of cotton quality from a very simple test. If differences in the weight per individual flat strip exist from card to card on the same blend of cotton, then it is reasonable to expect that differences in the settings or clothing between these individual cards exist. In a long-range program, the differences in the weight of these individual flat strips averaged for any particular blend of cotton can be compared to the weights obtained with other varieties or blends of cotton, and thus, any trend towards lower grade cotton or poorer quality can be indicated early. Because flat waste is correlated to other types of waste and to the total amount of waste, it is possible to get this indication without going into the expense and other difficulties associated with making regular card waste studies. In addition, cards with certain settings that are improper, as for example the settings of the flats themselves, will cause variation in the weight of these strips in a transverse direction; in other words, there would be a resulting variation across the width of the card, and the evaluation of individual flat waste will bring this condition to the attention of the technicians doing the work.

The test procedure involves the following steps: First, with the card in operation, the roll of strips on the card is removed and the waste removed from the machine at that spot; then the card is allowed to operate as before in a normal way; the individual flat strips are observed as they are formed to determine if there is any variation across the width of the card; this is continued until ten strips have been accumulated. These can be taken from the card without stopping any operations.

These strips preferably should be conditioned in a laboratory for upwards of one hour and then weighed and recorded. Obviously, the record should include the card number, the type of work being run, the date, and any special features referring to clothing, speed, or the like. The frequency of sampling can be less than that of the nep count, with once a month being a normal requirement. In addition, this test should be made after grinding or whenever card settings are changed; in fact, it should be considered as a "must" for such a time.

Sliver Lap and Ribbon Lap Control

The degree of efficiency of combing as well as the control of the combed sliver size depends upon a supply of guaranteed even-running comber laps.

One of the factors contributing to this desired evenness of lap is the great number of doublings. For example, one conventional type of sliver lap machine uses twenty ends of sliver to make one lap, and in turn, the ribbon lap machine doubles four of these sliver laps to make one ribbon lap. Thus, each finished lap represents the cross section of eighty cans of card sliver. It can be assumed, therefore, that if proper control of weights has been attained at the finisher picker and at the cards, and if the proper gearing has been established on the sliver and ribbon lap machines, there will be little requirement for further change. For this reason, a spot check of sliver lap and ribbon lap weights once a week should be enough. The failure of the results of a spot check to meet specifications indicates that investigation is in order.

The method of making the test is similar to that for card sliver. Three to five 1-yard lengths of lap are measured from each machine, weighed on a grain scale, and the results are tabulated. The results of weighings are expressed in grains per yard.

Comber Control

Sliver Size and Noil Per Cent

The combing process is used to obtain yarns of a quality higher than can be obtained with the less costly carding process. Combing is an extra process introduced after carding and designed to remove the short fibers that are a part of the array. As a result of the elimination of these short fibers, yarns that are smoother, more even, more lustrous, and freer from imperfections than carded yarns can be spun. These combed yarns are used for thread yarn and fine fabrics such as the better broadcloths, lawns, balloon cloth, and the like. Obviously, the control of the combing process from a quality standpoint must be centered around the efficiency of removing the short fibers; from a cost point of view, variations in the amounts of waste removed are serious also.

The short fibers removed are classed as comber noil. The percentage of short fibers is usually from 12 to 16 per cent of the weight fed, although for the very long Sea Island cottons (up to 2-inch staple), the waste per cent must be raised to as high as 22 per cent, and at the other extreme for semi-combed work, this value might be dropped to a minimum as low as seven per cent.

There is a definite correlation between the amount of waste removed and the ultimate single yarn strength. This relation is approximately as follows: For each 2 per cent change in the amount of noil removed by the comber, there is a one per cent change in the strength of single yarn. Furthermore, material costs are related closely to the per cent of waste extracted. It should be obvious that the greater the amount of noils removed, the greater the cost of combed sliver; although noils have a definite resale value, this cash value is less than that of raw cotton. There is also the cost involved in processing these noils from the raw stock through the combing. When a standard percentage of noil removal has been established, based on a full consideration of the quality of the work and costs, this standard must be maintained within close tolerances.

Another aspect that should be noted is that any deviation from the standard per cent of noil removal will have an immediate effect on the weight of the sliver produced. A change of 1 per cent in the waste removed in normal practice (60-grain sliver, 16-per cent noil) changes the sliver weight by more than 1 per cent. Because of the very close relationship between sliver size and the amount of waste removed by the comber, it is advisable to coordinate the two tests. A suggested schedule of routine testing for each comber is one series of tests a week, with a maximum of ten days between tests.

The procedure for the two tests, sliver size and comber waste, is as follows:

Before sampling the sliver and noil and while the machine is in operation, check to see that each lap is feeding properly and that the individual slivers are feeding properly into the draw box.

Stop the comber, remove the can of combed sliver, and select five 1-yard samples from the top of the can. These 1-yard samples are placed in a wire basket for conditioning and weighing in the laboratory. The results of these weighings constitute the sliver size control. They indicate average weight only.

The following portion of the procedure applies directly to determination of waste per cent. Break off the sliver at the bite of the calender roll. On combers equipped with aspirators or other built-in systems of collecting waste at one central point, the web of waste is broken off at that point nearest the condensing section. On combers in which each individual head feeds the waste through to its own receptacle, these individual webs of waste should be broken off as near to the web-forming zone as possible.

What has been accomplished by the above procedure is the clearing of the machine in preparation for a trial run. The trial run consists of running the comber for about fifty to sixty nips. At the completion of this run, the sliver and noil that have been simultaneously produced are collected separately by breaking the material at the same position as was done when the machine was cleared. These materials provide the means for determining the comber waste per cent by one of two methods.

The waste per cent is calculated by the following:

$$\text{Waste \%} = \left(\frac{\text{Noil weight}}{\text{Noil weight} + \text{sliver weight}}\right) 100$$

One quick method of determining comber waste per cent is by the use of a waste-per cent indicator. This instrument is a quadrant scale designed and calibrated so that by placing the sliver on one pan and the waste on the other pan, the indicator will point directly to the per cent waste.

If a comber shows a waste percentage out of line with standard, a correction on mechanical operation is made, and then a recheck

Comber Waste Per Cent Determination

Standard Per Cent________

To________________ Date________________

Comber Number	Stock	Waste Per Cent	Weight	Corrected Per Cent	Sliver Weight	Approval

Fig. 19.3. Control form.

is made and entered on a sheet. It is a good idea to make the test for sliver weight on the comber at the same time as the waste determination. Figure 19.3 shows one type of form used.

On those combers without central waste collecting systems, it is possible to examine the waste produced at each individual comber head. A visual examination will show cloudiness of web or other troubles that can be traced to mechanical difficulties. However, waste tests taken at each head will give a better picture of the performance of the machine. It is advisable, therefore, to supplement the usual waste tests on the comber with the more complete test of evaluating the weight of noil at each head. It is not essential to get the sliver weight at each head, but it does make a much more complete test to do it this way. Thus, an individual head comber noil determination test is made in the same general manner as for the previously described test, but each head is treated as an entity; the sliver and waste simultaneously produced are collected, conditioned, and weighed. From the data, the individual head waste per cent is calculated. Discrepancies of a serious nature should call for immediate remedy at the comber. This test is much more time-consuming than the regular comber test. However, if the individual heads are kept in control, the comber will be in control. A test of this nature once every two weeks is better than a regular comber test every week, but a good procedure is to make the regular test three weeks in a row, and then on the fourth week to make the individual head test. Thus, one-fourth of the combers in a department would be checked weekly.

Comber Sliver Evenness. It should be emphasized that the two series of tests previously outlined are basic in nature and are designed to help guarantee that the yarn to be spun shall be consistent in yarn size. Thus, both the combed sliver weight and the amount of waste removed are important in predicting if the spinner can successfully produce yarns of a given count without abnormal differences existing. Under more serious consideration, it shall be seen that these are all factors of long-term variation or evenness, and the relationship between long-term and short-term variation must be understood in any quality control program. As the name infers, a long-term variation is one that might be visualized as existing between one bobbin and another or between the beginning of a bobbin and the end of the same bobbin of spun yarn. It could be the variation between the sliver weight on a comber at the top of a can and at the bottom or run-out of the same can of sliver, or between the sliver weight produced by one comber in comparison with another comber. For example, in Table 19.3, there is a definite long-term variation existing between Comber 1 and Comber 3, and to some extent existing also between

Comber 2 and Comber 3. Likewise for Comber 4, in which the weighings have been made at intervals of 10 yards, there is a long-range variation existing which progressed from light weight sliver to heavy and then back to light. It can be visualized that if all combers in an area began to produce sliver heavier than the 54-grain-per-yard standard given in the example, then there would be a long-term variation arising, and the final result would be that the yarn spun from this material would rapidly approach a heavy

TABLE 19.3
Comber Control, Sliver Evenness

Comber Size, Grains Per Yard, Standard 54-Grain			
Comber 1	Comber 2	Comber 3	Comber 4
53.2	54.2	57.1	52.6 (1st yard)
54.5	54.5	56.5	53.5 (10th yard)
55.0	53.9	55.8	53.8 (20th yard)
53.6	54.7	56.2	54.9 (30th yard)
52.9	55.2	56.8	55.3 (40th yard)
Ave. 53.8	54.5	56.5	54.1 (50th yard)
			51.6 (60th yard)

condition. This would result in the necessity of making changes in the gearing on drawing or spinning to correct for average yarn size. At the same time, certain of the rovings in the creel would have been produced from the light-weight sliver and others from the heavy-weight sliver, so a high degree of variation would exist in yarn count numbers.

It is obvious that the constant control of the average weight of the material produced at each comber becomes a preventive measure, for the elimination of variation at the comber means that if other conditions are controlled as effectively, the yarn spun will be suitable. However, it will be noted that 1 yard lengths or possibly longer than 1-yard lengths have been tested. The question arises as to what the variation would be within the 1-yard length, and it is here that we approach short-term variation factors. Short-term variation alludes to variations that occur from inch to inch or at least within very short lengths of textile material, and this control or lack of control of evenness over very short lengths is just as critical a factor in determining the quality of the final yarn as is the long-term variation.

Short-term variation of sliver evenness can be evaluated at the comber by measuring the evenness of the sliver produced. This can be done by use of any one of the instruments previously described. First of all, there is the Saco-Lowell sliver tester which has been described in its application to the card sliver. The use for comber sliver is identical to that for card sliver, and the results of such tests give a good indication of the comparative rating of the evenness of material produced by the given mill.

Comber sliver can also be tested for evenness by use of the Uster Evenness Tester, the Pacific tester, or the Brush Uniformity Analyzer.

The degree of evenness that can be expected in any manufacturing process is a controversial subject. However, certain standards have been established through usage of the various testing machines, and the tabulation of these standards is given in the section discussing evenness testing.

Drawing Frame Sliver Control

Sliver Size

The control of sliver size at the drawing frame is of utmost importance; in fact, it has become almost a standardized procedure for all cotton and staple spinning mills to use drawing sliver size as a key control point for yarn evenness. With a system properly set up and maintained, it has been found that very little, if any, gear changing need be done subsequent to the drawing process except with changes in production schedule or organization.

One item of contention that arises frequently is whether the control should be exercised on the first process or on the second process of drawing in cases where there are two processes, and likewise whether it should be done on the third process in those few cases where three processes are employed. While there might be advantages in controlling on the first process of drawing, generally it is conceded that there are better reasons for controlling on the final process of drawing. For example, inasmuch as control of sliver size is a control of long-range variation, it seems logical that the first process of drawing in providing as it does the doubling of six or more ends of sliver, supplies material to the second drawing having characteristics that give a better opportunity to get average values in results. Therefore, this sliver when fed to the next or final process of drawing should show less long-term variation and consequently make the control of sliver size more effective.

The question arises frequently as to why the finisher drawing frame is preferable to roving or spinning frames as a final gear change point for control of evenness. An analysis of the situation will reveal the following advantages: First, material is sampled more easily at the drawing frame from a physical point of view, for the sliver to be sampled can be taken at or between doffs. Secondly, damage to the material can be avoided when sampling, and it can be done without disturbing the operation of the machine to any appreciable extent. Compare this to the roving frame or spinning frame when samples are taken during the operation of the frame. Thirdly, gear changes can be made on the drawing frame without disturbing quality. In this respect, it should be noted that it is very difficult to change draft gears on roving or spinning frames without disturbing the exact position of the drafting rolls, for generally it is necessary to move the steel rolls slightly in order to get the gear teeth to mesh properly, and this results in either a thick or thin spot in all ends on that particular side of the spinning frame or roving frame. Fourth, it is possible to get a very close control of change on a drawing frame by using the crown gear rather than the draft gear as a change gear. The reason for this is quite simple, for usually draft gears on drawing frames are of about 40 to 50 teeth in size and therefore a change of 1 tooth gives at least a 2 per cent change in sliver size. On the other hand, by changing a crown gear of 100 teeth, a change of 1 tooth gives a 1 per cent change in sliver weight. Many mills have found it advantageous to paint their crown gears in accordance with the number of teeth; for example, a 99-tooth gear could be painted red, a 100-tooth gear could be painted blue, and a 101-tooth gear could be painted yellow. Any change outside of the limits of these three gears would be taken care of by making a change of the draft gear. Fifth, there is very little loss of time in making a change on the drawing frame, and therefore efficiency can be maintained at high levels. Sixth, there is no difficulty in transporting the samples from the processing area to the testing laboratory, and samples can be reprocessed easily without any great loss of material. Seventh, the samples represent a much larger proportion of the work being processed than a similar amount of testing would provide at the spinning. Finally, the better the control of the work being fed to the roving and the spinning, obviously the less changes that are required at those particular processes. It is less work to change a few heads of drawing frames than it is to change fifteen or twenty spinning frames or three or four roving frames. Experience has shown that with this control at the drawing frame, gear changes to control either roving or spinning size at those particular processes can be reduced substantially or eliminated entirely.

Test Procedure, Sliver Size. The procedure for making the test is very similar to the methods described previously for cards and combers. Generally speaking, one good method of sampling is to take 5 yards of sliver at each delivery on each drawing frame, and the preferred method is to take these samples in 1 yard lengths rather than in the 5-yard lengths in order to give a better picture of yard-to-yard variation. When possible, the sampling should be done between doffs of the cans on the drawing frame. The procedure is to remove the full can at each delivery from the drawing frame and to start and let the drawing sliver feed through the coiler until it reaches the floor. This gives a little over a yard of material which has not been subjected to any stresses by being curled in the can. The sample at each drawing delivery is then laid on a 1-yard template and the 1-yard lengths measured and cut off and the individual samples twisted or rolled to form a little skein-like ball. These then can be placed in baskets containing individual squares or units which can be premarked to identify the particular delivery head and drawing frame. When all samples have been collected by the laboratory technician, they should be taken to the laboratory for proper conditioning.

If the baskets in which the samples are conveyed from the card room to the laboratory are constructed of open mesh wire, then the time necessary for these samples to reach moisture equilibrium would probably be no more than two hours and possibly less. In

DRAWING SIZE CONTROL

Draw Frame No. 1 Time: 10 AM Date: ______

Regain Indicator 9 %

Gears On: Draft 45 Change To: Draft 45

Crown 100 Crown 99

Delivery	#1	#2	#3	#4
	Sliver Weight, Grains			
Standard	55	55	55	55
Actual 1	52.1	53.6	55.2	54.6
2	56.1	52.8	54.6	57.1
3	54.8	54.7	55.8	55.0
4	55.2	55.2	55.2	55.3
5	55.6	55.7	56.1	54.8
Average Delivered	54.8	54.4	55.4	55.4

Average Head 55.0

Average Corrected to 7 Per Cent Regain 54.0

Calculations.

$$55\left(\frac{107}{109}\right) = 54.0$$

Fig. 19.4. Drawing size control.

any event, the samples should be exposed to the proper atmospheric conditions (70° F — 65 per cent R.H.) for a length of time sufficient to give moisture equilibrium. The samples can then be weighed and the results recorded on a proper form, similar to the one shown in Fig. 19.4.

Moisture Correction for Drawing Sliver

If the mill does not have a central control laboratory, then it is necessary to make corrections for changes in the moisture present in the atmosphere. One method of doing this is to use an instrument similar to the Aldrich Regain Indicator. This indicator gives the amount of regain in a typical sample of cotton so that if the drawing sliver samples are weighed in the area adjacent to the regain indicator, it is possible to correct the sliver size to a standard regain of say 7 or $7\frac{1}{2}$ per cent. If this is done, it is necessary to make the correction for regain only on the average weight for each drawing frame, for it is obvious that the variations that exist between yard lengths can be examined for the weight of the material as is. However, all the deliveries for any given drawing frame, which means for any particular set of draft gearing, should be obtained in order that the necessity of changing draft or crown gears can be determined.

In Fig. 19.4 is also illustrated the method of recording sliver weight and of calculating average sliver weight corrected to standard regain, all based on the assumption that the average must be corrected for regain. In the event that a controlled laboratory is available, obviously the information on the regain indicator and the correction for regain can be omitted, all other portions remaining the same. In making out this control form, the procedure is as follows: The technician writes in the draw frame number, the time and date, and then the reading of the regain indicator. After sampling the individual deliveries from the drawing head, the samples are weighed and the results entered on the form as indicated under the actual weight column for each delivery. These are then averaged and the grand average for the drawing head is obtained. This average is then corrected to the standard regain used, 7 per cent in this case, for which the calculation is shown on the sheet. In this particular case, the corrected sliver weight is 54 grains as opposed to a standard of 55 grains.

Drawing Frame Gear Changes

The question arises as to whether or not a change should be made in the draft gearing to correct the 54-grain sliver nearer to the standard of 55. There are different systems in use to aid in making a decision. Generally, the mill sets up a standard difference beyond which a change in gearing is made. This standard should not be less than 1 grain per yard and probably should not exceed $1\frac{1}{2}$ grains per yard. For example, if the corrected average weight were 54.3 grains per yard rather than the 54 as given on the control form, under this system no change would be made in the gearing. However, if the change limit were plus or minus 1 grain per yard, then a change in the gearing would be made, and it is assumed that such was the case and a change of gearing recommended. Another means of avoiding excessive change of gear is to set the limit of actual difference at a high level, say a variation of plus or minus $1\frac{1}{2}$ grains per yard, and then to change only on the second sizing of the material. For example, some mills check drawing sliver size twice each shift, and no change is made in gearing until the second sizing indicates that that particular frame or drawing head continues out of control. In other words, they wait to see that a definite trend away from standard has been in effect and then the change is made. Except in extreme cases where obviously something is in error, a change in crown gear only is made. This results in a 1 per cent change. In the example being discussed, assuming that changes are made for a plus or minus 1 grain-per-yard deviation, the technician recommends a change of 1 tooth in the crown gear, from a 100-tooth to a 99-tooth gear. Theoretically, this will change the sliver size from 54 to about $54\frac{1}{2}$, but the idea of making the change of 1 tooth only is to avoid extreme fluctuation in sizing. If this same drawing frame shows on the next sizing that the sliver is yet out of line, then a second change is made.

If only three crown gears are used for all changing, such as the 99, 100, and 101-tooth gears, conditions arise where it will be obvious that a gear larger than 101 teeth or smaller than 99 teeth shall be required to effect the change necessary. Assume, for example, that the drawing frame has a 45-tooth draft gear and a 99-tooth crown gear in operation, and it is required to change the material to a heavier weight (remember that the crown gear is a driven gear in the draft gearing and that if a larger draft gear would normally be required, then a smaller crown gear would have to be used). With a 45-tooth draft gear and a 99-tooth crown gear, in order to correct the drafting to make the work heavier by 1 per cent, a 98-tooth crown gear would be necessary, but as has been stated, 99 is the smallest available. In this case, therefore, the draft gear would be changed from a 45 to a 46-tooth gear. In changing from a 45 to a 46-tooth draft gear, the equivalent crown gear would be a 101-tooth gear (within very close limits). Thus, we might say that a 45-99 combination is equivalent to a 46-101 combination. Therefore, a 46-tooth draft gear can be used, and the crown gear can be changed from the equivalent level of 101 down to a 100-tooth gear. This entire change gives approximately 1 per cent increase in the weight of the sliver being delivered.

Sliver Size Records. It is a good practice for the laboratory to maintain a gear change control form for drawing. This control might be termed a running inventory of the size of the gears on

GEAR CHANGE RECORD

	Head Number									
	1		2		3		4		5	
Date	D	C	D	C	D	C	D	C	D	C
5/2	45	100	45	101	46	99	45	100	45	101
5/3	x	x	x	x	46	100	x	x	x	x
5/4	45	99	x	x	x	x	x	x	x	x
5/5	x	x	x	x	x	x	45	99	45	100
5/6	45	100	45	100	x	x	46	100	x	x
5/8	Etc.									

Fig. 19.5. Gear change control, drawing.

each drawing frame, so that it is not necessary to count the teeth on the drawing frame each time a change is contemplated. Figure 19.5, Gear change control, drawing, illustrates the type of gear inventory that can be kept. The cross marks on the form indicate that no change in gearing has been made that particular day. If gear changes are to be made, then the laboratory technician should put through a written request to the card room stipulating what frames are to be changed and what gears are to be replaced. Figure 19.6, Gear change order, is the type of written instruction that should be sent through to the card room. It will be noted that two

GEAR CHANGE ORDER

Copies: Card Room (2) Date: ____________
Superintendent
File

Please make the following changes on drawing frame gearing:

Frame No.	Gears On		Gears to Put On	
	Draft	Crown	Draft	Crown
1	45	100	45	99

Fig. 19.6. Gear change order.

copies would go to the card room, one for the overseer of the department and one for the third hand or fixer who is to do the changing, and in addition, a copy should be sent to the superintendent and one kept for the laboratory file.

Summary, Drawing Sliver Size. In summary, drawing frame sliver sizing should be performed at least once each shift of operation and if possible, twice each shift of full operation. At least five 1-yard samples should be taken from each delivery and averages corrected to a standard regain if the sizing is not done in a conditioned laboratory. Gear changes should be made only when the sliver deviates from the standard by a predetermined amount, and this variation should not be less than 1 grain per yard plus or minus nor more than $1\frac{1}{2}$ grains per yard plus or minus. In addition, if the sizing is done twice each shift, gear changes can be deferred until the second test indicates that the trend, either heavy or light, continues and that a gear change is in order. Before any changes are made, the fact should be established that the change in weight is not due to variation in moisture content. A record should be kept of the draft and crown gears on each drawing frame as a running inventory, and any gear changes should be specified by the laboratory with possibly the approval of the overseer of the department in writing, and copies kept for future reference. Crown gears should be changed, and it is recommended that they be painted to identify the size.

Drawing Frame, Evenness. Just as the drawing frame is a prime process for checking the sliver weight and thus controlling long-rate variation, so also it is a prime target for control of short-range evenness. As previously explained for comber sliver, a test can be made on any one of the mechanical or electronic instruments designed for this work, such as the Saco-Lowell Graphic Sliver Tester, the Pacific Evenness Tester, the Uster Evenness Tester, or the Brush Uniformity Analyzer. This subject is covered later.

Roving Control, Hank Size

Each individual hank roving that is produced should be sized as a standard quality control measure. The frequency of the tests should be a daily sizing for each hank roving being made. The number of samples to be sized should be a minimum of eight to twelve or an average of ten bobbins as a minimum for each type of hank roving.

Table 19.4 provides one standard for determining the number of roving bobbins to size, depending upon the number of roving frames in operation on each particular hank roving.

If the drawing frame is used as a gear change control area, then it should be found that very little if any gear changes need be made on the roving frame except as a result of a long-term trend and in order to control the average size at some new given standard. In other words, with changes of cotton or of raw material, it is obvious that gear changes on the roving frame might be required, but daily changes should be avoided if possible. If the mill finds that it is necessary to use the roving frame as a gear change area, then it is wise to use the crown gear as a change gear in much the same manner as has been described for the drawing frame. However, it is reiterated that if the drawing frame is used as a control unit, and if this is proceeded by proper opening, picking, and carding, then very little, if any, changing should be required as a daily quality control measure on the roving process.

TABLE 19.4

Sampling Roving Frames
No. of bobbins to sample

Frames on	Total	Per frame
1	8	8
2	12	6
3 or 4	16	4 to 6
5 or 6	18	3 to 5
Over 6	18 and up	2 or 3

Because of time lost, interference with operations, and other normal difficulties, it is an annoying procedure to stop the roving frame during the ordinary operation of the machine in order to sample roving bobbins. It is recommended, therefore, that full bobbins be selected from the frames right after the doff. It can be argued that variations occur from the beginning to the completion of a doff and that by sampling all full bobbins, this variation would be neglected. However, this is not the case, for if the extent of variation that might exist from beginning to end of a doff is to be evaluated, this should be done as a separate research investigation, and it should involve primarily the measure of short-range variations. By making all weighings to obtain roving size on the full bobbins, the mill is assured that tests are made in an identical manner for all roving types. Thus, from the point of view of roving size control, variables are eliminated from the results, and efforts to maintain proper roving sizes are easier. It should also be noted that some variation may exist between the roving produced on the front line as opposed to the back line of spindles on the roving frame, particularly on older machines. Again, it is believed that if random selections of roving bobbins are made, the results should be consistent insofar as any effect of this nature is concerned.

In making the actual test, the laboratory technician should select roving bobbins in the department at some predetermined time each day so that arrangements can be made to get completed full roving bobbins. In order to avoid variations that occur from spindle to spindle, one good method is to have the roving frame operator set aside full roving bobbins from certain predetermined spindles each day. For instance, if four rovings are to be selected from each of four frames, then two bobbins from the front line of spindles and two from the back line of spindles from each of the four frames can be set aside by the tender. The spindles that are selected or sampled by the tender are previously marked for selection, and the sampling can be done on the same spindles every day. The roving should be conditioned in a laboratory for a sufficient length of time to insure moisture equilibrium. In order to hasten the conditioning period, the roving bobbins can be reeled immediately on receipt in the laboratory and the reelings then put aside in a wire basket exposed to the conditioned atmosphere, and thus the roving being in an open state will reach equilibrium quicker than if packed on the roving bobbin. In this connection, it is possible to design a box or cage through which a continual flow of air is drawn by an exhaust fan, thereby giving a quick reaction to actual laboratory atmospheric conditions.

The sample length taken from each roving bobbin is 12 yards obtained by reeling the roving on a standard reel with a 1-yard (or $\frac{1}{2}$-yard) circumference. Care should be taken when reeling the roving that it is not stretched and the size changed. Weighings can be done on the usual grain scale. The weight in grains can be converted into hank size by use of the usual chart which provides for the conversion of grains per 120 yards into yarn size. The only difference that should be noted here is that these charts are made for 120 yards as opposed to the 12-yard reeling for roving, and therefore the roving size is obtained by moving the decimal point one place to the left from readings on the chart. These charts are based on the formula that yarn size is equal to 100 divided by the grains per 120 yards.

DAILY ROVING SIZE REPORT					
To: Department Head Superintendent Laboratory			Date: ______		
Hank Roving					
Stock					
Sizings	1				
	2				
	3				
	4				
	5				
	6				
	7				
	8				
	9				
	10				
	11				
	12				
Average					
Maximum					
Minimum					
Range					

Fig. 19.7. Daily roving size report. (Regain reading or regain correction factor to be included if controlled laboratory not available.)

It is also possible to use direct reading scales for obtaining the roving size. These quadrants are calibrated to give hank size directly, rather than weight. These are discussed more fully under spinning. More recently the Shadowgraph scale has been developed to give yarn or roving size directly. These scales are ideal for mill and laboratory use, for they are quick and easy to use.

Results should be tabulated on a standard form showing the date, the standard size of the roving, the stock from which the material is made, and a space on the form for each individual weighing, as well as the average of the hank size. Figure 19.7 shows how this information can be tabulated.

In summary, roving should be sized daily, and the number of samples to select from each roving frame depends upon the number of frames in operation. The sample length is 12 yards; samples should be conditioned before sizing. In the event that controlled laboratory conditions are not available, then corrections in the average roving size should be made depending upon the amount of regain in the sample. This can be done as previously discussed by the use of regain indicators or by making the roving weighings in an area in which the relative humidity can be measured during the test period and corrections then made on the average based on regain tables. Gear changing should be avoided as much as possible on the roving frame except when new rovings are started or other major changes made in organization; this recommendation is based on the condition that the mill uses the drawing frame process as the prime control spot for gear changing. Short-range uniformity is discussed later.

CHAPTER XX

Yarn Numbering

In the spinning room, increased efficiency of operation and improved yarn quality are two major benefits to be derived from effective quality control throughout the preceding operations.

The important features of yarn which are evaluated to determine its quality are yarn number, strength, twist, appearance, and evenness. The variability of each of these properties, within and between yarn packages, and within and between shipments or lots, is also of major importance in subsequent processing. For the sale yarn mill, spun yarn is the finished product, and quality standards must be maintained to meet customer specifications. If the control system in the carding department has failed in its objectives, then certainly the yarn that is to be sold will not meet the the customer's specifications. The point that becomes more evident as quality control programs are studied is that these same programs must be all-inclusive and not limited to one section or area of the plant. In the past, many mills have been content to study the spun yarn, generally using the simplest of tests. However, with the increase in statistical methods of analysis and with the tighter specifications required by the armed services as well as by industrial users, such methods are too weak for continued success, and therefore the well-rounded program must be followed. The tests that are listed in this section are important in themselves and as part of the entire picture of quality control.

Yarn Numbering Systems

There exists an unfortunate condition in that so many systems are in use for numbering yarns. Despite the activity of national and international groups who are endeavoring to correct this situation by the adoption of a universal system, it is feared that it shall be some time before any relief materializes.

All yarn numbering systems depend on one of two relationships, either weight per unit length, or the reciprocal relationship of length per unit weight. Beyond this, variations from one system to another are confined to variations in the units used for weight and length.

Cotton English System

The English system used for numbering cotton yarns is the one in widest use, although it is not necessarily the most effective. It is a system based on length per unit weight. However, the system is limited in application to roving and yarn, so in measuring lap and sliver weights, recourse is made to direct or weight per unit length methods. These have been previously elaborated upon; in summary for review, they are: (*1*) ounces per linear yard for picker laps (generally accepted as being of a 40-inch width, although occasionally 45-inch laps are used on special machinery, particularly for waste); (*2*) grains per linear yard for sliver, sliver laps, and ribbon laps.

For yarn or roving in the English system, the yarn number is measured by the number of 840-yard hanks in one pound of the material. Expressed as an algebraic relationship, this is:

$$\text{Yarn number or roving hank size} = \frac{\text{Hanks}}{\text{Pound}}$$

in which the ratio of yards to hanks is 840.

The basic formula can be converted to any other units that might be desired by the substitution of proper units. A few fundamental conversions are given herewith to illustrate the approach.

1. To convert to units of yards and pounds, let N = yarn number or roving hank size; Hk = length in hanks; Yd = length in yards; Lb = pounds

$$N = \frac{Hk}{Lb} \tag{1}$$

But,

$$\frac{Yd}{Hk} = 840, \text{ and so } Hk = \frac{Yd}{840}$$

Substituting $Yd/840$ for Hk, the result is

$$N = \frac{Yd}{840(Lb)} \qquad (2)$$

2. To convert to units of yards and grains, use formula (2) as a start. Let Gr = Grains.

$$\frac{Gr}{Lb} = 7000, \text{ and so } Lb = \frac{Gr}{7000}$$

$$N = \frac{Yd}{(840)(Lb)} = \frac{(Yd)(7000)}{(840)(Gr)} = \frac{(8.33)(Yd)}{Gr} = \frac{8.33^1}{Gr/Yd}$$

In making any conversion, it is helpful to set up a formula with the conversion units written down step by step so that like units can be canceled out of the equation, leaving the final units showing. To illustrate this principle, the factor for converting yarn number (English system) to units of centimeters and grams is shown. The second column of Table 20.1 gives the units being introduced at each step, the third column the numerical value of these units, and the last column of the table shows the final units *at each step*. (In = inches, Gm = grams, cm = centimeters.)

TABLE 20.1
Converting Yarn Number from English System to Metric System

Step	Conversion	Numerical value	Units at end of each step
1	$\left(\frac{Hk}{Lb}\right)$		
2	$\left(\frac{Yd}{Hk}\right)$	840	$\frac{Yd}{Lb}$
3	$\left(\frac{Lb}{Gr}\right)$	$\frac{1}{7000}$	$\frac{Yd}{Gr}$
4	$\left(\frac{In}{Yd}\right)$	36	$\frac{In}{Gr}$
5	$\left(\frac{Gr}{Gm}\right)$	15.43	$\frac{In}{Gm}$
6	$\left(\frac{cm}{In}\right)$	2.54	$\frac{cm}{Gm}$

[1] Read as "eight and one-third."

To get the conversion factor as a numerical value, substitute the known numerical value for each element. For example, take this for the first two steps above to get yarn number in terms of yards and grains (as illustrated before).

In doing this, the yarn number, N, which is equal to Hk/Lb by definition, is substituted for Hk/Lb, and the following relationship exists:

$$N\left(\frac{Yd}{Hk}\right)\left(\frac{Lb}{Gr}\right)=\frac{Yd}{Gr}$$

$$N(840)\left(\frac{1}{7000}\right)=\frac{Yd}{Gr}$$

$$N=\frac{8.33\ Yd}{Gr}$$

In a like manner, apply the same system to the entire equation to get units in terms of centimeters and grams.

$$(N)(840)\frac{1}{7000}(36)(15.43)(2.54)=\frac{cm}{Gm}$$

Therefore,

$$N=\frac{cm}{Gm}(0.059)$$

where 0.059 is the conversion factor.

Thus, the yarn number can be converted to units of centimeters and grams by a calculated constant; conversely, if a yarn length is measured in centimeters and its weight is found in grams, the yarn number can be obtained. By knowing any two of the three elements — yarn number, centimeters, or grams — the third can be calculated.

Denier System

The yarn numbering system used for all continuous filament synthetics as well as the staple synthetics in the staple form is the denier system. This is a direct numbering system, inasmuch as the units are in terms of weight per unit length. The definition of the denier system states that the denier is the number of grams per 9,000 meters, and this is also stated by some authorities to be the number of 0.05-gram weights per 450 meters; obviously, both of these indicate the same thing. This system has been in use for many years, originally being set up in France for use with silk. It lends itself to universal use except for one thing, and this is that the length unfortunately was chosen at 9,000 meters rather than 10,000 or 1,000 meters. Other systems, the Grex and Tex systems, are similar, but substitute 10,000 and 1,000-meter lengths respectively as the basis.

Being a direct numbering system, fine yarns or fine deniers have small numbers, and coarse or heavy yarns have larger numbers. For example, 100 denier is one-third the weight of 300 denier, two ends of 100 denier would form 200 denier, and so forth. It will be noted that this is the opposite of the cotton English system in which coarse yarns have small numbers and fine yarns have large numbers. Since many textile mills today process spun yarn both from cotton and synthetics as well as the continuous synthetics, and furthermore, as many fabrics are made from combinations of these different yarns, it is imperative to be able to convert readily from one system to the other.

The basic relationship for denier is as follows:

$$Den = \frac{Gm}{L}$$

in which Den = Denier number; L = Length in 9,000-meter units; M = Meters; Gm = Grams.

Expressed as an equation, the value of L is:

$$\text{Meters per unit length} = 9{,}000$$

or

$$\frac{M}{L} = 9{,}000$$

and

$$L = \frac{M}{9{,}000}$$

The resultant expression for denier then becomes:

$$Den = \frac{Gm(9{,}000)}{M}$$

EXAMPLE. A 450-meter length of yarn weighs 5 grams. What is the denier of the yarn?

$$Den = \frac{5}{450}(9{,}000) = 100$$

EXAMPLE. A spool of 50-denier nylon weighs 35 grams (yarn only); what length of yarn is on the spool?

$$50 = \frac{35(9{,}000)}{M}$$

$$M = 6300$$

To convert the basic formula to units other than grams and meters, follow the systems previously described for cotton yarn.

EXAMPLE. Convert denier to units of pounds and yards (necessary in cloth calculations), and also into cotton number. (See Table 20.2).

TABLE 20.2

Converting Denier to Pounds and Yards and to Cotton Number

Step	Conversion	Numerical value	Units at end of each step
1	$\left(\frac{Gm}{L}\right)$		
2	$\left(\frac{L}{M}\right)$	$\frac{1}{9000}$	$\frac{Gm}{M}$
3	$\left(\frac{M}{Yd}\right)$	0.914	$\frac{Gm}{Yd}$
4	$\left(\frac{Lb}{Gm}\right)$	$\frac{1}{453.59}$	$\frac{Lb}{Yd}$
5	$\left(\frac{Yd}{Hk}\right)$	840	$\frac{Lb}{Hk}$

If the work is carried only to the units Lb/Yd in the fourth step the first relationship (denier in pounds and yards) can be determined. Note that Den is substituted for Gm/L.

Thus,

$$(Den)\left(\frac{1}{9{,}000}\right)(0.914)\left(\frac{1}{453.59}\right) = \frac{Lb}{Yd}$$

or,

$$Den = \frac{Lb}{Yd}(4464483)$$

or,

$$\frac{Yd}{Lb} = \frac{4464483}{Den}$$

To convert denier to cotton number, carry the computation through to Lb/Hk in the fifth step, which is the reciprocal of the yarn number in the English system.
Thus,

$$(Den)\frac{1}{9{,}000}(0.914)\frac{1}{453.59}(840) = \frac{Lb}{Hk} = \frac{1}{N}$$

or,

$$\frac{D}{5315} = \frac{1}{N}$$

So,

$$Den = \frac{5315}{N}$$

and

$$N = \frac{5315}{Den}$$

Worsted System

The worsted system, like the English cotton system, is an indirect yarn numbering method. In place of the 840-yard hank length of the English cotton numbering system, the hank length for the worsted system is 560 yards. The basic formula for the worsted system is:

$$\text{Worsted number} = \frac{\text{Length}}{\text{Pounds}}$$

in which the yards per unit length equal 560. Thus, the algebraic relationship to use in the worsted system is:

$$\text{Worsted number} = \frac{Yd}{560(Lb)}$$

Having this basic relationship, it is possible to convert into other units in exactly the same way as has been demonstrated for the cotton count system with the one exception that 560 is used in place of the 840-yard length. The use of this shorter length means that the worsted number is numerically higher than that for an identical weight of cotton yarn expressed in the English system. For example, a 30s worsted yarn would be equivalent to a 20s cotton yarn, inasmuch as both would have the same number of yards per pound.

Woolen Run

There are two systems of numbering woolen yarn that are in common usage in the United States, these being the woolen run and the woolen cut. The woolen run system is again an indirect numbering method inasmuch as the run is equal to length per unit weight. In the woolen run system, the unit of length is 1600 yards as opposed to the 840 for the cotton English system. The formula to use for the woolen run system is:

$$\text{Woolen Run} = \frac{Yd}{1600(Lb)}$$

Because a 1600-yard length is used, it results in giving a number smaller in proportion to cotton yarn for the same material. Thus, a 10-run wool would have 16,000 yards per pound, and this would be equivalent to 19s cotton yarn. The handling of this relationship of run in converting to other systems is done in the same way as previously discussed.

Woolen Cut System, Linen Lea

The woolen cut system, which is identical also to the linen lea, is based on length per unit weight just as for the cotton, worsted, and woolen run systems, but in this case the unit of length is 300 yards. The basic formula is:

$$\text{Woolen cut} = \frac{Yd}{300(Lb)}$$

A 10s cut woolen or linen yarn would be equivalent to 3.6 cotton count; both have 3,000 yards per pound.

Typp System

A few years ago a concerted effort was made to introduce a system labeled the Typp system (abbreviation of thousands of yards per pound). This proposed system has met with very little success, despite the fact that it was advocated as a universal system by many organizations. Its weakness lies in the fact that it is not based on metric units and also in that it is an indirect system of the ratio of length to weight. For these reasons, it is fortunate that the system has not been adopted for universal use, as many of the ills present in the cotton system continue with the Typp system.

The basic formula for the Typp system is:

$$\text{Typp number} = \frac{Yd}{1000(Lb)}$$

In comparison with the cotton system, a 10s cotton yarn would be 11.9 Typp yarn. The treatment of the basic formula for conversion into other units is affected in the same way as previously demonstrated for the cotton system.

Tex System

The Tex system is a direct system based on weight per unit length and differs only from the denier system in that a 1,000-meter length is selected in place of the 9,000-meter length. Thus, obviously this system is metric and decimal in character and lends itself to the ideal type of universal system, for it is possible to identify materials from a lap to an individual fiber by using a metric unit. For example, a single fiber could be identified for size as being of a given Tex and by the same token a picker lap can be identified the same way, and thus the relationship between the two obtained by a simple comparison. Compare this to the present system in use for cotton fibers, laps, sliver, and yarn. In the first place, cotton fibers are described as being so many "micrograms per inch of length"; then the picker laps are ounces per linear yard, the sliver is grains per linear yard, and the yarn size is length in hanks per pound. In effect then, the cotton system becomes involved with four distinct measurements, none of which lend themselves very readily to interchange. On the other hand, with the Tex system, comparisons can be made readily, inasmuch as the units of measure are the same all the way from fiber to yarn.

The basic formula is as follows:

$$Tex = \frac{1{,}000(Gm)}{M}$$

The conversion of Tex to denier is relatively simple, provided that care is taken in multiplying Tex (not dividing) by the conversion unit of 9 to get denier. If it is remembered that the Tex uses a length of 1,000 meters and that denier uses a length of 9,000 meters, then the ratio of Tex to denier is one of 1,000 to 9,000, or of 1 to 9.

$$\frac{Tex}{Den} = \frac{1}{9}$$

$$Tex = \frac{Den}{9}$$

Therefore,

$$Tex \times 9 = Den$$

Another way to remember this is that the Tex number is always relatively smaller than the equivalent denier because a shorter length is used. For example, 100 Tex is equivalent to 900 denier.

Conversion Constants

In Tables 20.3 to 20.6 are given conversion factors for the various systems discussed. Information of the type shown in these tables is most valuable where speed in obtaining information is required, but it should be emphasized that an understanding of the derivation of these relationships is just as important.

Yarn Numbering, Spun Yarn

For mills spinning yarn from cotton or synthetic staples, the yarn number on the cotton English numbering system is obtained by the use of a 120-yard skein. The source of supply for the measurements would be spinning frame bobbins, tubes, cones, or

TABLE 20.3
Yarn Numbering Conversion Table

Using	Do This	To Get This	
Cotton yarn number with 840-yard hank	Mult. by 1.50	Worsted	560-yard hank
	Mult. by 0.525	Woolen run	1600 yard/run
$N = \frac{Hk}{Lb} = \frac{Yd}{840(Lb)}$	Mult. by 2.80	Woolen cut	300 yard/cut
	Divide 5315 by yarn no.	Denier	Gm/9000 meters
(Also used for spun silk)	Divide 590.6 by yarn no.	Tex	Gm/1000 meters
Denier with 9000-meter length	Divide 5315 by den	Cotton no.	840-yard hank
$Den = \frac{9000(Gm)}{Meters}$	Divide 7972 by den	Worsted	560-yard length
	Divide 2790 by den	Woolen run	1600-yard length
	Mult. by $\frac{1}{9}$	Tex	Gm/1000 meters
	Divide den by 637.8	Grains/yard	
Worsted with 560-yard length	Mult. by $\frac{2}{3}$	Cotton no.	840-yard hank
$N = \frac{Yd}{560(Lb)}$	Mult. by 1.87	Woolen cut	300-yard length
	Mult. by 0.35	Woolen run	1600-yard length
	Divide 7972 by worsted	Denier	Gm/9000 meters
	Divide 886 by worsted	Tex	Gm/1000 meters
Woolen run with 1600-yard length	Mult. by 5.33	Woolen cut	300-yard length
	Mult. by 1.90	Cotton no.	840-yard hank
	Mult. by 2.86	Worsted	560-yard length
$Run = \frac{Yd}{1600(Lb)}$	Divide 2790 by woolen	Denier	Gm/9000 meters
Woolen cut Also linen lea with 300-yard length	Mult. by 0.36	Cotton no.	840-yard hank
	Mult. by 0.54	Worsted	560-yard length
$Cut = \frac{Yd}{300(Lb)}$	Mult. by 0.19	Woolen run	1600-yard length
Typp with 1,000-yard length	Mult. by 1.19	Cotton no.	840-yard hank
	Mult. by 1.79	Worsted	560-yard length
$N = \frac{Yd}{1,000(Lb)}$	Mult. by 0.63	Woolen run	1600-yard length
	Mult. by 3.33	Woolen cut	300-yard length

Table continued

TABLE 20.3 (*Continued*)

Using	Do This	To Get This	
Tex with 1,000-meter	Divide 590.6 by Tex	Cotton no.	840-yard hank
length	Mult. by 9.0	Denier	Gm/9000 meters
$Tex = \frac{1000\ (grams)}{meters}$	Divide 885.8 by Tex	Worsted	560-yard length
	Divide 310 by Tex	Woolen run	1600-yard length
Grains per yard	Mult. by 637.8	Denier	

TABLE 20.4
Summary of Other Relationships

1. $$\text{Cotton number} = \frac{\text{hanks}}{\text{pounds}} = \frac{\text{yards}}{840\ (\text{pounds})} = \frac{8.33\ (\text{yards})}{\text{grains}}$$

2. $$\text{Denier} = \frac{9000\ (\text{grams})}{\text{meters}} = \frac{9843\ (\text{grams})}{\text{yards}} = \frac{4{,}464{,}483\ (\text{pounds})}{\text{yards}} = \frac{637.8\ (\text{grains})}{\text{yards}}$$

3. $$\text{Worsted number} = \frac{\text{lengths}}{\text{pounds}} = \frac{\text{yards}}{560\ (\text{pounds})} = \frac{12.5\ (\text{yards})}{\text{grains}}$$

4. $$\text{Woolen run} = \frac{\text{lengths}}{\text{pounds}} = \frac{\text{yards}}{1600\ (\text{pounds})} = \frac{4.375\ (\text{yards})}{\text{grains}}$$

5. $$\text{Woolen cut} = \frac{\text{lengths}}{\text{pounds}} = \frac{\text{yards}}{300\ (\text{pounds})} = \frac{2.33\ (\text{yards})}{\text{grains}}$$

6. $$\text{Typp} = \frac{\text{lengths}}{\text{pounds}} = \frac{\text{yards}}{1000\ (\text{pounds})} = \frac{7.0\ (\text{yards})}{\text{grains}}$$

7. $$\text{Tex} = \frac{1000\ (\text{grams})}{\text{meters}} = \frac{1094\ (\text{grams})}{\text{yards}} = \frac{496{,}054\ (\text{pounds})}{\text{yards}}$$

similar large supply packages. The use of the 120-yard length skein, which is also called a lea, serves a dual purpose, for this skein or lea is used in order to obtain the strength of the yarn. In quality control work, the skein is tested on a pendulum-type strength tester, and immediately following that, the yarn number is determined.

The 120-yard skein is made up of eighty wraps of $1\frac{1}{2}$ yards each.

TABLE 20.5
Summary, Yards per Pound

For cotton system, 840-yard hank	840 × (yarn number)
For denier system, 9,000-meter length	4,464,483 ÷ (denier)
For worsted system, 560-yard hank	560 × (worsted number)
For woolen system, 1600-yard length	1600 × (run)
For woolen, 300-yard length	300 × (cut number)
For Typp, 1,000-yard length	1000 × (number)
For Tex, 1,000-meter length	496,054 ÷ (Tex)

TABLE 20.6
Conversion of English and Metric Units

Weight		Length	
Grams per pound	453.59	Inches per meter	39.37
Grams per grain	0.0648	Inches per centimeter	0.3937
Grams per ounce	28.35	Meters per foot	0.3048
Grains per pound	7000.00	Meters per yard	0.9144
Grains per gram	15.43	Centimeters per inch	2.54
Grains per ounce	437.50	Feet per meter	3.281
Ounces per grain	0.00229	Yards per meter	1.0936
Ounces per gram	0.03527		

The simplest form of reel is hand operated. The reel itself is made up of six arms carried about the center with a circumference of $1\frac{1}{2}$ yards. The crank mechanism to drive the reel is arranged with gearing so that forty turns of the crank result in eighty turns of the reel for a complete reeling of 120 yards. Usually four ends, although sometimes six ends or more, can be reeled simultaneously. Each individual end has a pigtail to give tension and to serve as a guide. During the formation of the skein, the pigtail traverses from an initial starting point to one side and then back to the original point again with a total traverse of about one and one-half inches. This traverse prevents the yarn from building up upon itself in any one spot, and in addition makes a flat skein for use on the strength testing machine. A yardage device indicates the yardage reeled at any time, and in addition, the machine is provided with a warning bell that is actuated just before the completion of the

120-yard length. The speed at which the reel is to be cranked is one of experience, although the speed limits specified by the American Society for Testing Materials are from 100 to 300 yards per minute. In this respect, if the machine is operated so that a 120-yard length

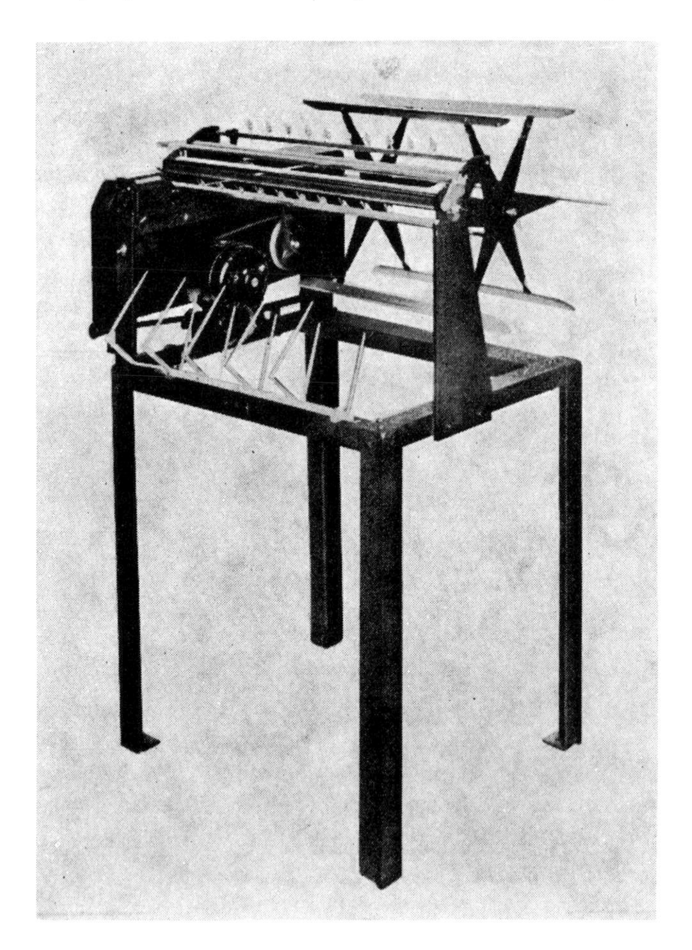

Fig. 20.1. Suter skein reel.

is reeled in half a minute, then the speed can be considered to be within the limits specified. The number of wraps of yarn on the pigtail together with the speed at which the skein is wound determine the tension of the yarn. The proper tension is largely a matter of experience, for unfortunately no standards have been set up by any organization specifying just what the proper tension should be. In any event, excessive tension means that the yarn size will be on the light side and possibly erroneous results will be obtained in measuring skein strength as well as yarn number. An experienced operator can generally ascertain the proper tension with which to wind these skeins using a hand-driven reel by feeling the skeins

Fig. 20.2. Fidelity skein reel.

when they have been completely wound. The skeins that are almost impossible to remove from the reel obviously have been wound with excessive tension, and skeins that slip off too readily usually are too softly wound.

On completion of the formation of the skein, it is wise to tie the leading and trailing ends together, doing this in a manner that binds the skein together with a single loop, using a square knot. By doing this, more accurate results are obtained during strength testing, and this single loop of yarn tied around the skein itself serves to keep the skein intact when several are mounted side by side on holders.

Power-Driven Reels

Power-driven reels are valuable in that they save time and provide greater control of tension during the reeling operation. There are two models made specifically for laboratory work: one sold by Alfred Suter of New York, (Fig. 20.1) and one by the Fidelity Machine Company of Philadelphia, (Fig. 20.2). On the older model reels, provision is made to reset the original setting of the guide bar so that an additional ten skeins can be made after the completion of the first ten. A third setting of the guide bar allows a final ten skeins to be formed so that a total of thirty skeins can be made before they are all removed from the machine. It should be noted that this system is satisfactory only in those cases where ten bobbins are used for each of the three skeins. On the newer model of the Suter machine, only ten skeins can be made at any one time. The tensioning of the yarn is obtained by an adjustable tension device which is made up of a series of three polished rods extending the length of the machine over which are located in alternate positions three other rods, and these three top rods can be raised or lowered as a unit to give the washboard tension effect. This series of six rods is preceded by a single plush-covered roll to give initial tensioning and to prevent horizontal movement of yarn over the tension bars. The relative position of the upper three rods is controlled by a small cam which has a calibrated index marker. Thus, having once established a proper position of the cam, the same position can be obtained at a later date by using the same index reading. These reels run at a constant speed of 300 yards per minute (or less if so arranged). Therefore, the tension of the yarn is dependent upon the package on which the yarn is wound and the setting of the tension rods. As for the hand-wound skeins, a properly wound skein is a matter of judgment, for no standards have been established controlling this. As a starting point, it will be found that a yarn tension of 10 to 20 grams will serve to give a properly wound skein for 20s down to 5s yarn; it should be understood that these are approximate tension readings only and that the exact readings should be verified by the mill itself. One of the main points to keep in mind in this respect is that the tension used from day to day should be maintained at the same level; the control of variation of yarn number becomes very difficult if the yarn is wound one day with a high tension and the next day with a low tension.

The traverse of the yarn for the power-driven reels is quite similar to that for the hand-driven type, averaging about 1 inch in width. In order to remove a skein from the older model Suter reels, one reel arm of the reel is folded over, thus effectively reducing the circumference of the reel. By collapsing the reel arms, the skeins can be moved lengthwise on the reel. Unfortunately, on this older model both ends of the reel are supported by bearings, and it is necessary, therefore, to use a rather troublesome device to get the skeins from the reel and around and off of the bearing supporting the end of the reel. In the newer model machines, however, this trouble has been eliminated by making the reel open-ended, and in place of having a single arm fold over, alternate arms are moved simultaneously to form a three instead of a six-armed reel; therefore, the circumference is effectively reduced so skeins can be removed quite readily over the end of the reel. Tensioning on the newer model reels is similar in principle to that on the older types in that a metallic rod type of tension is used with the exception that on the newer reels fewer rods are employed.

The stop motion on the earlier model Suter machines was arranged to give either 30, 60, or 120-yard lengths. The newer model machines use the preset electric counter which can be set for any yardage desired.

Procedure for Preparation of Skein

1. Place the supply package in the creel of the reel and place the yarn through the proper tension devices.
2. Attach yarn to the reel under the holder provided for this purpose.
3. Adjust tension devices to give proper tensions as previously outlined.
4. Set yardage indicator or automatic knock-off motion for power-driven reels.

5. Wind skein.
6. Cut or break yarn from the supply package and tie ends of skeins together so that the skein is bound together. The knot should be tied so that the skein will be held together but not be bound at the knot.
7. Remove the skein from the reel. This can be facilitated on most power-driven reels by collapsing one or more arms of the reel.
8. Place skeins on skein holder.

Skein Lengths. As has been pointed out, a 120-yard skein is considered as standard for use in obtaining the yarn number for singles spun yarn; this includes not only cotton yarns, but also spun synthetic staple yarns. The same length is used also for yarns numbered on the worsted system, although some mills have employed an 80-yard length rather than the 120-yard length. For ply yarns, shorter lengths are used in obtaining the yarn number if the recommendations of the American Society for Testing Materials are followed. These are shown in the following tabulation. It should be added that some mills spinning very coarse yarns frequently drop the 120-yard length for 10/1 and coarser yarns to a length of 60 yards.

Equivalent single number	Yards to reel for test
20 and above	60
3 to 20	24
Below 3	12

Skein Conditioning. Once the skeins are complete, if they are to be used for strength tests, then they should be mounted on a suitable holder to keep the skein in as near its original shape and condition as possible. Holders for skeins are made with parallel pins spaced just under 27 inches. The skeins should be conditioned in the laboratory until moisture equilibrium is obtained. Moisture equilibrium is defined by the American Society for Testing Materials, Committee D-13, as follows:

It shall be considered that moisture equilibrium is reached when, after free exposure to air in motion, two successive weighings not less than fifteen minutes apart do not differ by more than 0.1 per cent of the total weight. Moisture equilibrium shall be approached from the dry side. Where samples require drying before conditioning, no heat shall be applied in excess of 125°F.

Moisture equilibrium can be attained, generally speaking, by four hours exposure to a standard conditioned atmosphere. By agitation or movement through the atmosphere, equilibrium can be attained in a much shorter period. There are two common methods for decreasing the time necessary for the yarn to reach moisture equilibrium. One method is to rotate the skein holders at a speed of approximately twenty to thirty revolutions per minute. This gives sufficient movement of the yarn in the surrounding atmosphere for the yarn to reach moisture equilibrium in a period of three to four hours as compared with eight to twelve hours for stationary conditioning.

Another method for decreasing the conditioning time is to use a conditioning box. The box is constructed with a wire mesh top and an exhaust fan in the bottom. The skeins can be placed directly on the wire mesh, or they can be placed in a special type skein holder and the holder laid on the mesh. With the skeins in this position and the exhaust fan running, the air passes over the yarn at a high rate of speed, and moisture equilibrium is reached in three to four hours.

Some laboratories have found it necessary to precondition the yarn so that the regain of all yarns tested will be the same; that is, the regain of all yarns will be the absolute regain at standard conditions. In preconditioning the yarn, it is placed in an atmosphere of high temperature and low relative humidity so that some of the moisture will be driven from the yarn. Care should be taken in preconditioning yarn that the temperature does not exceed 120 to 130° F., and for best results it should not be lower than 90° F. The relative humidity should also be controlled to not less than 20 per cent and not greater than 40 per cent. The conditions are controlled within these limits so that the fibers will have less moisture in them when the preconditioning period is ended than they would have at standard conditions. In this way, the material will always be conditioned from the dry side with respect to the hysteresis curve. If the preconditioning is done in a drying oven, extreme care should be taken that all of the moisture is not removed from the fibers. The loss and gain of moisture is not a complete reversible reaction, especially if the reaction is taken to zero regain.

Fig. 20.3. Quadrant balance for yarn.

Quadrant Scales

The performance of routine tests can be time-consuming, and

therefore every effort should be made to use labor-saving instruments. The quadrant scale is made so that a direct reading of yarn number is obtained when a skein is hung on the quadrant arm. The scale readings generally cover two ranges of yarn numbers, and supplementary counterbalancing weights can be added or removed to provide the proper scale range for the particular yarn being run. In addition, special counterbalancing weights are provided for the ply yarns. These quadrants, similar to the type made by Alfred Suter and shown in Fig. 20.3 are designed to give the accuracy required in testing. The scales can be read to the nearest one-fourth of a yarn number, and despite some concern to the contrary, this degree of accuracy is sufficient for the numbering of yarn in routine control work.

The quadrant scales are constructed so that automatic dampening of the scale indicator is accomplished by means of magnetic action. This, together with the techniques that can be developed by the trained operator, allows weighing of skeins and the indication of the yarn number of the skein very rapidly. When used in conjunction with the skein-breaking test, one procedure which is very satisfactory is as follows: The operator places the skein in the testing machine until rupture; the machine is then reversed to bring it back to zero position. In this interval the technician records the strength of the skein (as will be described in a later section of this book). Then as soon as the strength tester has returned to a position that allows the skein to be removed, the tender can remove the broken skein and hang it onto the hook of the quadrant balance. The technician then introduces another skein into the strength-testing machine to get its break; after starting the machine into operation, he can check the yarn number of the previous skein, for by this time the quadrant pointer has come to rest. The technician then turns back to the strength-testing machine, observing the action until the rupture of this second skein occurs, at which time he repeats the procedure outlined above.

Procedure for Calibration and Use of Quadrant Balance

1. Balance cabinet until the pointer of the quadrant balance points to zero.

2. Use marked weights to check balance throughout range of scale.
3. Place proper counterbalance weight on balance.
4. Place a 120-yard skein on balance hook and allow indicator to come to rest.
5. Read and record yarn number directly from scale.

Calculations. No calculations are necessary when using the quadrant balance.

Other Scales; Shadograph. Recently a direct reading model

Fig. 20.4. Shadograph.

of the Shadograph scale has been developed for weighing yarn and roving skeins (The Exact Weight Scale Co., Columbus, Ohio). See Fig. 20.4. The dial can be calibrated in yarn number, weight, or both. The scales are accurate, quick-reacting, and extremely simple to use. Their use should save time and effort where full quality control programs are used.

Frequency of Test for Yarn Number. Each yarn being spun should be tested at least once each day of three-shift operation. The number of bobbins to test is rightfully a function of the relative production of a particular yarn, as well as the number of frames in operation on each type of yarn. Whereas some mills designate certain spinning frames as test frames, it is generally conceded that a more randomized selection is preferable, for statistical control is based upon random selection throughout the entire population. A minimum of eight individual bobbins of yarn should be selected for each yarn number being spun; preferably ten samples should be selected in order to give a better indication of the average. The bobbins should be selected at random from frames in operation and in such a way that the maximum number of frames is represented in the final sampling. A good index of the number of bobbins to sample for testing is the use of the limits of 100 to 150 pounds production per sample. Table 20.7 giving the number of tests to make has been set up to illustrate the principles upon which test frequency can be based. This table should provide a good starting point for the mill interested in good quality control procedure. However, individual conditions that exist would necessitate some change from this standard. As is indicated, the table is based on 20/1 yarn, and therefore changes must be made in the mill practice for yarns other than 20/1. A further examination of the table will indicate that there is an overlapping of the items under the heading "Frames on". This allows for the exercise of individual judgment on the selection of the number of bobbins to test.

A survey of the industry would reveal that the majority of mills are satisfied in testing eight, ten, or twelve bobbins for each yarn number being produced, which indicates immediately that no consideration is given to the relative quantities of these materials being manufactured. For proper control, the rate of production should be a factor, and the suggestions incorporated above have taken this into account. In those cases where more than sixteen or twenty bobbins are tested, it is also recommended that the selection of bobbins be made at different times during the operating period. By this practice any trends can be discovered and steps

TABLE 20.7

Number of Tests [a]

Frames on	Bobbins to test
Up to 4	8
4 to 7	12
6 to 10	16
8 to 12	20
10 to 16	24
Over 16	28

[a] Based on three-shift operation of 20/1 yarn with approximate limits of 100 to 150 pounds production per test. For other yarn numbers, use in proportion of production. For example, on 10/1 the frames would be cut approximately in half so that for a 16-bobbin test the frames would be 3 to 5; on 40/1 the frames for a 16-bobbin test would be 12 to 20.

taken to prevent trouble much more quickly. For example, if a mill is operating thirty frames, the test samples should be collected at least twice and preferably three times per shift, thus affording a much higher factor of safety in controlling trends.

Universal Yarn Numbering Balance

The Universal yarn numbering balance is a quick method for determining the approximate yarn number of a 1-yard length of yarn or of several short lengths totaling 1 yard. The determination of yarn number based on a single 1-yard length is not generally recommended for quality control purposes; however, this particular type of balance is very convenient for fabric analysis where only short lengths are available, for separating mixed yarn, and for getting a quick estimate of yarn size.

The balance itself is of the torsion type, which accomplishes the balancing of the load by changing the degree of torsion on the spring within the instrument. To use, the sample of yarn is hung on a hook on the beam, which is located within a weighing chamber to eliminate the effect of air drafts. A magnetic damping device insures the beam coming to rest quickly. The balancing load is applied by an external lever, which also carries the pointer and

Fig. 20.5. Universal yarn numbering balance.

vernier scale to indicate the load on the dial. Fig. 20.5 shows one of these precision balances. This same type of balance is used in conjunction with certain of the cotton fiber tests.

Procedure for Preparation of Yarn Sample

1. Hold one end of yarn with the left hand at the proper mark on the ruler. For cotton and spun synthetics, take a 1-yard length. For filament synthetics, take 0.9 meters.

2. With the right hand, place sufficient tension to remove the crimp, but not enough tension to stretch the yarn.
3. Cut the sample with the stationary knife attached to the end of the ruler.
4. Form yarn into a sufficiently small loop so that it will hang from the balance hook and not touch any part of the balance cabinet. Do not handle the test specimen any more than is necessary to prevent moisture pickup from the hands.

Procedure for Calibration and Use of Balance

1. With front adjusting screws, adjust balance until the spirit level indicates the balance level.
2. Check zero of balance by releasing the beam clamp on the left hand side of the balance cabinet. If the balance pointer is not at zero, adjust the knob at the rear of the cabinet (in center of back plate) until the pointer indicates zero.
3. Lock beam and place the sample on the balance hook.
4. Close the door to the weighing chamber and make sure the sample does not touch the cabinet at any point.
5. Release the beam clamp and adjust the index lever until the pointer indicates zero.
6. Read and record the yarn number or denier, depending upon the sample being tested.
7. Lock the beam, remove sample, and return index lever to zero position.

Calculations. For cotton, spun synthetics, and continuous filament synthetics, read the yarn number or denier directly from the scale, and no calculations will be necessary. For yarns on the cotton system that are heavy enough to carry the pointer off the scale, use 18, 12, or 9-inch lengths and divide the readings by 2, 3, and 4 respectively. For yarns on the denier system, use .45, .225, or .1125 meters, and multiply by 2, 3, and 4 respectively.

Number and Frequency of Tests. The number and frequency of tests depends upon how the test results are to be used. For example, if the tests are to be used to give an indication if the yarn is a 20/1 or 30/1, then only one test is necessary. However, if the balance is used in connection with fabric analysis, then several tests should be made, depending upon the amount of yarn available for analysis.

Yarn Number from Skeins using Yarn Number Tables

The yarn number, N, on the cotton system is calculated from the following formulas:

$$N = \frac{(Yd)(8.33)}{(Gr)}$$

or

$$N = \frac{1{,}000}{Gr/120 \text{ yds}}$$

Tables have been prepared by different companies as a service to the industry which simplify the determination of the yarn number. These tables give the yarn number directly for the corresponding weight in grains of the standard 120-yard skein length. Thus, all that is necessary is to get the weight of the skein in grains and then from the table read off the equivalent cotton yarn number.

Grain Scales. In making the weighings of the skeins, the type of balance ordinarily used is a simple beam balance calibrated in grains and with a sliding beam weight to give a total of 20 grains in tenths. The scales should be accurate to 0.1 grain. In cases where the skeins are used for yarn break, the weighings should be made immediately after the strength test is made.

If continuous filament yarns or yarns to be numbered in the denier system are to be measured, the calculation for the denier is according to the following formula:

$$\text{Denier} = \frac{\text{Grams}\,(9{,}000)}{\text{Meters}} = \frac{\text{Grains}\,(637.8)}{\text{Yards}}$$

In cases where a 120-yard skein length is used, this formula simplifies to:

$$\text{Denier} = (\text{Grains per 120 yards})(5.313)$$

When 450 meters are used, then

$$\text{Denier} = (\text{Grams per 450 meters})(20)$$

If the Tex system is used, then the formula for yarn number using a 120-yard skein is

$$\text{Tex Number} = (\text{Grains per 120 yards})(0.5905)$$

Yarn Size on Beams

The true average size of yarn can be determined very accurately by analysis of the section beam data. The formula for determining the average yarn size on a section beam is as follows:

$$\text{Yarn Number} = \frac{(\text{Yards})(\text{Ends})}{840(\text{Weight, lbs.})}$$

Inasmuch as standard mill practice requires the careful measurement of beam yardage as well as the accurate determination of the net weight of yarn on a beam, it is natural that by use of these figures in the above formula, a very accurate measurement of the average number being produced in the mill can be obtained. This value is of service in determining long-range or season-to-season trends and also in determining the efficiency of the quality control system in use. The following example shows the application of the formula:

A 40,000-yard section beam of 40/1 yarn is made with 500 ends. The net weight of the yarn on the beam is 600 pounds. Determine the average number.

$$N = \frac{40{,}000\,(500)}{840(600)} = 39.68$$

Novelty Yarns

The term "novelty" is given to those yarns produced in a manner not normal to regular production and where the yarn has an unique texture characteristic. Some novelty yarns are spun as single yarns; particularly in this category are slub and Himalaya-type yarns which are characterized by heavy places of varying lengths. Such yarns can be treated in a normal way insofar as the measurement of the yarn number is concerned. The majority of novelty yarns, however, consists of two or more individual yarns in a ply. Within the final yarn, individual yarns may vary in length. For example, in *seed* or *rice* yarns, a single end of yarn may be wound intermittently around a core yarn (usually of 2-ply construction itself) to form the *seeds*. In turn, a second twisting process of this base with another end called the binder is performed to bind the seeds into a stabilized construction. The problem of yarn size, together with that of determining the per cent of each component in the final yarn, becomes a little involved because of the different lengths of the components together with the different amounts of contraction at each stage of processing.

The basic formula for determining yarn size of a multi-ply novelty yarn is:

$$\frac{L_1}{N_1} + \frac{L_2}{N_2} + \frac{L_3}{N_3} + \dots \frac{L_n}{N_n} = \frac{L_x\left(\frac{100 - \%C}{100}\right)}{N_x}$$

where L_1, L_2, L_3, and L_n are lengths of individual yarns N_1, N_2, N_3, and N_n (English system), L_x is the length of the product yarn N_x, and C is the contraction. Note that L_x must be the same as the shortest length L.

The equation can be applied for each process of twisting. The relative amount of any individual component N_1 is represented by the ratio of L_1/N_1 to L_x/N_x for L/N is a measure of weight.

EXAMPLE. The first process of a novelty yarn is made by twisting in the Z direction a length of 32 inches of 40/1 dyed yarn around a 10-inch length of a core yarn of 20/2s. The 40/1 yarn is fed intermittently to form seeds at every $\frac{1}{2}$ inch. There is a contraction of 6 per cent in this first process.

A second binding process of twisting is performed: In this, a binder of 30/1 yarn is twisted around the first process yarn with an S direction. Equal lengths of the 30/1 and the first process yarn are fed on the twister. There is an actual extension of the yarn of 2 per cent; that is, instead of contracting, the final yarn has relaxed by this amount, so that a 10-inch length becomes 10.2 inches. The problem is to find the final calculated yarn size and per cent of each component.

1. For first process yarn numbers:

$$\frac{32}{40} + \frac{10}{20/2} = \frac{10\left(\frac{100 - 6}{100}\right)}{N_x}$$

$$0.8 + 1.0 = \frac{9.40}{N_x}$$

$$N_x = \frac{9.4}{1.8} = 5.22 \text{ equivalent single yarn number}$$

2. For first process per cent:

$$0.8 + 1.0 = 1.8$$

$$\frac{0.8}{1.8}(100) = 44.4\ \% \text{ of } 40/1 \qquad \frac{1.0}{1.8}(100) = 55.6\ \% \text{ of } 20/2$$

3. For final yarn number:

$$\frac{9.4}{5.22} + \frac{9.4}{30} = \frac{9.4\left(\frac{100 + 2}{100}\right)}{N_y}$$

$$1.8 + 0.313 = \frac{9.59}{N_y}$$

$$N_y = \frac{9.59}{2.113} = 4.54 \text{ equivalent single final yarn size}$$

4. For per cent in the second process and in the final yarn:

$$1.8 + 0.313 = 2.113$$

$$\frac{1.8}{2.113}(100) = 85.2\ \% \text{ of first process yarn}$$

$$\frac{0.313}{2.113}(100) = 14.8\ \% \text{ of } 30/1$$

85.2 % of 44.4 =	37.8 % of 40/1 in final yarn
85.2 % of 55.6 =	47.4 % of 20/2 in final yarn
	14.8 % of 30/1
Total	100 %

CHAPTER XXI

Yarn Strength

Strength has long been accepted by many as one of the most vital characteristics of yarn. Over the years, strength together with appearance has always been in the foreground when yarn has been evaluated, whether in the mill or in the market, whether for a consumer product or an industrial product. Appearance of yarn has been a matter of judgment, a somewhat nebulous characteristic, but strength has always been a measurable property of the yarn. It has been something that could be evaluated, described, bragged about, or blamed. The spinner congratulates himself for producing high strength yarn and denounces the cotton buyer for low strength yarn. The weaver denounces both for poor weaving and weak fabric.

Some contend that the importance of yarn strength has long been overemphasized. Whether or not this is justified, in yarns for consumer use, for military use, and for industrial use, the yarn strength remains as the criterion most frequently established and usually the most difficult to attain. Just as for cotton, the yarn strength or tenacity of synthetics usually is the first property determined.

Factors Affecting Yarn Strength

In considering yarns spun from staple, there are many inherent variables that influence the final strength of the spun yarn. Some of these variables for cotton and staple synthetics are listed herewith together with the resulting effects.

Staple Length. Longer staple cotton gives higher strength. With synthetics where much longer staple lengths than cotton are available, the increase levels off after the optimum length.

Fiber Fineness. Fine fibers give greater yarn strength than coarse fibers when spun into a given yarn size: this is due to the greater internal friction provided by the more numerous fine fibers.

Fiber Strength. Logically, a strong fiber produces a stronger yarn than a weak fiber: compare a nylon yarn with a wool yarn of similar weight.

Twist. For any single spun yarn, there is always a twist that gives maximum strength. A twist less than or greater than this optimum amount results in a yarn of lower strength. The amount of twist for maximum strength is related to the twist angle, and in general for any given fiber the twist angle for maximum strength remains nearly constant over the range of yarns that can be spun from this fiber. Uneven twist results in a variation in strength, and the effects of this variation are more noticeable in single end tests than in skein tests. In ply yarn, uneven twist distribution or uneven tension applied to the individual components during twisting cause corkscrew yarn, and the strength of such a yarn is always less than that of normal yarn of the same size. Obviously, twist has a very definite and important bearing on yarn strength, both for single and ply yarns.

Evenness. The greater the uniformity of a spun yarn, the higher is its strength, and the more uneven a yarn, the lower is its strength. For example, a yarn having a variation of plus or minus 50 per cent means that the weight per unit length has been reduced by 50 per cent from the average in thin spots. The strength of this fine section of yarn would be correspondingly lower. In addition, non-uniformity of the yarn itself is reflected in a similar lack of uniformity in the distribution of twist, which in turn has a deleterious effect on the strength. Investigations of the relationship between evenness and strength indicate that they are very closely related.

Fiber Length Distribution. Variations in the distribution of fiber lengths will cause a variation in yarn strength. The greater the percentage of short fibers, the lower the strength of the yarn. It is possible for two samples of fibers to have the same average length and yet produce yarns with different strengths due to the fact that one sample might have a greater preponderance of short fibers. The purpose of the Fibrograph, Suter Webb Sorter, and similar instruments is to detect such conditions.

Fiber Finish. The type and amount of chemical finish applied to fibers, particularly the man-made fibers, has a very definite effect on the strength of the yarn, as well as on the processing characteristics of the staple.

Other Causes. There are undoubtedly many other variables that have a very definite effect on the strength of yarn. One such element would be chemical treatments which have tendered the yarn. Apparent differences in yarn strength could be due to the moisture content of the material. Another variable that has been the subject of recent investigation is that of the position and individual elongation characteristics of the fibers. For example, if two fibers that have the same strength are twisted together so that one is loose and the other is tight, the tight fiber assumes the load and the strength of the pair is that of the single fiber. The same type of condition may exist with a spun yarn, except that many fibers are involved: it has been estimated that only 60 to 70 per cent of the fibers in a yarn break under a tensile load.

The Two Approaches: Multiple End and Single End Tests

There are in general two approaches to obtain the strength of a yarn. The first of these is the multiple end test, and the most common in this category is the skein test in which several ends are tested simultaneously. The second type is the single end test.

The skein test, which has been in international use for many years for determining the strength of cotton yarn, is made by applying a tensile load to a standard skein of yarn. These skeins are 120 yards in length made up of 80 wraps of $1\frac{1}{2}$ yards each as described in the previous chapter on yarn numbering. In making the test, the skein is placed on the testing machine over small spools about 27 inches apart. Because of this positioning, it will be observed that with the 80 wraps in the skein, a total of 160 ends are subjected to the tensile load. In the single strand method of testing, on the other hand, a single length of yarn is held by appropriate clamps, generally 10 inches between the jaws in the United States, and up to 20 inches between the jaws in other countries. The results of the skein strength tests are most frequently reported in units of pounds of strength, whereas for single strand tests, the strength is generally reported in pounds, ounces, or grams.

Although there are 160 ends subjected to the load in the skein test, it does not follow that the results of this test are 160 times greater than the strength obtained on the same yarn by the single strand method. An analysis of the two testing methods shows that with the single strand test with 10 inches between the jaws, there is a reduction of non-uniformity occurring in the selected length. However, with a 120-yard length of yarn composed, in effect, of 160 ends, there is a very high probability of maximum non-uniformity or unevenness in the sample to be tested. In the skein test, as the load is applied and any one strand is broken because of non-uniformity or because of variation in tension or elongation, then there are only 159 ends available to support the stress being applied. Furthermore, because one continuous length of yarn is used in the skein test, the rupture of any single strand will result in slippage of the terminal ends of this strand so that actually the 160 ends originally supporting the load are now reduced to less than 159 ends, possibly to 157 or less. As a result of this reaction, following the rupture of one or two individual ends, very little if any further load is sustained by the skein. It is for this reason that the strength of yarn as obtained by the skein test method is always relatively lower per strand than that obtained by the single strand method.

There are several variables such as twist, yarn number, and evenness which influence the relationship of the strength as measured by the single strand test to the equivalent single end strength as calculated from the skein test. For any given singles yarn, if the single end strength is multiplied by a constant ranging from a minimum of 90 to a maximum of 130, a fair approximation of the skein strength will result. Theoretically, if the single strand strength of a yarn is multiplied by 160, the total number of ends in a skein, the result will be the skein strength. For ply yarns, the relationship varies from approximately 115 to 145. It is somewhat higher than for singles yarns since the ply has comparatively fewer weak spots.

The results of the skein tests are generally expressed in pounds. These results are used to calculate the count-strength product or break factor, which is a product of the skein strength in pounds and the yarn number. This index of strength is described in greater detail later. In addition, when strength tests are made, it is possible to measure the actual amount of elongation. This also will be amplified later.

Another multiple type test that has been used by some laboratories is the Seriplane type of test. In this test the yarn is wound in much the usual manner on a reel with as many ends as are desired in the final test. Instead of removing the skein from the reel, masking tape is used to bind these ends of yarn together, generally 10 inches apart. The sample of yarn is then cut from the balance of the skein and the 10-inch Seriplane type of sample can be placed into the strength testing machine in much the same manner as fabric and the tensile test made. It is claimed that by this test a smaller variation in yarn strength in comparison to the single yarn test is possible, with results that more nearly represent the true average strength of the yarn. However, because of the time necessary in preparing the samples and because of the questionable superiority of this test over the other accepted methods, it has not been adopted to any wide extent, and in fact is very rarely used.

In summary, the skein test is a quick easy method of obtaining an index of the strength of yarn. It will detect changes in the level of performance of yarn with respect to strength, even though it does not measure the single strand strength as such. Another point in favor of the skein test is the ease with which the yarn number can be determined in connection with the test. For comparable accuracy, the single end test is more time-consuming and more expensive to make unless the new type of automatic machines are used. However, it does give the actual strength of the yarn and at the same time also gives some indication as to the weakest spots that may be expected in a yarn. Since the results of this test do show the variations in yarn strength, the data will have more variation, as shown in Table 21.1, than the skein test. This means that more single end tests must be made from the same yarn than skein tests for the average to have the same reliability. Since the single end test is more expensive if performed on non-automatic machines, mills frequently perform fewer single end tests, which means that the average is much less reliable than the skein test.

TABLE 21.1

Approximate Values for the Coefficient of Variation in Per Cent for Strength and Elongation of Yarns

Test	Staple synthetics		Cotton	Continuous filament	
	Singles	Ply	singles	Dry	Wet
Skein strength	5 to 10	4 to 7	10 [a]	—	—
Single end strength	14 to 19	8 to 13	15 [a]	1.5 to 3	1.5 to 3
% Elongation	10 to 14	5 to 10		2 to 6	2 to 5

[a] ASTM, D-13 Handbook.

Types of Testing Machines

Pendulum Type. What might be considered as the original method of making tensile strength tests is performed on the pendulum type of strength tester. In the pendulum type of tester, one of the two jaws is traversed at a constant rate of speed so as to exert a tensile load on the test specimen. The second jaw of the machine holding the other end of the test specimen works against a pendulum appropriately weighted. Because the actuating jaw of the pendulum tester moves at a constant rate, the test specimen is elongated at a constant rate, except for the effect of a small movement of the secondary jaw that actuates the pendulum. The rate of loading depends upon the extensibility of the test specimen so that in effect the actual rate of loading varies for all materials and is not constant.

The pendulum principle of loading is used in testing machines designed for use with fibers, yarns, cords, fabrics, and other textile materials. It is used not only for multiple end skein testing, but also for single end testing, in both manually operated and automatic instruments. One example of the pendulum single end tester is the Suter single end tester used extensively for the continuous filament yarns; examples of the automatic type are the Moscrop and the Autodinamografo testers. The following discussion is based on the use of the Scott type of pendulum tester.

Pendulum testers are in the simplest terms weighing devices with the force supplied by an actuating jaw translated through the material being tested to the second jaw. A pendulum is used to provide the load, and a dial gives a means of recording the extent of the load. Pendulum testers are made for yarns with normal elongation, and therefore for certain of the synthetic yarns with very high elongation characteristics, the machines must be modified to some extent. Pendulum testers are made in many capacities and in both vertical and horizontal models; they are heavy-duty type machines capable of continuous use in the laboratory. At the same time, however, there are certain errors that influence the accuracy of the results. For example, the inertia of the pendulum device, overthrow due to momentum forces, friction, and in general, the inaccuracies in the weighing device itself all can cause errors. In spite of such errors, this type of tester remains very popular in the textile industry because of the low cost of the machine, the relatively low cost of making tests, and because the accuracy of the machine is sufficient for routine quality control test work.

Inclined Plane Type. The second basic type of strength tester is the inclined plane instrument, which has been predominant for use in making single strand tests in the synthetics industry. In the inclined plane type of machine, the load is supplied by gradually increasing the angle of slope of the plane carrying a track or tracks upon which the load is free to roll.

Inclined plane testers of the type manufactured by Henry L. Scott, Incorporated, of Providence, Rhode Island, are of the autographic type, for both the load and elongation characteristics are autographed directly onto a chart. The elongation can be measured directly in inches or other units of length, for the pen movement is controlled by the actual elongation of the yarn. The amount of load is determined by the weights that are used on the rolling carriage, so that a range of load capacities is possible. Strength testers of this type are used for single yarn, ply yarn, cable yarn, cord yarn, continuous filament yarns, and in any other application where actual strength is of importance. In general, this type of machine is more expensive than the pendulum types, but it is more accurate, and shows the weakest spot in any given sample of yarn. The normal gauge length used in this country is 10 inches, although this can be varied at will. In this respect, the strength of the yarn varies to some extent with the test length, with the longer test lengths giving the lower average strength.

The charts reveal the stress-strain relationship of the yarn and lend themselves to analysis of the characteristics associated with strength and elongation, as for example, elastic limit and toughness. In addition, these machines generally can be set for repeated stress-strain loading so that other results are obtainable to give a better understanding of the yarn characteristics. This is particularly valuable with the continuous filament yarns.

The same principle of the inclined plane method of loading has been adapted for use with the Uster single strand automatic yarn tester which is described in more detail later.

Strain Gauge Type. A third type of tensile tester utilizes a constant rate of movement cross head for applying the load, and a strain gauge for measuring the amount of tensile load supported by the test sample. In this country this principle is exemplified by the Instron Tester built by Instron Engineering Corporation.

Pendulum Testers

Principles of Operation. In pendulum testers, the sample to be subjected to a tensile load is placed between the two jaws. In the case of yarn, the skein is looped over small spools fastened to these jaws, and in the case of a single strand of yarn or fabric, the material is clamped in each pair of jaws. (It should be noted that these machines are designed for use for both yarn and fabric, and that most test details discussed are applicable to both.) One of these jaws, referred to as the lower or bottom jaw in vertical machines, is the actuating jaw. This moves at a constant rate of traverse away from the top or reaction jaw at a speed which should be adjusted to 12 $\pm$ 0.5 inches per minute. The sample, whether it is skein or fabric, is held firmly in place by both jaws, and the movement of the lower jar away from the top jaw through the movement derived by the motor gearing and screw thread device imparts the load to the specimen. This load is transmitted through the fabric or the yarn to the upper jaw, which in turn transmits the load to the head of the machine by means of a chain and pulley or drum device. Attached to the drum is the pendulum which opposes the pulling force, and the magnitude of the force is then determined by the arc through which the pendulum swings. By reference to

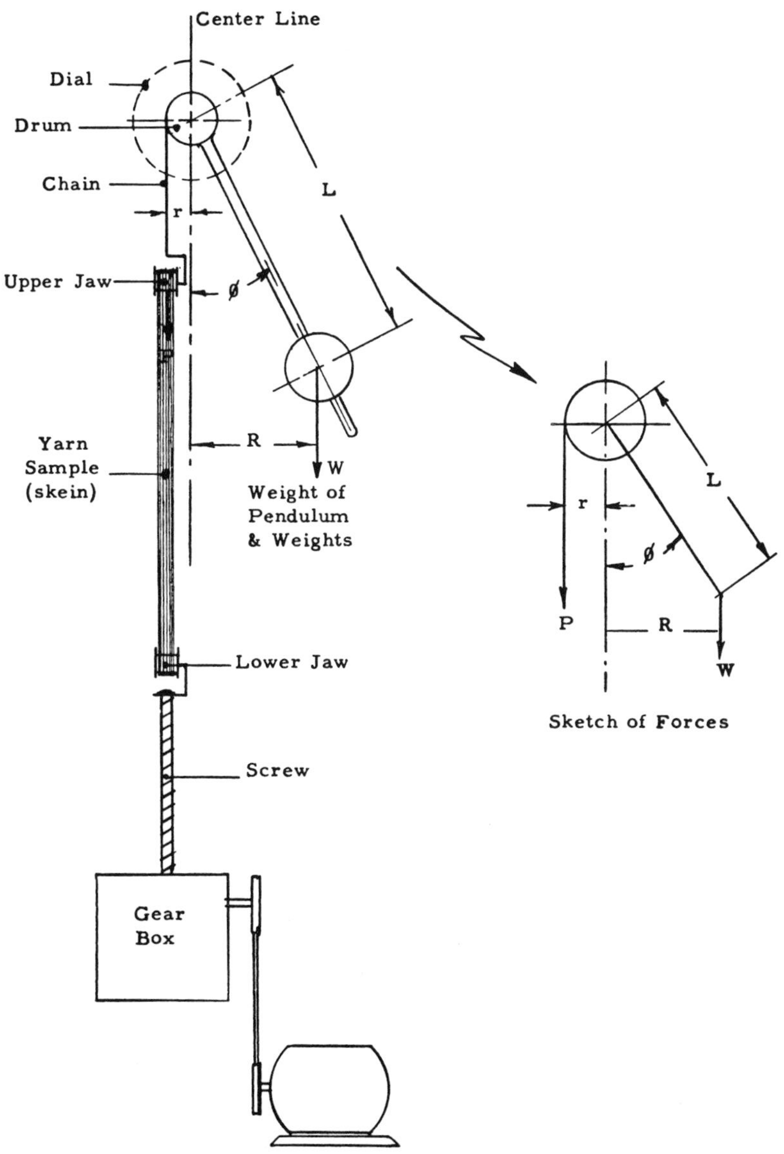

Fig. 21.1. Schematic diagram of a pendulum-type tester.

Fig. 21.1, the following relationship may be derived, letting W = weight of pendulum and arm:

$$Pr = WR$$
$$R = L \sin \phi$$

Therefore,

$$Pr = WL \sin \phi$$

With r, W, and L constant, P varies as $\sin \phi$, or

$$P = k \sin \phi$$

If a non-extensible material is used for a test, then the load would increase as $k \sin \phi$. The sine curve approaches a straight line between the angles of 9° and 45°, which are the two limits designated as being the acceptable extremes. That is, results of tests on a pendulum tester are inaccurate if the break occurs when the angle of the pendulum to the vertical is less than 9° or greater than 45°. Inasmuch as the sine curve approaches a straight line within these limits, the rate of loading for a non-extensible material would approach a constant rate. However, textile materials elongate under load, some to an extreme amount. Therefore, the upper jaw does not move in proportion to the lower or actuating jaw when testing a textile material, which means that the pendulum type machine cannot be considered as having a constant rate of loading.

It would be found wise to put limit marks at 9° and 45° on laboratory machines in order that technicians performing routine tests may keep within these limits.

It is also possible to get a measurement of the elongation of the sample by using an attachment that is available for most models. However, the measurement of elongation is generally confined to use with fabric. With elongation attachments, the platen is moved vertically at the same rate as the lower jaw, and then the swing of the pendulum controls the horizontal movement of the autographing pen to indicate the load. Special charts must be used on these instruments because of the fact that the normal curve described by a material having zero elongation would be at a small angle from the horizontal, due to the movement of the upper jaw coordinated with the movement of the pendulum away from the vertical. These general principles apply to the pendulum system of loading, whether fibers, yarns, or fabrics are to be tested.

Calibration of Pendulum Testers. No attempt will be made to give complete instructions for calibrating all types and makes of pendulum testers. However, the following procedure is given for

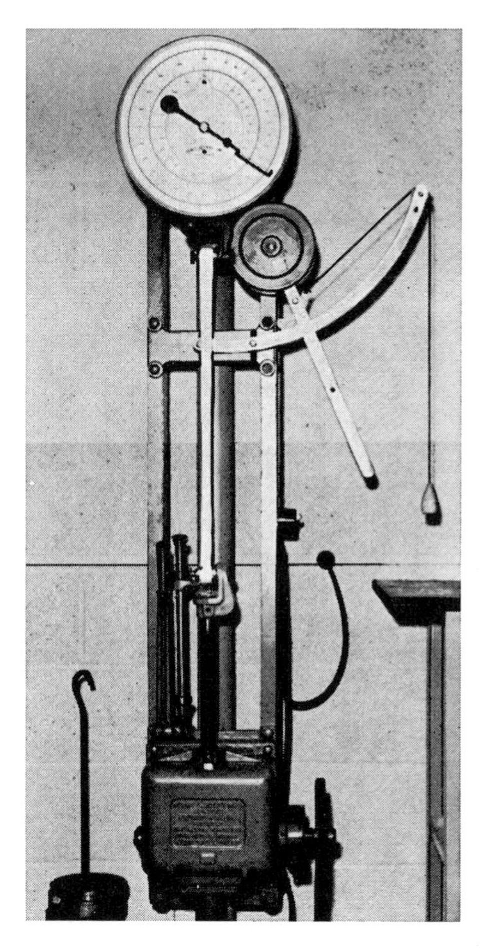

Fig. 21.2. Pendulum tester (Scott).

the standard type tester as manufactured by Scott Testers. This particular pendulum tester has been selected very widely for laboratory use in the textile industry. These instructions follow very closely to the procedures recommended by Scott Testers. See Fig. 21.2.

Before calibration, it is necessary to have available different weights of known amounts, and the exact weight in each case should be determined on scales that have been themselves calibrated for accuracy. In computing the amount of a known weight, the weight of all supplementary apparatus used to attach it to the clamp should be included. The first step is to balance the machine by allowing the pendulum to swing to its normal position at dead center. With the pendulum in this position, adjust the dial pointer to zero, thus giving the correct zero position. Next, one of the known weights is attached to the upper clamp. Then the weight is allowed to exert its load on the weighing head by manaully lowering it gradually until it reaches equilibrium. It should not be dropped rapidly, and it should not be lowered too slowly. Instead, an attempt should be made to lower it at a speed consistent with that under which the stress is applied during actual test work. When at rest in this new position, the dial pointer should register the exact amount of the known weight, plus the attaching apparatus. In this first calibration step, the pawls must be in the same position on the quadrant rack as in actual testing; in other words, they should be in action. Furthermore, the weight used to calibrate should be attached in such a way that there will be no binding or friction caused by contact of the weight with other parts of the tester.

It is advisable to test the machine over the entire range within which it is to be operated, with special attention being given to the limits within which the materials to be tested are expected to break. This means that more than one weight must be available to get proper calibration. It is possible to select weights so that two or more can be used together to give several calibration levels.

If a high reading results from checking the recording head, there are three possible causes: (*1*) On earlier model testers the weight on the pendulum was held by a set screw, and this may have been raised closer to the pivot of the pendulum. (*2*) Mass may have been removed from the pendulum weights. (*3*) The pendulum weight may have been removed and put back on the pendulum upside down.

If a high reading cannot be corrected by adjusting the position of the weight, it is advisable to send the head of the tester to the manufacturer for recalibration.

Usually any error found in calibration is a low reading, generally caused by undue friction within the head mechanism. First, check the bearings holding the head shaft. These are lubricated with a special grease, in which dirt and lint may accumulate after years of service. To clean, remove bearings from the head and wash carefully in gasoline or commercial solvent. After washing, cup the bearing in the palm of the left hand so that the inner race does not rub against the skin; place the right thumb against the inner race with moderate pressure, revolving it slowly. Any particles of rust or foreign matter can thus be felt. To produce a good calibration, the bearings must be entirely clean of any material that would impair free rotation.

After cleaning, lubricate the bearing with a very small amount of white vaseline. Never use gear lubricants or machine oil, which tend to oxidize upon exposure.

In replacing ball bearings in the head, leave about 1/64th inch end play in the shaft; avoid tightening the check nuts too much.

Another cause of low reading may be in the pointer gears. Their teeth, also the bearings of the pinion shaft in the head, should be thoroughly cleaned, If gear teeth show excessive wear, the gears should be replaced. In reassembling the head, a very small amount of oil should be placed in the shaft of the pinion and on the teeth of the gears.

Undue friction may be caused by the condition of the chain supporting the upper clamp. This chain should be thoroughly cleaned and reoiled with a few drops of light machine oil. It should also be checked for undue wear on the points that contact the head drum. The drum should be inspected to make sure that the chain has not caused excessive wear on its surface.

In servicing the head, the teeth of the quadrant should be cleaned. Inspection may reveal that quadrant teeth have become bent, requiring straightening. The points of the pawls should be inspected to see if they are bent or dulled. A special case exists in the head of those testers having a guide for the upper clamp; friction can occur where the clamp rod passes through, if the guide has become bent or if the machine is not installed precisely vertical.

If discrepancies still remain after correcting the tester head, the difficulty may lie in the clamps.

On flat grip clamps for testing fabric or yarn, the gripping surfaces must be absolutely flat and parallel. Place a sheet of white paper and very soft pencil carbon paper between the clamps and close them tightly. If carbon is not deposited uniformly over the white paper, it will indicate that the surfaces have become worn by testing hard material, or distorted by excessive pressure on the closing screw. Such clamps should be returned to the factory for resurfacing and realignment.

To determine parallelism of the long edge of the gripping surface, move the lower clamp down about one inch; measure the distance between the edges of the gripping surface at both ends. If not parallel, adjust the clamp which is out of line by straightening the connection joining this clamp to the testing machine.

Cord, twine, or similar materials may require drum-type clamps. If so, the drum surface must be absolutely smooth, while the gripping surface of the clamps must apply uniform pressure throughout.

Procedure for Skein Strength Tests on Pendulum Testers

1. Prepare skein as described in the yarn numbering section.
2. Equip the tester with spools of not less than 1 inch in diameter and not less than 1 inch in length. One spool should rotate freely after being placed on the tester.
3. The gauge between the spools should be set so that just sufficient distance is allowed for placing the skein around the spools in a flat band. Care should be taken to prevent disturbance of the yarn in the skein.
4. With the skein properly placed over the spools, hold the yarn under tension with one hand by pulling the front side of the skein toward the operator and initiate movement of the lower spool by pulling the release handle with the other hand. Hold

the yarn under slight tension until the machine applies tension on the skein; this is done in order to keep the skein in its undisturbed condition with all strands as parallel as possible.

5. Allow the lower spool to continue the downward movement until the skein breaks. (The skein will not break completely, but a sufficient number of ends will rupture to cause slippage of the yarn. When this occurs, the pendulum will stop.)
6. Press the return lever to reverse the machine, which will bring the lower or actuating jaw back to its pre-set position. When many tests are to be made on a yarn, the collar on the return lever can be set to a proper position, so that reversal is automatic
7. Record the strength in pounds from the proper scale on the dial. (If only the top weight is on the pendulum, read the inside scale. If both weights are on the pendulum, read the outside scale.) Note: If the break does not occur when the pendulum is between a 9° and 45° angle with the vertical, change the capacity of the instrument or use another tester with the proper load range.

When making skein tests for both strength and yarn number, the following procedure should be used:

1. Place the skein on the tester and start the machine.
2. Allow the skein to break and reverse movement of lower spool either manually or automatically.
3. As soon as the tension is released on the skein, remove the skein and place it on the weighing hook of the quadrant balance.
4. Read and record the strength from the dial of the tensile tester.
5. Release the pendulum pawls and return the pendulum to zero position.
6. Place the new skein on spools of the tester and start the downward movement of the lower spool, keeping manually applied tension on the skein until the machine picks up the load.
7. Read and record the yarn number on the skein from the quadrant balance.
8. Remove and discard the skein from the balance.
9. Repeat steps as outlined above for all skeins to be tested.

The object of this particular procedure is to reduce the waiting time of the operator to a minimum. While one skein is being tested, the yarn number of the skein on the balance is being read and recorded. While the lower spool is returning to the starting position, the skein just broken is being removed from the strength tester and placed on the balance. The strength of the skein is recorded and the pendulum returned to the zero position. In following this system, the waiting time of both the machine and operator is reduced to a minimum.

Calculations. The average skein strength for the skeins of cotton yarn tested may be corrected to a specified yarn number by the following formula:

$$S_2 = \frac{C_1 S_1 - (C_2 - C_1)(21.7)}{C_2}$$

where S_2 = Adjusted strength; S_1 = Observed strength; C_2 = Specified yarn number; C_1 = Observed yarn number.

A detailed discussion of the formula for getting the strength at a given yarn number from the actual strength at some actual yarn number is discussed in Chapter 11. It has been pointed out that the correction factor of 21.7, which is approved by A.S.T.M., can be replaced by an adjusted value of 18.27 to give more accurate results, so that the formula becomes:

$$S_2 = \frac{C_1 S_1 - (C_2 - C_1)(18.27)}{C_2}$$

In effect, the correction value 18.27 means that for each unit change in yarn number, the count-strength product increases by 18.27 units (for coarser yarn) or decreases by 18.27 units (for finer yarn). The correction value of 18.27 applies to use only for American Uplands cotton in carded yarns, and is an average value only. Individually, yarns from different cottons will reflect in different slopes. A rounded value of 18.3 will be found to be sufficiently accurate for mill use.

For staple synthetics, it will be found that other correction values will be found. In general, these values are smaller for the fine deniers, and larger for the coarse deniers. The authors have found this variation to extend from as low as 16 on certain $1\frac{1}{2}$ denier to over 30 for 3.0 denier, to nearly 60 for $5\frac{1}{2}$ denier on one make of viscose staple. These values are given solely to indicate the trend to be expected with different deniers; they will be found to vary considerably with the different types of synthetic fibers.

The most common method of expressing skein strength of cotton and staple synthetic yarns is in the form of the count-strength product, which is also called "break factor" and "product, yarn number and pounds lea strength." This is calculated by multiplying the average skein strength by the average yarn number. Either the adjusted skein strength and the specified yarn number or the observed strength and number may be used for the calculations, whichever is of greater interest.

EXAMPLE.

$$\text{average skein strength} = 70.6 \text{ lbs.}$$
$$\text{average yarn number} = 31.2$$
$$\text{Count strength product} = (70.6)(31.2) = 2202.7$$

The yarn strength for 30/1 yarn is estimated to be

$$S = \frac{(70.6)(31.2) - (30 - 31.2)(18.3)}{30}$$
$$= \frac{2202.7 + 22.0}{30} = 74.2$$

The count-strength product which is normally not associated with any units is actually the "breaking length" with units of hanks.

$$\text{Count Strength Product} = (\text{Yarn Number})(\text{Pounds})$$

Since

$$\text{Yarn number} = \frac{(\text{Hanks})}{(\text{Pounds})}$$

then

$$\text{Count strength product} = \frac{(\text{Hanks})}{(\text{Pounds})}(\text{Pounds})$$
$$= \text{Hanks}$$

Thus the count-strength product is actually the number of hanks the yarn can support from the standpoint of strength. For example, if a 20/1 yarn has a count-strength product of 2000 hanks, it means a skein of the 20/1 yarn could support the weight of 2000 hanks of the same yarn.

Frequency and Number of Skein Tests. The frequency of skein strength testing should be daily on each shift. The number of bobbins to select per yarn number per shift depends upon the amount of each yarn number produced with reference to the total production of the spinning department. The sampling should be stratified with respect to the proportion of the production represented by each yarn number; however, once the number of bobbins per yarn number has been established, they should then be selected at random. At least three and not more than five breaks should be made per bobbin. Either four or five breaks per bobbin are generally preferred since that forms a convenient subgroup size when using the statistical control chart technique. It is also preferable to make four or five breaks from several bobbins than to make fifteen or twenty breaks from a few bobbins. In gathering the bobbins to test, it is recommended that the sample be collected at least twice per shift and preferably three times in preference to doing all of the sampling at one time. This is discussed in detail under "Number of Tests" in the yarn numbering section.

Pendulum Testers: Suter Types

Two types of pendulum testers are produced by Suter: one is motor-driven and the other is activated by a dead-weight system. See Fig. 21.3.

The motor-driven tester is equipped with a drum-type recorder and can be adapted for testing cord, skeins, and fabric. The maximum capacity for this particular type of tester is 400 pounds. In general, the procedures for testing given for the Scott machines cover the operation of these.

The dead weight type of Suter tester is used primarily for single strand yarn strength and elongation and has been used very extensively by the manufacturers of continuous filament yarns for quality control work. The instrument is available in several different capacities with the maximum load being equal to 20 pounds or the equivalent in grams. The older models were adjusted for a sample length of $18\frac{3}{16}$ inches, and in many cases these models were reworked for a 20-inch sample of yarn. The later models are generally designed for a gauge length of 10 inches.

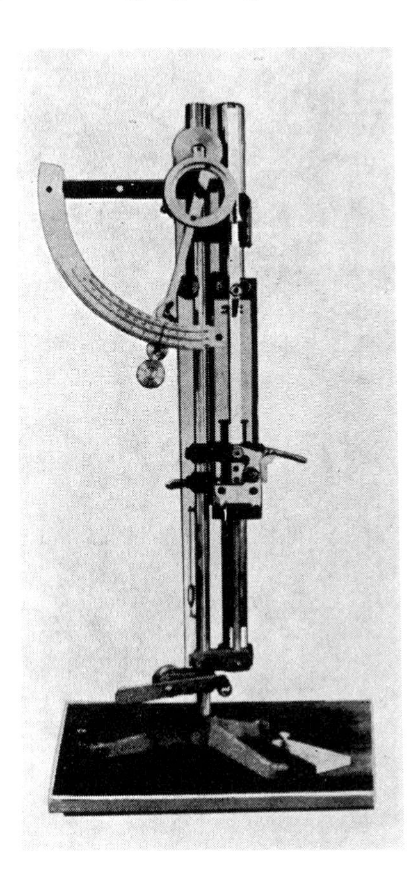

Fig. 21.3. Suter single end pendulum tester.

These instruments are primarily manual type because the motion of the actuating jaw is by a dead-weight system, rather than by a motor drive. The weight that carries the lower actuating jaw moves in a vertical axis, and during this movement it forces a piston which is attached to the weight into a cylinder containing a lightweight oil, the entire mechanism acting as a dash pot. The rate of descent of the weight, together with the lower jaw, can be controlled by a knurled screw adjustment which controls the rate of flow of the oil in by-passing the piston.

One feature of this instrument that is found useful in many

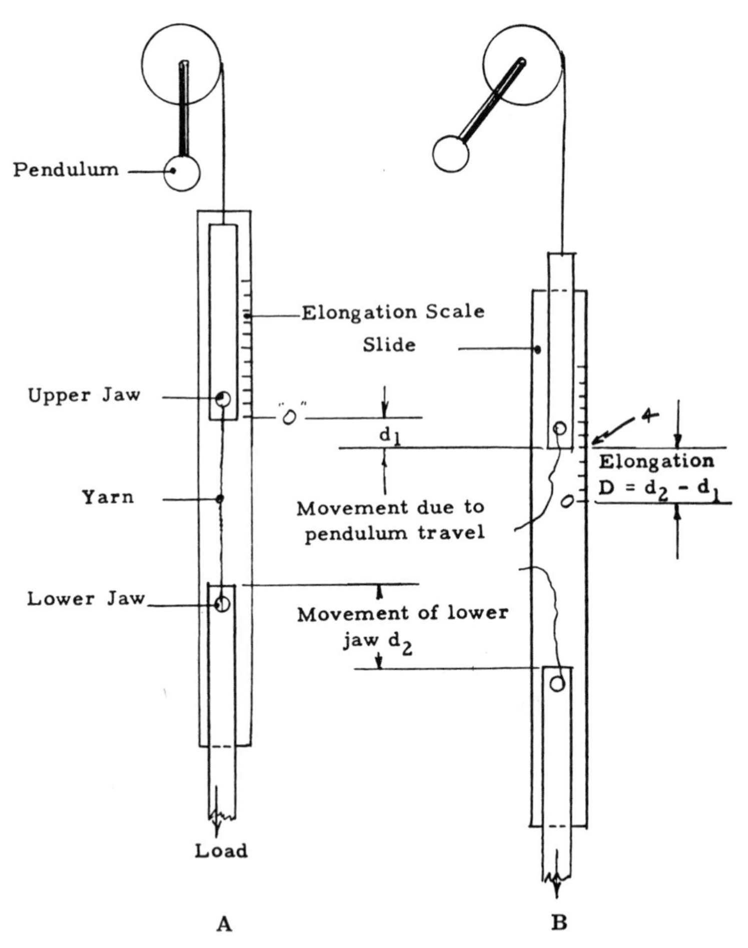

Fig. 21.4. Suter single strand strength and elongation tester. A: relative positions at start of test; B: relative positions at yarn rupture.

laboratories is the automatic elongation scale which gives a direct reading of the per cent elongation as well as actual elongation.

The calibration of the per cent elongation scale is based on some predetermined gauge length, and therefore care should be taken to be sure that the gauge length corresponds to the scale being used. The operation of the elongation scale is illustrated diagrammatically in Fig. 21.4, in which the relative positions of the jaws and elongation scale slide are shown for the start of a test and for the end of a test at yarn rupture. The elongation scale is a counterweighted balanced slide which is positioned at the start of a test so that the lower part of the upper jaw indicates zero. When the lower jaw is released so that the load is applied by movement of the weights, the elongation scale is carried intimately downward with it, so that during the test the elongation scale and the lower jaw move precisely the same amount. Immediately at rupture of the yarn, a toggle arrangement releases the connection between the lower jaw and the elongation scale so that the scale comes to rest, whereas the lower jaw can continue to move downward until it reaches its maximum traverse position. As a result of this arrangement, the movement of the upper jaw due to the pendulum swing is compensated for, and the net elongation shown at D is equal to the total movement of the lower jaw d_2, minus the movement due to pendulum travel d_1. Obviously, this feature is very valuable in routine quality control work.

One of the main disadvantages of this particular instrument is that the manual operation poses quite a burden on the technician, for at the completion of each individual strength test, the lower jaw with its weighting system must be returned to its normal zero position by hand, either by lifting directly or by cranking it up by an arrangement provided on certain models.

Pendulum Tester: Moscrop

The Moscrop tester, manufactured by Cook and Company, Manchester, England, also uses the pendulum system of loading. This instrument is designed so as to break six ends of yarn simultaneously and record the strength and elongation of each yarn separately. The tester will make eighty breaks from each of the six packages of yarn placed on the machine during one complete cycle. It will reject the broken ends and automatically place the new test sample in the jaws for testing.

The recording system consists of a platen for holding the recording paper and six pins, one for each testing position, which punches a small hole in the paper after each test.

The instrument is supplied with several sets of pendulum weights for changing the capacity. The maximum load that can be applied is 64 ounces.

A system was developed by Campbell and Field (1) for analyzing charts from the Moscrop, which gives a reliable estimate of the average and standard deviation without the use of any laborious or time-consuming methods. Their technique is based on the assumption that the results of the strength tests will form a normal distribution. On this basis, approximately 68 per cent of the results will be within the limits of the average plus and minus one standard deviation. This means that the remainder, or 32 per cent, will be outside these limits with 16 per cent of the eighty, or thirteen above and thirteen below. Therefore, if the two points are located on the chart above and below which, respectively, thirteen observations lie, the one sigma lines are located at these points In other words, the lowest thirteen points are counted and a line drawn at that point, and a corresponding line is drawn below the top thirteen points. The difference between these two points divided by two is the standard deviation, and the lower point plus the standard deviation is the mean or average line. With these two values, it is then possible to calculate additional values such as standard error and coefficient of variation.

By making the above calculations for each bobbin, it is also possible to evaluate statistically the average values for all six bobbins which constitute a complete test.

In applying these techniques, any impressions made on the chart that are erroneous due to such malfunction as yarn slippage in the jaws should be disregarded. In that case, the number of values which are in error are subtracted from 80 and the remainder divided proportionately to the figures discussed above.

Inclined Plane Type of Testers

Scott Machine

Principle of Operation. The principle of operation of inclined plane testers is shown in Fig. 21.5. With the tracks of the inclined plane tester in a perfectly level position, there is no load or strain being applied to the test specimen. However, as the track

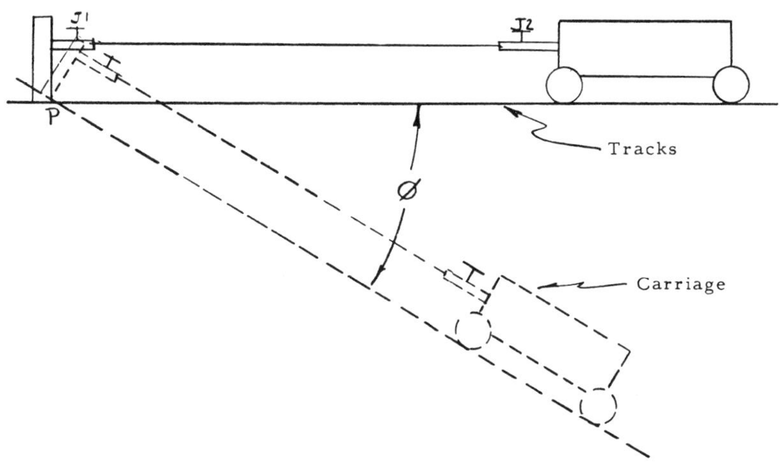

Fig. 21.5. Diagram illustrating the principle of loading of an inclined plane type of testing machine.

begins its movement through the angle ϕ, load is applied to the yarn, and neglecting friction, the load applied varies directly with the sine of the angle ϕ. For this reason, inclined plane testers are considered constant-rate-of-loading machines; however, for most materials this type of instrument is not actually a constant-rate-of-loading tester. This is due to the fact that most textile materials do not obey Hooke's Law for the complete test. (The strains produced when stresses below the elastic limit are applied are proportional to the stress intensities.)

The inclined plane type of testers have a small amount of inertia and friction to overcome, and in doing so, set up an oscillation in the test specimen. This oscillation is recorded at the very beginning of the stress-strain curve in the form of a diminishing sine curve; however, it is usually overcome within the first few small squares on the chart, and for this reason does not have any great influence on the test results.

The load on the yarn may be calculated by the following relationship:

$$\text{Load on yarn} = (\text{Weight of carriage})(\sin \phi)$$

with ϕ being the angle through which the carriage has passed. For example, if the actual weight of the carriage is 591.5 grams, and it has moved through a 20° angle, what load neglecting friction is being applied to the yarn?

$$\begin{aligned} \text{Load on yarn} &= (\text{Weight of carriage})(\sin \phi) \\ &= (591.5)(0.34202) = 202.3 \text{ grams} \end{aligned}$$

From the above relationship, it can be seen that the load on the yarn varies as the sine of ϕ since the weight of the carriage remains constant, and since the sine of ϕ increases at a constant rate with respect to time, the load is applied at a constant rate.

Calibration. The first step in the calibration of the Scott inclined plane machine is to weigh the carriage complete with all attachments that ride with it. For each range to a given maximum load, there is an individual weight to be attached to the carriage. These weights are stamped with the maximum capacity of load attainable; this is the rated weight or capacity in each case. This rated weight will not be the same as the actual weight of the carriage plus accessories, for the carriage moves through a maximum angle on the model IP-2 machine of 25°. Therefore, the maximum rated load is equal to the actual weight of the carriage multiplied by the sine of 25°. Listed in Table 21.1 are the rated carriage weights with the corresponding actual weights of the carriage plus accessories.

TABLE 21.1
Carriage Weights, Rated and Actual

Rated capacity, grams	Actual weight, grams
250	591.5
500	1183.1
1000	2366.2
2000	4732.4

On the IP-4, the maximum angle is 30°, which means that the rated weight of the carriage would always be exactly one-half of the actual carriage weight. With this exception, the same principles apply to the IP-4 as for the IP-2.

After determining that the carriage weight is correct, see that the rims of the wheels and tracks are smooth and free of all dirt, rust, and so forth. Place the carriage on the track midway of its run. If the carriage does not remain stationary in this position, it means that the tracks are not level. If the tracks are not level, place the carriage at the extreme left end of the track and place a spirit level on the exposed section of the track and adjust the position of the track until it is level.

Adjust the pen to rest in the zero horizontal on the chart. Then start the plane inclining. The line drawn will start vertically indicating combined starting friction and inertia, but should move away from the vertical within the first two small spaces in the chart to indicate a satisfactory calibration. If it does not, proceed as follows:

1. See that the pen point is in good mechanical condition and sliding freely.
2. With commercial solvent and a soft rag, clean foreign material from the wheels and track.
3. Check tracking of the wheels.
4. Remove the wheels and wash ball bearings; repack per instructions.
5. Plain-bearing wheels: Check condition of pivots and indentation axle and point in frame.
6. In replacing either type bearing, take care not to restrict the rotation of the wheel.
7. Check the track alignment; tracks must be parallel and in the same plane.
8. Make sure that the carriage is in a perfectly vertical position.

Procedure for Use of Scott IP-2 Type of Testers

1. Set the jaws with 10 inches between the inside of the jaws. This is a critical measurement since it is the basis for calculating elongation.

2. Insert the pen in the carriage and place the chart in the proper place. The chart should be aligned with stop pins at the bottom and left side of the platen. The pen should be adjusted to the bottom horizontal line and the vertical line on the extreme left of the chart. This is an important step; if the chart is not in the proper position, the strength and elongation measurements will both be in error. To check the stop pins, place a chart in the position as described above and push the chart along the track without starting the machine. The line drawn by the pen should be superimposed on the bottom line of the chart. To check the vertical line, put the carriage in the starting position, and the pen should draw a line superimposed on the vertical line of the chart.
3. Place the proper weight on the carriage. The weight should be selected so that the specimen will break when the pen is between the seventh and ninth heavy line from the bottom of the chart. When the proper weight is selected, record the rated weight at the top of the chart.
4. Adjust the weight on the tension arm for the desired pre-test tension load. This is necessary so as to have uniform pretensioning of all test specimens.
5. Take the yarn from the side of the package and fasten between the clamps of the pretensioning device. Do not disturb the twist of the yarn.
6. Pull the length of the yarn through both jaws, and just far enough to pull the clamp on the tension device to a vertical position.
7. Fasten the movable jaw first and then fasten the stationary jaw. The sample is now ready for testing.
8. Place the machine in operation by turning the control knob to the right. After the machine starts, turn the knob back to the original position. This will give the carriage a quicker return action.
9. After the break occurs, pull the lever to the left. The machine will return to the original position.
10. If a series of tests are to be made on the same chart in the original starting position, it may be found helpful to number the

curves at the breaking point to facilitate reading the chart. It has also been found helpful to record the strength in grams and the stretch in inches at the time of the break to reduce the time necessary to read the values from the charts.

11. If the break occurs in the area immediately adjacent to the jaw, it should be classified as a "jaw break" and be recorded as such. If jaw breaks occur frequently, the condition of the jaws should be examined or possibly replaced; however, if only an occasional jaw break is encountered, it can be disregarded. Since a strand of yarn will always break at its weakest point and there is no method for controlling the position of the weak spots with respect to the jaws, the law of probability will account for a certain number of breaks in the vicinity of each jaw. Based on this assumption, if the break is within the prescribed limits, it is reasonable to assume that the jaws did not affect the break and the result can be accepted.

Uster Fully-Automatic Tester

The Uster automatic yarn strength tester, shown in Fig. 21.6 and developed by Zellweger Limited of Switzerland, has been designed to make a series of tests on any one package automatically. The machine records graphically the elongation and strength for each individual test, totals the individual elongations and strengths, records the number of tests performed, and gives the means to provide a graphic picture of the strength distribution.

Principle of Operation. The Uster automatic tester uses the inclined plane principle of loading, in which the pull on the specimen sample increases proportionately with time. With this instrument, the rate of loading can be set so that the maximum load is reached in from ten to ninety seconds.

The gauge length of the sample tested is 500 mm., which is just shy of 20 inches in length. In contrast, the usual standard for American practice is for a sample length of 10 inches. However, the use of this long length should not be detrimental for the analysis of results from either routine testing or from research programs. The machine can be set so that there is an automatic pretensioning of the sample before the load is applied, with a pretensioning within the limits of 0.5 to 10 grams. This range is sufficient for any normal yarn and has been designed to meet the European practice of applying a pretension equivalent to the weight of a 100 to 500-meter length of the yarn under test.

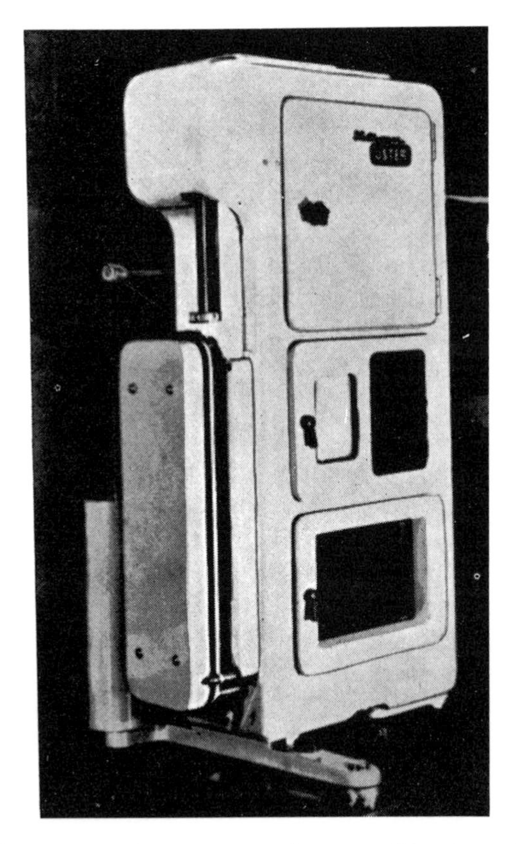

Fig. 21.6. Uster automatic yarn strength tester. (Courtesy of Uster Corporation.)

The instrument can be set so that it will make any one of several different series of tests automatically. At the completion of the final test in the series, the machine stops automatically. The number of different tests that can be made are 20, 40, 60, 80, 100, etc., up to 200. At the completion of the final test, a repeat button can be actuated to give a complete repetition without resetting any of the controls. A 4-digit counter totals the number of tests.

The machine can be adjusted for six load capacities, which are attained by using different combinations of weights on the inclined plane carriage. These load capacities are from zero in each case up to 200, 400, 600, 1000, 1500, and 2000 grams.

The breaking strengths for each individual test are totaled on a 4-digit counter so that at the end of a series of tests the sum of the breaking loads can be tabulated. In order to take care of the differences in load capacity, a reading of 10 on the 4-digit counter gives the standard equal to the load capacity selected. For example, if the machine is set for a load capacity of 600, then a reading of 10 would indicate a load of 600 grams, or a reading of 6 would indicate a load of 360 grams. In addition to giving the totals of the strength, the breaking strength for each individual test is recorded on a ruled paper strip with the length of the line indicating the amount of the load. See Fig. 21.7.

In a manner similar to that for strength, elongation is also recorded. Another 4-digit counter totals the sum of the elongations for the series of tests, with a reading of 1 indicating a 10 per cent elongation. The maximum capacity of the machine for elongation at rupture is 40 per cent. Yarns with an elongation at rupture greater than this will cause the machine to stop automatically. Elongation is indicated also for each individual test by the autographed line on a ruled paper strip. In practice, it is wise to use different colored inks for the elongation and strength autographs in order to distinguish between them.

The grips that hold the yarn sample are designed to hold the yarn firmly without slippage and at the same time prevent damage to the yarn. These grip or jaw heads have changeable packings that accommodate different types of yarn. For fine yarn and synthetics, soft rubber packings are used, and for the coarse or high-strength yarns, either fiber or steel packing is used.

One of the most interesting features of the machine is the frequency distribution device. At the completion of each individual test a small ball is fed into the proper vertical groove in a plate

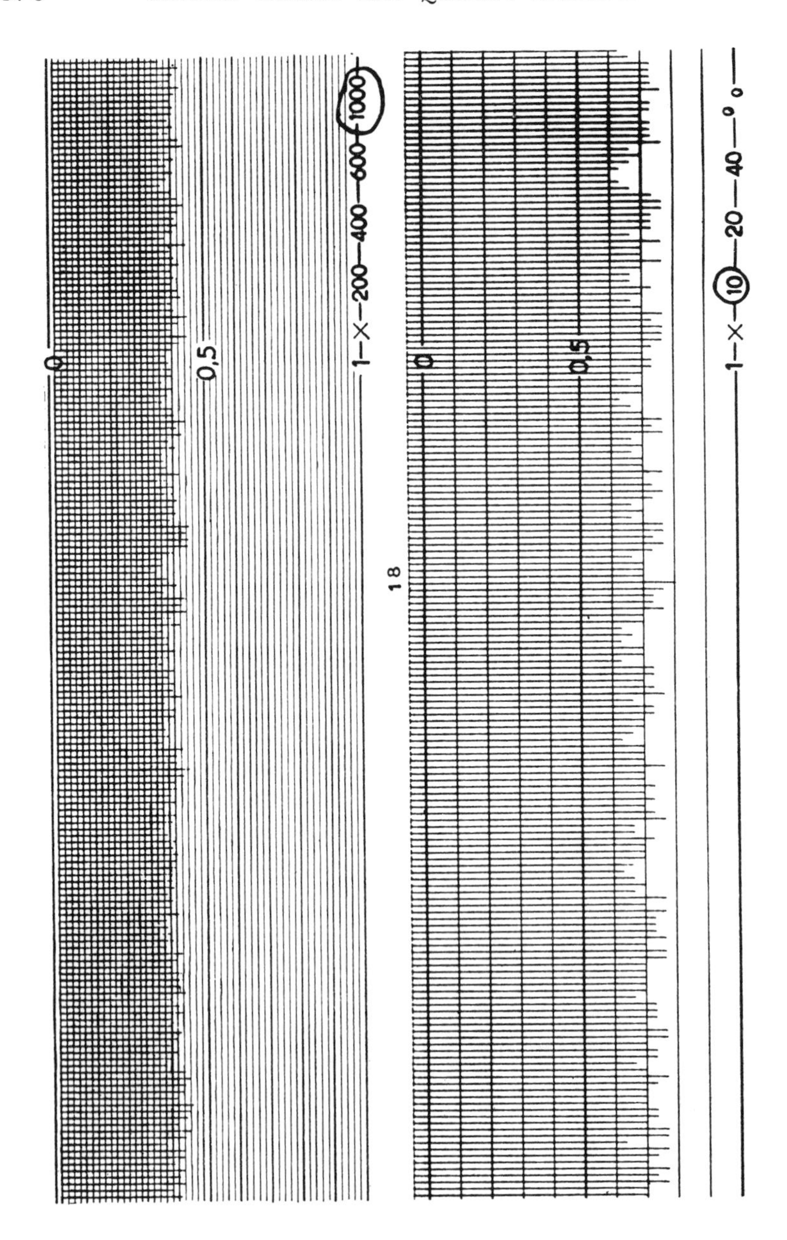

Fig. 21.7. Breaking strength and elongation autographs from Uster automatic yarn tester.

corresponding to its proper breaking strength. There are 100 grooves provided so the load interval is $\frac{1}{100}$th of the range being used. At the completion of any series of tests, the frequency distribution diagram formed by these balls in their respective grooves can be transferred onto a ruled chart and the diagram analyzed to give several valuable measurements.

Operation of Machine. In making a test, the bobbin is placed on the spindle provided and the end properly threaded. Then the machine is set to give the desired number of tests so that the machine will stop automatically on completion of these. Then either knowing or being able to estimate the average amount of elongation, the desired measuring range for elongation is set. Elongation is set in ranges of 0 to 10, 0 to 20, and 0 to 40 per cent. The rate of increase of load is set. Next, the load capacity of the machine is established. With no additional weights, this is 200 grams. Therefore, additional weights must be added to give the proper range desired. All three of the counters should be reset to zero, and the frequency distribution chart should be clear of any balls. The main switch is thrown to an indicator marked "I". This starts the motor and simultaneously causes a buzzing sound. The noise is stopped by pushing the starting knob of the machine. If anything goes wrong during the test, this buzzing sound gives notice of trouble. For example, the machine will stop automatically if no yarn is rethreaded through.

At the completion of the desired number of tests, the machine stops automatically, at which time the operator should cut the current. The readings on the different counters for the number of tests, the sum of the breaking strengths, and the sum of the elongations should be recorded. The diagram strips for both elongation and strength are removed from the machine. Finally, a copy of the frequency distribution diagram is made. To do this, the cover is opened and the recording plate is swung out into an inclined position. Then a diagram sheet is placed properly over the plate, clamped in place, the glass cover plate raised temporarily, and a copying stone rubbed over the surface of the paper. By doing this, the arrangement of the individual balls is transferred onto the paper so that it becomes visible as the frequency diagram.

Following this, the glass plate is returned to its position by a technique described by the manufacturers, the small balls removed from the plate and returned to the machine proper for re-use.

Calculations

Average Breaking Strength and Elongation

The calculation of the average breaking strength and the per cent elongation can be performed from machine readings in the following way.

ΣB = Summation of breaks as read from counter = 749
ΣE = Summation of elongation as read from counter = 143
N = Number of breaks = 200
R = Range of breaking load scale in grams = 1000
R_e = Range of elongation scale in per cent = 10
K = Breaking load constant = 2.45
e = Elongation constant = 0

1. Average breaking load, grams $= R\left(\frac{\Sigma B}{10N} + \frac{K}{100}\right)$

$$= 1000\left(\frac{749}{2000} + \frac{2.45}{100}\right)$$

$$= 1000(.3745 + .0245)$$

$$= 1000(.3990) = 399 \text{ grams}$$

2. Average elongation, % $= \frac{10\Sigma E}{N} + e$

$$= \frac{10(143)}{200} + 0$$

$$= \frac{1430}{200} + 0 = 7.15\ \%$$

Short Method for Determining Standard Deviation using Plastic Ruler ($8\frac{1}{4}$" × $1\frac{3}{4}$") With Knob

1. From the bottom of the frequency diagram Fig. 21.8 mark off 5 % of the balls from the left side and 5 % of the balls from the

right side with a red pencil. Measure the distance between the two red lines in units from the red (5 %) scale of the ruler.

2. Mark off 10 % of the balls from the left side and 10 % of the balls from the right side with a blue pencil. Measure the distance between the two blue lines in units from the blue (10%) scale of the ruler.

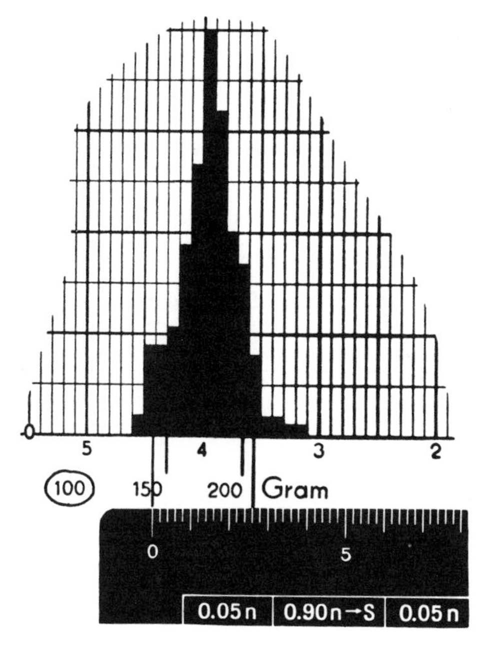

Fig. 21.8. Frequency diagram and analysis ruler. Uster automatic yarn tester.

3. Add together the units measured from the red (5 %) and blue (10 %) scales and obtain an average. This average is standard deviation expressed as a per cent of the range, indicated as X in the following equation.

$$\text{Standard deviation in grams} = \left(\frac{X}{100}\right) R$$

Where X = Measured base of chart, i.e., standard deviation in per cent of breaking range; R = Range of breaking load scale.

$$\text{Standard deviation, grams} = \left(\frac{2.6}{100}\right) 1000 = 26 \text{ grams}$$

Coefficient of variation of breaking load

$$= \frac{\text{Standard deviation in grams}}{\text{Average break}} (100) = \frac{26}{399} (100) = 6.52\ \%$$

Standard Deviation for Breaking Strength

The standard deviation of the breaking strength can be calculated by the grouped frequency method. Table 21.2 shows the method of tabulating the data. The nominal average used in the tabulation is zero. D = deviation from zero; F = frequency of deviation; W = grams per interval (10 grams for 1000-gm. carriage).

TABLE 21.2

Standard Deviation for Breaking Strength

D	F	FD	FD^2
−8	1	−8	64
−7	1	−7	49
−6	2	−12	72
−5	2	−10	50
−4	8	−32	128
−3	17	−51	153
−2	20	−40	80
−1	32	−32	32
0	40	0	0
1	27	27	27
2	19	38	76
3	11	33	99
4	9	36	144
5	9	45	225
6	2	12	72
	$\Sigma = 200$	−1	1271

$$\sigma = \sqrt{\frac{\Sigma FD^2}{n} - \left(\frac{\Sigma FD}{n}\right)^2}$$

$$\frac{\Sigma FD^2}{n} = \frac{1271}{200} = 6.355$$

$$\left(\frac{\Sigma FD}{n}\right)^2 = \left(\frac{-1}{200}\right)^2 = (0.005)^2 = 0.000025$$

$$\sigma = \sqrt{6.355 - (0.000025)} = 2.52$$

$$\sigma \text{ in grams} = \sigma W = 2.52(10)$$
$$= 25.2$$

$$\text{Actual average} = \text{nominal average} + \frac{\Sigma FD}{n}(W)$$
$$= 400 + \left(\frac{-1}{200}\right)(10)$$
$$= 399.95 \text{ grams}$$

$$\text{Coefficient of variation, } V = \frac{\sigma}{\bar{X}}(100)$$
$$= \frac{25.2}{399.95}(100)$$
$$= 6.30\ \%$$

Standard Deviation for Per Cent Elongation

The calculation by the grouped frequency method is shown below. Data are given in Table 21.3, in which the nominal elongation is zero. D = deviation; f = frequency of deviation; W = units of elongation per interval (0.3 % in this case).

TABLE 21.3

Standard Deviation for Per Cent Elongation

% elongation	D	F	FD	FD^2
6.1	−3	1	− 3	9
6.4	−2	11	−22	44
6.7	−1	30	−30	30
7.0	0	62	0	0
7.3	1	54	54	54
7.6	2	41	82	164
7.9	3	1	3	9
		Σ = 200	+84	310

$$\sigma = \sqrt{\frac{\Sigma FD^2}{n} - \left(\frac{\Sigma FD}{n}\right)^2}$$

$$\frac{\Sigma FD^2}{n} = \frac{310}{200} = 1.55$$

$$\left(\frac{\Sigma FD}{n}\right)^2 = \left(\frac{84}{200}\right)^2 = (0.42)^2 = 0.1764$$

$$\sigma = \sqrt{1.55 - 0.1764} = \sqrt{1.3736} = 1.172$$

$$\begin{aligned} \sigma \text{ in \% elongation} &= \sigma(W) \\ &= 1.172(0.3) \\ &= 0.3516 \end{aligned}$$

$$\begin{aligned} \text{Actual \% elongation} &= \text{nominal per cent elongation} + \left(\frac{\Sigma FD}{n}\right)W \\ &= 7.0 + \left(\frac{84}{200}\right)0.3 \\ &= 7.0 + (0.42)0.3 = 7.0 + 0.126 \end{aligned}$$

$$\text{Actual \% elongation} = 7.126\ \%$$

$$\begin{aligned} \text{Coefficient of variation, } V &= \frac{\sigma}{\bar{X}}(100) \\ &= \frac{0.3516}{7.126}(100) \\ &= 4.93\ \% \end{aligned}$$

Strain Gauge Testers

The recent introduction of the electronic strain gauge to the testing, quality control, and research phases of the textile industry has resulted in the measuring and investigating of certain phenomena previously beyond the scope of existing equipment. The strain gauge has been applied very successfully to such instruments as strength, impact, compression, and resiliency testers. The initial research in the use and application of the strain gauge was conducted at the Massachusetts Institute of Technology.

The gauge is based on the principle that metal wire changes resistance when placed under a tensile or compressional force.

This change in resistance is converted to a voltage change by the use of certain electrical circuits and recorded. The wire used must have a linear resistance-strain relationship and be stable with respect to temperature and relative humidity. The wires are covered with either paper or Bakelite for insulation and then cemented to a suitable weight bar. The member to which the wire is cemented is usually either a metal bar or brass tube. With force being applied to the weight bar, deformation occurs and a corresponding strain is placed on the wire in the gauge, thus a change in the electrical resistance of the wire. This system has been proven to be extremely sensitive to loads and load changes, and with only a negligible amount of inertia.

Instron Tester

The Instron tensile tester is one of the finest and most versatile instruments developed using the strain-gauge system of measuring load for use in textile laboratories. The floor model is shown in Fig. 21.9. More recently a table model has been developed and put on the market. The Inston is a valuable instrument for both the fundamental and applied research laboratory due to its versatility and the wide range of tests it can perform. However, because of the relative complexity of operation of the instrument and the high initial cost, the floor models are more suitable for research investigations than for routine quality control testing. It is very probable that the less complex table models will find wider application than the floor model for use in routine work.

The Instron floor model can be obtained with a number of different tensile load cells so that a wide range of textile materials can be tested. These load cells range as follows: 10 to 50 grams, 100 to 2,000 grams, 1 to 50 pounds, and 10 to 1,000 pounds. In addition, another cell is available for certain models with a range of from 100 up to 5,000 pounds.

The machine can be set so that different gauge lengths can be used. The rate of jaw separation can be controlled over a wide range, specifically from 0.02 up to 20 inches per minute, with rapid return of the lower jaw if desired. The machine can be set for repeated tests to give hysteresis and relaxation measurements between adjustable extension points. These points can be set so as to reflect a minimum and a maximum extension, or a minimum and maximum load, or combinations of the two. This equipment enables the instrument to be used for performing a wide variety of

Fig. 21.9. Instron tensile tester.

tests, such as strength with different gauge lengths and with different rates of loading, tearing, flexing, and repeated loading and unloading within any given range below the rupture point.

The instrument is equipped with a Leeds and Northrup high-speed graphic recorder with a special gear box for chart speed changes. Controls are provided to make it possible to have automatic reversal of direction of the Leeds and Northrup strip chart to follow the up and down motion of the moving cross head. With this extra equipment for the recording system, it is possible to plot conventional hysteresis diagrams. The machine can be set to use any one of three methods of operating the chart: first as a function of time, secondly to follow the cross-head motion, and finally, to follow the cross-head motion but always in the same direction. As with the autographed charts from the inclined plane testers similar to the IP-2, the chart can be analyzed for stress-strain characteristics.

Different types of jaws are available, depending upon the use desired. These jaws range from heavy units designed for fabric down to very light units with a 50-gram capacity used for testing small fibers, wires, and so on. In testing single fibers, it will be found that special techniques will have to be adapted in order to handle the fibers. This can be very exacting work, and time-consuming as well. One of the original techniques involved the aligning of the fibers on a piece of cardboard and scotch-taping them individually at top and bottom. Then the original fibers could be cut away from the group, the paper backing removed, and the test made. However, at the School of Textiles, N. C. State College, a vacuum system has been developed which greatly simplifies the handling of single fibers for testing on the Instron, as shown in Fig. 21.10. In this system, the first need is a vacuum pump from which a rubber tube leads to a small stand adjacent to the tester. An upper jaw addition made of a metal tube and hook arrangement is placed in the vacuum line holder. The technician picks up a single fiber with tweezers and holds the end of it near the open end of this tube so that the vacuum draws the end into the jaw. Then a small tapered plug can be inserted in the tube to lock the fiber in place. This unit can then be removed from the vacuum line and hung on a hook on the upper recording head on the tester, as shown. Attached to the lower jaw is a similar arrangement with a permanent suction hose. The technician teases the end of the single fiber near the opening of this lower jaw; it is then sucked into the jaw and locked in place with another small tapered pin. The instrument is then ready for the test.

An automatic integrator can be obtained to use with this tester. As applied to the Instron tester, the area under the load deforma-

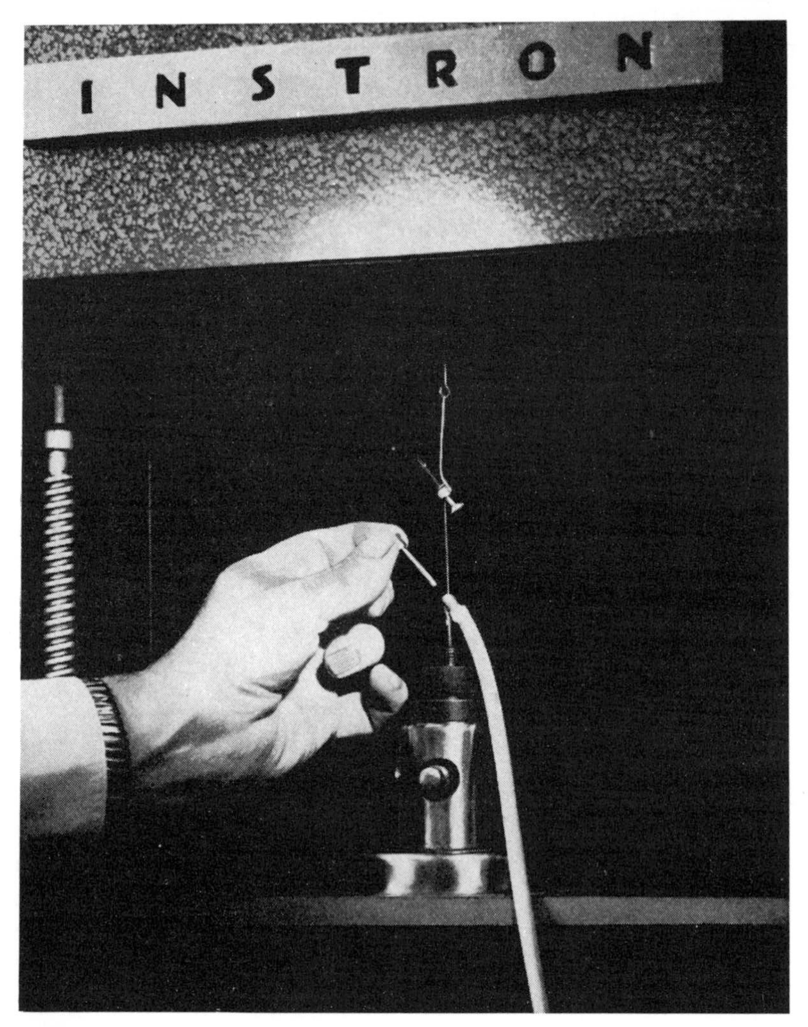

Fig. 21.10. N. C. State College vacuum jaws for the Instron tester.

tion curve, which is equivalent to the energy either absorbed or returned by the test sample, is available from the counter readings at the conclusion of each test.

Instron Table Model. The table model Instron has many of the features of the larger units described above, but is less versatile. Because of its compactness and relatively lower cost, it lends itself to routine quality and production control needs without losing its effectiveness for research work. As with all strain gauge testers, the inertialess weighing results in a rapid response to changes in load on the sample with the added accuracy that accrues to this method. In addition, the absence of mechanical friction so inherent to pendulum systems adds to the accuracy of results.

The full load capacity of this unit ranges from 10 grams to 100 pounds through choice of the load cells. Crosshead speeds can be varied over a wide range, with standard speeds from 0.2 to 50 inches per minute with eight different speeds or with low speeds (in alternate models) from 0.02 to 5.0 inches per minute. Rapid return traverse speeds over nearly the entire range are also available. The maximum crosshead travel of 33 inches makes these units suitable for nearly all types of textile tensile tests. As with the larger models, these table units have a high-speed graphic recorder, with chart speeds matching those of the standard speeds of the crosshead.

Other Strain Gauge Testers

Thwing-Albert Instrument Company. In addition to Instron models, there are also other suppliers for this type of equipment. The Thwing-Albert Instrument Company has developed an "Electro-Hydraulic Tensile Tester" that utilizes the electronic weighing cell. This is available with capacities ranging from one gram to 2,000 pounds with special models ranging up to 10,000 pounds.

The pulling force for the tester is provided by a hydraulic flow control system which is equipped for easy adjustment of gauge lengths of 12, 18, 24, and 30 inches. The rate of extension is also adjustable from .10 inch per minute to 20 inches per minute. Special controls are available for obtaining rates from 0.20 to 250 inches per minute.

Scott Testers, Incorporated. The Scott Testers, Incorporated, have available an electronic measuring head and recorder that is adaptable to their pendulum testers. The advantage of the type head is the response as compared to the pendulum system of loading.

References

1. Campbell, M. E. and Field, G. W.

CHAPTER XXII

Analysis of Strength Measurements

Methods of Expressing Strength

Tensile Strength. Tensile strength is defined as the force or load per unit area required to rupture a material. There are many different units used throughout the world in the textile industry to express strength. Basically, in English-speaking areas, the units most commonly used for fibers are pounds per square inch. Thus, the units used in the "bundle method" are pounds per square inch. In addition, the results of the Pressley test (p. 189) are converted to pounds per square inch. A cotton showing as a result of such tests a tensile strength of 80,000 pounds per square inch would be considered average.

At the same time, the strength of cotton can be expressed as having a certain "Pressley index" (p. 193) from which is calculated the pounds per square inch. For instance, a Pressley index of 7.8 corresponds approximately to 85,000 pounds per square inch. More recently a fiber strength index has been used to express the level of the cotton fiber strength, with an index of 100 indicating average strength.

The Clemson index has the same value as the Pressley index, the only difference being that the tests have been made on the Scott Clemson tester.

The strength of both spun and continuous filament yarns is not expressed as pounds per square inch. It is more accepted to use other measurements as described below.

Skein Strength. The strength of yarn in the skein form, with the skein being made up of 80 wraps of yarn totaling 120 yards, has been discussed in the previous chapter. The results are expressed as pounds in most English-speaking areas, although metric units are also used. Both cotton and spun synthetic yarns have their strengths expressed this way.

Count-Strength Product. It is customary to convert skein strength to a count-strength product in order to make comparisons between yarns more accurately. Over narrow ranges of yarn count, the count-strength product is constant. Over wide ranges the slope of the line representing the count strength can be found, or averages used so that predictions can be made. The slopes of these lines and the general relationships were discussed in Chapters 11 and 21.

Single Strand Strength. The strength of single stands of yarn is expressed in units of pounds, ounces, or grams in the United States. There has been a trend, particularly with the synthetics, to use single strand measurements in place of skein tests.

For fine yarns where the strength is less than 1 pound, it is general practice to report strength in units of either grams or ounces. For heavy yarns or cords, it is customary to express strength in pounds. In reporting results of strength tests, it is always good practice to report the yarn number as well as the strength.

Tenacity. Tenacity is defined as the force or load that is applied to a material per unit fineness. It should be pointed out that fineness is in effect the ratio of weight per unit length and is probably best illustrated by the denier system, in which the linear density is expressed as grams per 9000 meters. Thus, tenacity can be applied to any numbering system which is based on weight per unit lengths. However, the most common and accepted use of the term tenacity is in the units of grams per denier. The tenacity expressed as this load per unit fineness cannot be used for any yarn numbering systems that are based on the inverse ratio of lengths per unit weight. For instance, you cannot apply tenacity to staple yarns as such unless the yarn number is converted to a fineness unit similar to the denier system.

Tenacity is an excellent basis for comparing the strength of different types of yarns. The force required to rupture any given denier yarn can be reduced to a common basis of load per unit fineness, thus affording a very satisfactory and convenient method for comparison.

Breaking Length. One method of expressing yarn strength that is used extensively in Europe and has been introduced to a limited extent into this country is the *breaking length* (p. 363). This is the length of a yarn whose weight is equal to the breaking strength of the same yarn. As discussed under skein strength, the count-strength product is actually a breaking length. Since the count is expressed in units of hanks per pound and the skein strength is in units of pounds, the units for the product are hanks:

$$\frac{\text{hanks}}{\text{pounds}}(\text{pounds}) = \text{hanks}$$

For singles yarn, the break factor has the same meaning but is calculated in a slightly different manner. If a 20/1 has a single end strength of 0.75 pounds, the breaking length would be:

$$\begin{aligned}\text{Breaking length} &= (840)(\text{yarn number})(\text{pounds strength}) \\ &= (840)(20)(0.75) = 12{,}600 \text{ yards}\end{aligned}$$

(The yards per pound for the 20/1 are (840)(20), and if the yarn had an average strength of one pound, the breaking length would have been 16,800 yards. However, the strength was 0.75 pounds; therefore, the breaking length is (16,800)(0.75), or 12,600 yards. This means a single strand of this particular yarn could support 12,600 yards of the same yarn.)

For filament yarns the breaking length in meters is calculated from the basic relationship of denier; that is,

$$\text{Denier} = \frac{(9000)\,(\text{grams})}{\text{meters}}$$

$$\text{Meters} = \frac{(9000)\,(\text{grams})}{\text{denier}}$$

For example, if a 100-denier yarn has a breaking strength of 250 grams, the breaking length would be

$$\text{Breaking length} = \frac{(9000)\,(250)}{100} = 22{,}500 \text{ meters}$$

From the above, it becomes apparent that both the count-strength product and tenacity are measures of breaking length and, if properly expressed, are comparable. For example, suppose we compare 150-denier yarn having a total strength of 300 grams and a tenacity of 2 grams per denier with a 35/1 spun synthetic yarn having a single-strand strength of 10.6 ounces. The tenacity of the 150-denier yarn, expressed in units of meters length, is

$$\frac{9000(300)}{150} = 18{,}000 \text{ meters.}$$

The count-strength product for the single strand test, expressed in units of yards is

$$\frac{840\,(35)\,10.6}{16} = 19{,}477 \text{ yards}$$

and this is equal to 17,810 meters. Thus, the two yarns are approximately of equal strength, having nearly the same breaking lengths.

Comparison of Skein and Single End Test Results

For many years the skein test for yarn strength was used almost exclusively. With the advent of continuous filaments yarns, the use of single end testers became more prevalent. The introduction of the Moscrop, and later of the inclined plane testers, increased the usage of the single end method. The development of the Uster automatic tester and other fully automatic testers for single end work has given additional impetus to this trend to single end testing. Perhaps the reluctance against entirely deserting the skein test method stems from the fact that a great wealth of data have been accumulated over the years on skein tests.

The question most frequently asked concerning results from the two types of tests is "just what is the ratio of strength between the two tests?" Unfortunately, there is no specific answer that will hold true in all cases. With no consideration taken of the many factors influencing the test results, it could be said that the skein test results should average 160 times as high as those from a single end test on identical samples because of the 160 ends under test by the skein method. Other factors enter the picture to distort this relationship, such as twist, uniformity, sample preparation, and so forth. One rule always holds true: the skein test results are lower than single end results times 160. The basic reason is this: When a skein is subjected to a tensile load, some one weak spot in the 120-yard length breaks before the ultimate load is applied; this broken end represents a broken strand, and as the load increases, it may cause several slack strands that carry no load. Thus, with a steadily-increasing load due to the jaw separation, there are progressively less ends supporting the load. The result is that the ultimate load is reached very quickly after the failure of a few weak spots. With a single strand of yarn, the ultimate load reflects the strength of the weakest section of a relatively short length: 20 inches on the Uster and generally 10 inches in other U. S. practice.

Studies have been made at the School of Textiles, N. C. State College, to determine the relationship between skein strength and single end strength. Two relationships, one for carded yarns and one for combed yarns, have been found; these disregard differences in yarn number, staple length, twist, uniformity, and other factors. Therefore, they serve merely as an average indication of what to expect from skein strength. When the single end strength is known, or vice versa, these formulas are:

$$\text{For carded yarns: } S_s = -6 + 119.5 S_{se}$$

$$\text{For combed yarns: } S_s = -6 + 132.1 S_{se}$$

where S_s = Skein strength, pounds; S_{se} = Single end strength, pounds.

As a result of many tests over long periods of time, it has been found that for single yarns the skein strength equals 100 to 125 times the single end strength, and for ply yarns the ratio is 120 to 140.

$$\text{Thus, } S_s = (100 \text{ to } 125)\ S_{se} \text{ single yarns}$$

$$S_s = (120 \text{ to } 140)\ S_{se} \text{ ply yarns}$$

It becomes increasingly apparent that it is impossible to convert the wealth of information available from years of skein testing into comparable single end results with any degree of reliability. Averages can be predicted, but even these would be subject to error.

Comparison of Scott IP-2 with Uster Automatic Yarn Strength Tester

The steady development of quality consciousness and the potential economies to testing laboratories inherent in automatic single-strand testing machines have created a need for correlating the new type testers with the conventional Scott IP-2 inclined plane type. Since the new instruments, or at least the Uster automatic yarn strength tester, employ a 20-inch gauge length of yarn, and since the rate of loading on this type of machine can be varied, information on the differences, if any, that can be expected in results from the two instruments is important in maintaining mill standards.

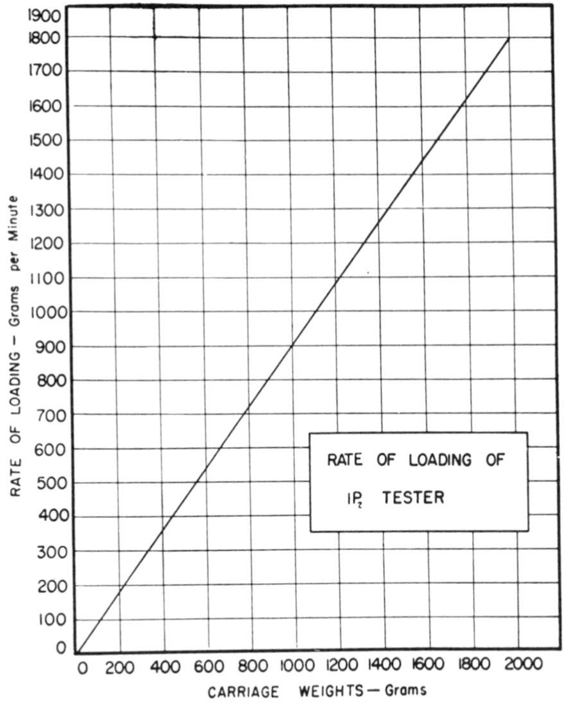

Fig. 22.1. Rate of loading on the Scott IP–2 tester plotted against carriage weight.

Recent studies made at the School of Textiles, N. C. State College, show an excellent over-all correlation between the two instruments. While yarn strength using the 20-inch gauge lengths runs slightly lower than for the 10-inch length used on the Scott IP-2, the difference averages only 22 grams. Differences in strength at various rates of loading were more pronounced. The accompanying charts illustrate the differences found in the experiments.

Fig. 22.1 charts the rate of loading on the IP-2 tester for various carriage weights. Fig. 22.2 shows the rates of loading for the Uster instrument with different carriage weights on various rates of jaw separation set by the rate of loading adjustment dial numbers of one through ten. (Note that the adjustment on the Uster when divided into 100 gives the seconds time for test from zero to the

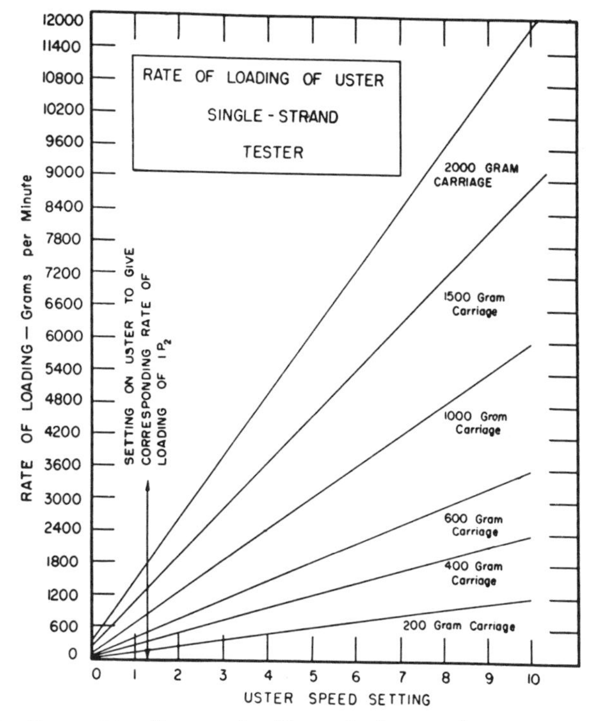

Fig. 22.2. Rate of loading on the Uster single strand tester plotted against Uster speed setting for various carriage weights.

total capacity being used.) The vertical line at the left of this chart indicates the theoretical setting on the Uster tester that would give a rate of loading corresponding to the IP-2.

Fig. 22.3 illustrates the differences in strengths for various yarn numbers as affected by rates of loading on the Uster. The higher rates of loading give higher strength values, and this tendency is more pronounced as the yarn count becomes heavier. One interesting point in addition to the change in strength as a result of tester speed is the change in slope when using different weights. The increase in slope with the increase in carriage weight follows the same general pattern as is shown in Fig. 22.2. It would not be expected, however, that the slopes for the corresponding carriages in Figs. 22.2 and 22.3 would be the same, for the slopes shown in

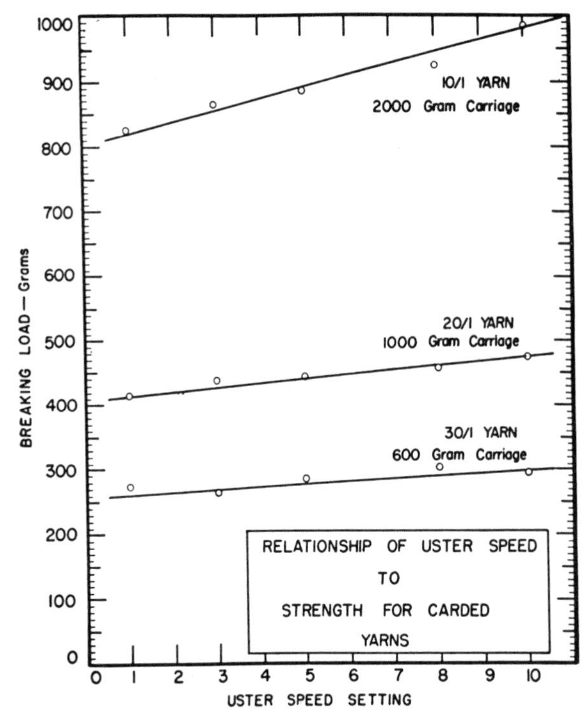

Fig. 22.3. Breaking load on the Uster tester plotted against Uster speed setting for various yarn numbers and carriage weights.

Fig. 22.3 would also be influenced by the rate of extension and the time to break of the material.

Fig. 22.4 shows the effect of gauge length on indicated yarn strength. As expected, the longer the length, the weaker the yarn appears to be. It clearly shows that the apparent strength is reduced by increasing the length of the test specimen; however, this set of curves should not be taken as representative for all yarns since the slope or rate of change in strength varies not only with the specimen length but also with yarn evenness and type of yarn.

Fig. 22.5 is a summary chart for carded, combed, and spun synthetic yarns, showing the correlation between results on the IP-2 and the Uster tester. The Uster speed for this particular study was set at position number 5, and a 10-inch gauge was used on the IP-2.

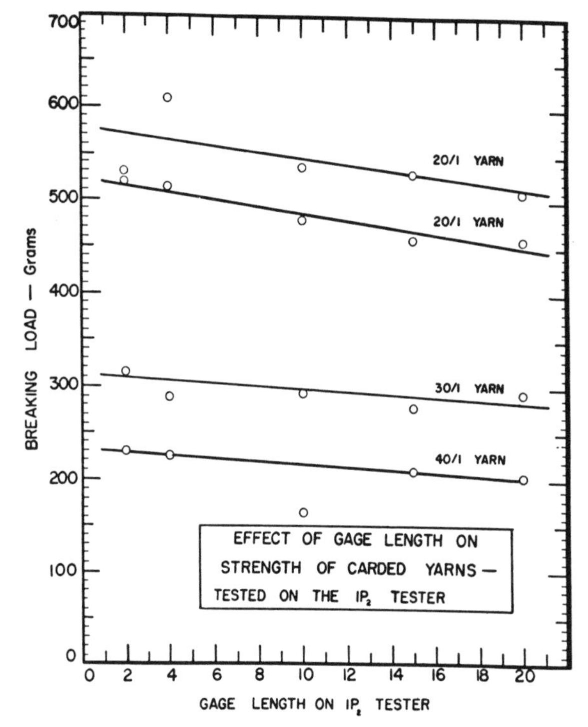

Fig. 22.4. Breaking load on the Scott IP–2 tester plotted against gage length for various yarn numbers.

Studies were also made using the Uster at speeds of positions 1, 3, 5, 8, and 10, and excellent correlation was obtained at these speeds between the Uster and the IP-2. However, the slopes for these respective speeds varied from the ones shown in this figure. This was due in part to the influence of machine speed and also to the fact that the rate of loading on the IP-2 remained constant. The slopes for the comparison between the testers varied from 0.97 to 1.2, depending upon the speed of the Uster tester. As shown, there is no appreciable effect resulting from the type of yarn tested or the testing instrument used. This is substantiated by the coefficient of correlation of 0.98. The equation for the least squares line was found to be

$$Y_c = 22.374 + 0.9785X.$$

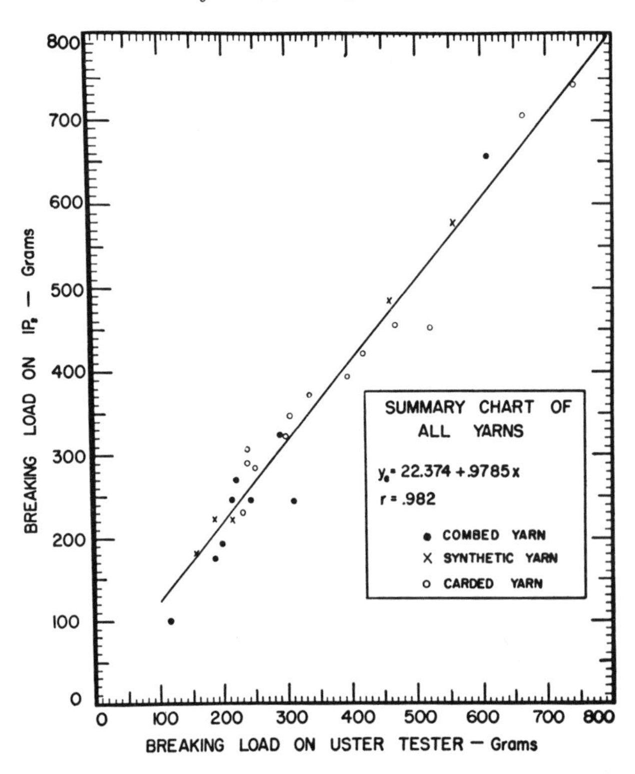

Fig. 22.5. Summary chart for carded, combed, and spun synthetic yarns showing the correlation between breaking loads on the IP–2 and the Uster testers.

Since the slope approaches unity and the intercept is approximately 22, for a quick approximation of IP-2 strength from Uster data, it is only necessary to add 22 grams to the average Uster breaking strength.

It was determined that with a coefficient of variation ranging from 8 to 13, 75 to 100 breaks on the Uster would give a very reliable average for each of the cones; however, in this particular study, all averages used were for a minimum of 200 breaks, with most yarns being tested with 400 breaks.

The methods recommended by the Uster Corporation for calculating the average strength, standard deviation, and per cent coefficient of variation were found to be very reliable for the sample sizes used. These methods are described in Chapter 21.

CHAPTER XXIII

Analysis of Stress-Strain Curves

When testing yarns on a routine basis, the only results usually recorded are strength and elongation. These everyday tests are the ones of importance to most quality control departments. In addition to strength and elongation, there are several other important properties which can be calculated from the stress-strain curve. Although these properties are more commonly associated with filament yarns, there are certain applications where they are of interest in connection with other types of yarn.

Type of Charts. The stress-strain or strength-elongation chart is actually a graph showing the reaction of a material to stress. With any stress there is always a corresponding strain in the material, and in the case of a tensile-type stress, the strain is measured as stretch or extensibility. For textile yarns, the stress is usually in units of grams, ounces or pounds, and the strain in units of inches or centimeters expressed as a per cent.

Some instruments are so arranged that the stress axis is the vertical axis of the chart, and on other instruments the stress is drawn on the horizontal side of the chart. The inclined plane type of machine recording units are so arranged that the stress or load side of the chart is the vertical axis.

When placing the load values on the chart, the maximum load which the machine can apply should be placed at the top of the chart. For example, on the IP-2, if the machine is set up with the 1000-gram carriage, the maximum load the machine could apply would be 1000 grams. Therefore, 1000 would be placed at the top of the chart. The bottom line will be zero, and the scale would then be calibrated in increments of 100. For the 250-gram carriage, 250 would be placed at the top of the chart and the scale increased in increments of 25 from the zero line. For the other capacities, the stress scale would be calibrated in a similar manner.

The strain axis is usually calibrated in units of actual strain, either in inches or centimeters. For most charts the units are inches.

Ultimate Tenacity. If the tenacity of a material is calculated at its ultimate strength, which would be that strength required to rupture the yarn, then we have the ultimate tenacity of the yarn. In speaking of the tenacity of synthetic yarns expressed in grams per denier, this particular figure or value is the one most commonly quoted, for it represents the ultimate strength of the material in question.

Ultimate tenacity in units of grams per denier is calculated by dividing the breaking strength in grams by the denier of the yarn being tested. By examining the stress-strain curve, Fig. 23.1, it can be seen that the breaking strength of the 100-denier yarn is 142.5 grams. The calculation is as follows:

$$\text{Tenacity, g.p.d.} = \frac{142.5 \text{ grams}}{100 \text{ denier}} = 1.43.$$

Note that the denier is always based on the original yarn.

Tenacity at Any Strain. Often it becomes desirable to compare two or more yarns for their tenacity at a given elongation. The result of calculating the tenacity at any such elongation gives the tenacity at any strain. This particular tenacity is desirable in many cases because very often it is necessary to know the load required to produce a given elongation. For example, for some industrial yarns it is necessary that the yarn stretch a given amount under certain loads, and the tenacity at this strain is used in making the calculations.

To calculate the tenacity at any strain, it is necessary to determine the load necessary to produce the given amount of strain. For example, in Fig. 23.1 what tenacity is necessary to produce an elongation of 15 per cent? The first step in this problem is to locate the point on the stress-strain curve which represents an elongation of 15 per cent. With 10 inches between the jaws, the actual elongation, x, will be:

$$x = 10\left(\tfrac{15}{100}\right) = 1.5 \text{ ins.}$$

This value of 1.5 inches is the point on the stress-strain curve

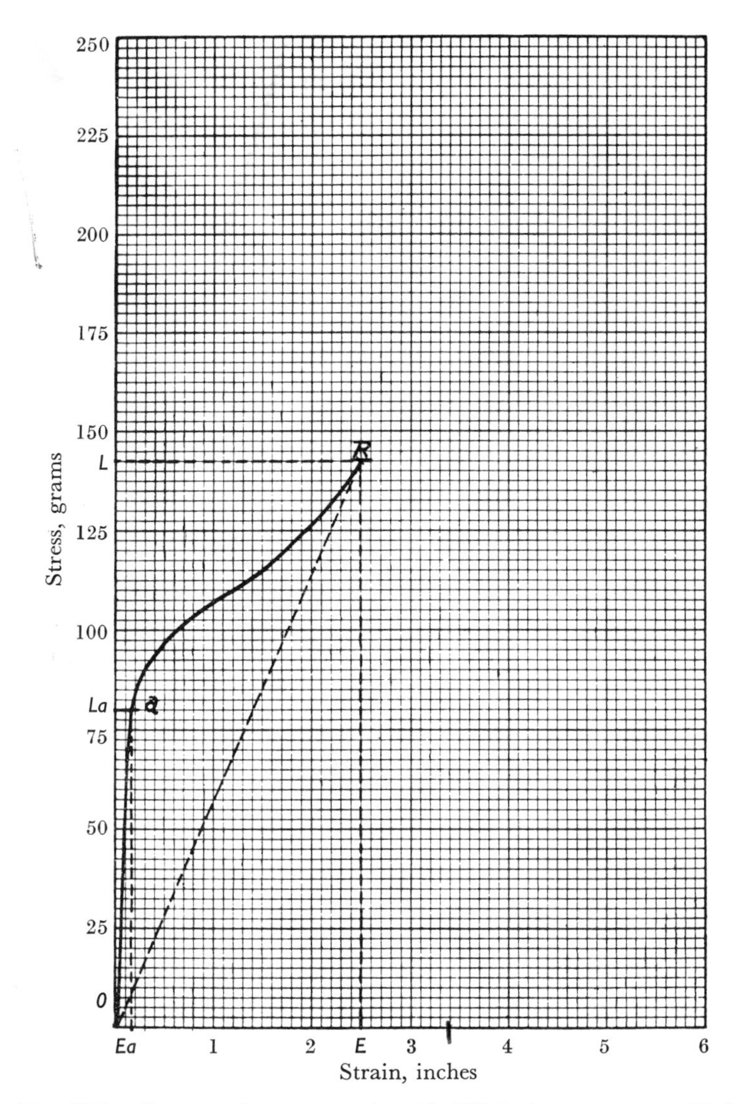

Fig. 23.1. Stress-strain curve made with 100-denier yarn on an IP–2 tester with a 250-gram carriage and 10 ins. between jaws. The chart size is approximately one-half actual. *L*: load necessary to rupture the material; *E*: ultimate elongation of the strand; *Ea*: strain or elongation at the elastic limit, or yield strain; *La*: point representing the stress or load at the elastic limit, or yield stress; *O–a*: the elastic stiffness; *O–R*: the average stiffness; *O–a–R–E–O*: area representing breaking toughness; *O–R–E–O*: area representing toughness index.

which represents 15 per cent elongation. By locating this point on the curve, it is found that a load of 115 grams, or 1.15 grams per denier, is required to produce an elongation of 15 per cent.

Elongation. Elongation is a measure of the extent of deformation along the axis of a material under a tensile stress and is expressed as a per cent change in length based on the original length of the test sample.

Ultimate Elongation. Ultimate elongation is the elongation at rupture. To calculate ultimate elongation, subtract the original length from the length at rupture, divide by the original length, and multiply by 100. The answer will be elongation expressed as a per cent based on the original length.

In Fig. 23.1 the original length, or the length of the test specimen, was 10 inches, and the sample stretched an additional 2.5 inches before rupture. The per cent elongation would then be calculated as follows:

$$\text{Elongation, } \% = \left(\frac{12.5 - 10}{10}\right) 100 = 25.$$

Elongation at Any Load. It is very desirable to know the elongation produced by any given load when designing a textile for a particular job. For example, in making a V-belt, the cords used in the belt should have a relatively low elongation within the working range of the belt. Therefore, most belt cord is tested on a machine which will automatically record in the form of a stress-strain curve the reaction of the cord to load. The curve is then analyzed for the amount of stretch or elongation produced by any given load.

To calculate the elongation of a material at any point on the stress-strain curve, follow the same method as for ultimate elongation, except use the length at the specified load in place of the ultimate length.

For the specimen curve, what would be the elongation produced by a load of 0.85 grams per denier? The load of 0.85 grams per denier represents a load expressed in grams of (0.85 g.p.d.)(100), or 85. Locating this point on the curve, it is found that the yarn stretched 0.2 inch. This represents an elongation of 2 per cent as calculated below:

$$\text{Elongation, } \% = \left(\frac{10.2 - 10}{10}\right) 100 = 2.$$

Elasticity. Elasticity is the tendency of a material in a deformed size or shape, as a result of stress, to return to its original shape on the removal of this stress. The usual quantitative measure of the elasticity of a material is the load which it will withstand without any permanent deformation.

Elastic Limit. One definition of elastic limit is: "The elastic limit of a material is the limit of stress intensity above which the intensity of stress ceases to be proportional to the strain."

For yarns or fibers that have been stretched beyond their elastic limit, the term "degree of elasticity" is used to express the ratio of recoverable strain to total strain at any given stress.

The elastic limit is usually expressed in terms of the yield stress, which is the load at the yield point. The yield stress is expressed as

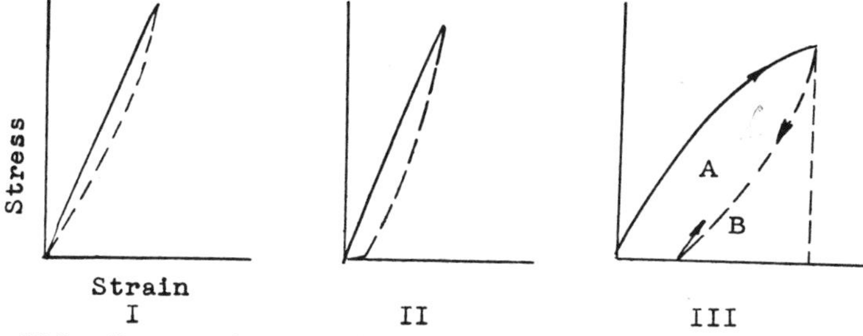

Fig. 23.2. Stress-strain curves showing: I, immediate elastic recovery; II, delayed elastic recovery; III, non-recoverable deformation. Area $(A + B)$ represents work performed. Area A represents work lost; area B, work recovered.

grams per denier, the grams being read directly from the stress-strain curve at the yield point.

The elastic limit may also be expressed in terms of the yield strain, which is the strain at the yield point. The yield stress and yield strain represent the limit within which the material is completely elastic.

The exact location of the elastic limit is difficult, if not impossible, to locate. A reasonably accurate approximation is usually satisfactory for textile materials since most working stresses are practically always well below this point.

In the specimen curve, assuming the elastic limit to be at the 80-gram line, the yield stress would be calculated by dividing the grams stress by the denier as shown below:

$$\text{Yield stress, g.p.d.} = \tfrac{80}{100} = 0.80$$

The yield strain is calculated in the same manner outlined previously for per cent elongation

$$\text{Yield strain, \%} = \left(\frac{10.15 - 10}{10}\right) 100 = 1.5$$

Resilience. Resilience is the ability of a fiber to absorb work without permanent deformation, or in other words, the ability to recover. Resilience may be thought of as the area under the stress-strain curve up to the elastic limit. Once the elastic limit has been exceeded, the ability of the material to recover has been reduced, and in some cases the material will never completely recover. Repeated loading and unloading of a material within the elastic limit will have very little if any effect on the ability to recover; however, if this is continued over a long period of time, failure will occur as the result of the prolonged loading.

To calculate the modulus of resilience, multiply the grams per denier at the elastic limit by the elongation at the elastic limit expressed as a decimal and divide by 2.

$$\text{Modulus of resilience, } \frac{\text{gm. cms.}}{\text{den. cm.}} = \frac{(0.80)(0.015)}{2} = 0.006$$

Stiffness. Stiffness is the ability of a material to resist deformation. In the case of a yarn subjected to a tensile force or pull, stiffness is the ability to resist elongation. The units for stiffness are grams per denier per unit elongation.

Average Stiffness. Average stiffness is the stiffness of a material from its original state to breaking point. It is the ratio of unit breaking stress to unit breaking strain and thus indicates the average stress for the entire range of extensibility of the material per unit increase in strain. Average stiffness is an indicator of the general character of a material with regard to the stiffness quality.

To calculate the average stiffness, divide the grams per denier at rupture by the ultimate elongation expressed as a decimal. From the specimen curve, the values would be:

Average stiffness, unit stress per unit strain or g.p.d. per unit

$$\text{elongation} = \frac{1.4}{0.25} = 5.6$$

Elastic Stiffness. Elastic stiffness is the stiffness of a material within its elastic limit and is indicated by the general slope of the stress-strain curve between the origin and the yield point. In engineering terms, the elastic stiffness is called modulus of elasticity or Young's modulus.

It is important to know the elastic limit and elastic stiffness of a material when selecting a fiber or yarn for a definite use. Since a fiber has little if any ability to return to its original shape once the elastic limit has been exceeded, the material is of no value. For example, in designing a diaphragm, if the textile used did not have a stiffness or elastic limit capable of withstanding the pressure to which the diaphragm was to be subjected, the yarn or fabric would be of little or no value as a reinforcement member after the first expansion of the diaphragm. The same would hold true for textiles used in the clothing or in any other industry.

The elastic stiffness from the specimen curve would be calculated as follows:

$$\text{Elastic stiffness, g.p.d. per unit elongation} = \frac{0.80}{0.015} = 53.4$$

Stiffness at Any Stress. It is also possible to measure the stiffness at any given load or at any specified point on the stress-strain curve. This specific stiffness rating may be calculated by two methods: first, by dividing the load, expressed as grams per denier, at the specified point by the corresponding elongation expressed as a decimal; second, by calculating the slope of the curve at the given point.

The two methods described above for stiffness at any stress will give two answers which are not analogous. The first gives only the average stiffness up to the point chosen, whereas the second method (the "slope") gives the instantaneous stiffness of the material at the point selected. It is difficult to say which of the two methods is more useful or correct for certain uses.

Since stiffness values are used mostly for comparison purposes, it is recommended that the average stiffness and elastic stiffness be used. Since the average stiffness will give an over-all picture of the material from the origin to the rupture point, it will give a satisfactory basis for general comparisons. As previously mentioned, the elastic stiffness is the stiffness of a material within its elastic limit and can be used as an excellent index for comparing the ability of fibers to absorb strain without permanent deformation.

Toughness. The thougness of a material is a measure of its ability to absorb work and gives an indication of durability of the material. The units used to describe toughness are gram centimeters per denier centimeter.

Toughness Index. The toughness index is the toughness of a material from origin to breaking point, assuming the stress-strain curve to be a straight line. This value is a useful index for comparing the toughness of one material with another; however, the value calculated in this manner will be in error due to the assumption that the stress-strain curve is a straight line.

To calculate the toughness index, simply multiply the grams per denier at break by the elongation, expressed as a decimal, at break, and divide by 2. This method assumes the curve to be a straight line and finds the area of the resulting triangle.

From the specimen curve

$$\text{Toughness index}, \frac{\text{gm. cms.}}{\text{den. cm.}} = \frac{(1.4)(0.25)}{2} = 0.18.$$

Average Toughness. Average toughness is the actual work per unit mass required to rupture yarn. To calculate the breaking toughness, it is necessary to measure the actual area under the stress-strain curve. This can be done with a planimeter; however, an alternate method is as satisfactory. To measure the area under the curve, count the number of squares under the curve and divide this value by the number of squares along the elongation axis of the chart. From the specimen curve, the total squares are 1001, and there are 25 squares along the elongation axis of the chart. By dividing the total number of squares (1100) by the number of squares along the elongation axis of the chart (25), the result is 44.

This number (44) is the number of squares which is set off on the gram axis of the chart. This calculation actually makes the area under the stress-strain curve into the form of a rectangle to facilitate the calculating of the area. With the number 44 set off on the gram axis of the chart, calculate the grams per denier load at this point. Multiply the grams per denier load by the ultimate elongation expressed as a decimal, and the answer will be the breaking toughness expressed as gram centimeters per denier centimeter. A summary of this calculation is as follows:

Total number of squares (approx.) = 1100
Number of squares on elongation axis = 25

$$\frac{1100}{25} = 44.$$

By counting off 44 squares on the vertical axis, the equivalent strength is found to be 110 grams.

$$\frac{110}{100} = 1.10 \text{ g.p.d.}$$

$$(1.10)(0.25) = 0.275 \frac{\text{gm. cms.}}{\text{den. cm.}}$$

Summary

Tenacity

$$\text{Ultimate tenacity, g.p.d.} = \frac{\text{load at rupture, gm.}}{\text{yarn denier}}$$

Tenacity at any strain, g.p.d.

$$= \frac{\text{tenacity at selected elongation, gm.}}{\text{yarn denier}}$$

To locate on the stress-strain curve the tenacity for the selected elongation, get chart location x on the elongation axis.

$$x = \text{gauge length} \frac{\text{selected elongation, \%}}{100}$$

Elongation

Ultimate elongation, %

$$= \frac{\text{length at rupture} - \text{original length}}{\text{original length}} (100).$$

Elongation at any load,

$$\text{Elongation, \%} = \frac{\text{elongation}}{\text{original length}}\,(100)$$

Calculate the load for the tenacity selected, (tenacity, g.p.d.) (denier) and on the stress-strain curve, get the corresponding elongation.

Elastic Limit

$$\text{Yield stress, g.p.d.} = \frac{\text{load at yield point, gm.}}{\text{denier}}$$

$$\text{Yield strain, \%} = \frac{\text{elongation at yield point}}{\text{original length}}\,(100)$$

Resilience

$$\text{Modulus of resilience, } \frac{\text{gm. cm.}}{\text{den. cm.}}$$

$$= \frac{(\text{load at yield point, gm.})(\text{elongation at yield point, \%})}{2(\text{denier})(100)}$$

Stiffness

Average stiffness, g.p.d. per unit elongation

$$= \frac{(\text{load at rupture, gm.})(100)}{(\text{denier})(\text{elongation, \%})}$$

Elastic stiffness, g.p.d. per unit elongation, or modulus

$$= \frac{(\text{load at elastic limit, gm.})(100)}{(\text{denier})(\text{elongation, \%})}$$

Toughness

$$\text{Toughness index, } \frac{\text{gm. cm.}}{\text{den. cm.}}$$

$$= \frac{(\text{load at rupture, gm.})(\text{ultimate elongation, \%})}{2(\text{denier})(100)}$$

Average toughness, $\frac{\text{gm. cm.}}{\text{den. cm.}}$, equals the area under the stress-strain curve.

$$= \frac{(\text{average strength, gm.})(\text{ultimate elongation, \%})}{\text{den.}(100)}.$$

TABLE 23.1

Classification of Engineering Quality Meanings of Textile Materials [a]

Performance characteristics in terms of people's requirements	Characteristics in relation to textiles	Words describing characteristics
Ease of bending	Flexibility	Pliable to stiff
Ease of squeezing	Compressibility	Soft to hard
Ease of stretching	Extensibility	Stretchy to non-stretchy
Ability to recover from deformation	Resilience (may be flexural, compressible, extensible, or torsional)	Springy to limp
Weight per unit volume	Density	Compact to open
Divergence of the surface from planeness; resistance to slipping offered by the surface	Surface contour: surface friction	Rough to smooth, harsh to slippery
Apparent difference in temperature of the fabric and the skin of the observer touching it	Thermal character	Cool to warm
To carry a dead load	Ultimate strength	Strength
To carry a load without deformation	Modulus of elasticity	Stiffness
To undergo deformation and return to original shape after cessation of deforming force	Elastic limit	Elasticity
To absorb shock without permanent deformation	Modulus of resilience	Resilience
To endure large, permanent deformation without rupture	Ultimate resilience	Toughness

[a] Suggested by the United States Testing Company.

CHAPTER XXIV

Yarn Grade and Appearance

Until recent years no means were available for either the qualitative or quantitative analysis of yarn for grade and appearance. The purchaser of yarn depended upon the integrity of the supplier and on his own experience to determine grade. When he dealt with a new supplier, one of the few things he could do in order to get a yarn of a known quality was to insist that the yarn delivered should be equal to that delivered by some recognized yarn mill. There was no definite method of specifying the grade of yarn for commercial transactions or for manufacturing and technical use, except by direct comparison. Furthermore, it has been long recognized that the general appearance of yarn is one of the prime qualities affecting its commercial value, for it is this very practical and aesthetic characteristic that influences to a large degree the character of the end product.

Technically, the determination of yarn grade and appearance in a yarn mill is analogous to fabric inspection in a weaving mill. Just as the weaving mill, in order to satisfy its customers, must inspect all fabrics produced, so should the yarn mill inspect its product for grade and appearance. Today practically all knitting yarns are sold or purchased with specific reference to a given grade standard, and it is this grade standard for visual comparison that has permitted the qualitative analysis of yarn for grade and appearance. For cotton yarns the grade designations are A, B, C, and D, based on standard yarn boards for each of five ranges of yarn numbers. Of course, it is possible to evaluate actual grade and appearance in intermediate grades such as B+ ,C+, B—, and so forth, for yarns that do not clearly fit into one of the broader categories, or which seem to miss one of the main categories in certain respects.

Cotton Yarn

The grade of any sample of yarn is determined by visual comparison with a photograph of a standard sample within the same range of yarn numbers. The standard yarn boards for comparison were originally developed by the United States Department of Agriculture and later adopted and distributed by the American Society for Testing Materials, Committee D–13. A complete set of boards consists of five boards covering five ranges of yarn numbers wrapped as shown in Table 24.1 with each board containing the photographs of four different standard samples of yarn graded A, B, C, and D.

TABLE 24.1

Wraps Per Inch for Grading Yarns

Yarn number	Wraps per in.
3.0 to 7.0	16
7.0 to 16.5	20
16.5 to 32.0	26
32.0 to 65.0	38
65.0 to 125.0	48

In order to determine the grade of yarn, it is first wound onto a cardboard with a dull black finish having the approximate dimensions of $5\frac{1}{2} \times 9\frac{1}{2}$ inches, as shown in Fig. 24.1. The yarn is wound with the longer dimension of the board and with sufficient tension to hold the yarn snugly and evenly in place, but not sufficiently high to bend the board or distort the yarn. The number of wraps of yarn per inch should be the same as that on the standard to be used for comparison. This is shown in Table 24.1. For example, if the sample of yarn to be compared is 20/1, then it should be wound with 26 wraps per inch. Instruments for the wrapping of these yarn boards are available commercially.

After the boards have been prepared, they are then compared visually with the cotton yarn appearance standard boards, and a grade designation such as A, B, C, or D, with intermediate classification of plus or minus, can be determined. It is wise in making the comparisons to have more than one technician make a comparison to get average grade values.

Fig. 24.1. Winding yarn.

These grades have been given numerical values, which is an aid in analyzing the results of evaluation of several different yarns and gives a reasonably reliable basis for establishment of the average grade of a lot of yarn. The recommended values for these grades are shown in Table 24.2. These index numbers are those used by the United States Department of Agriculture and are not necessarily industry standards. Some mills have adopted their own numerical values, while others for commercial transactions use the letter designations previously described.

In order that results of these visual comparisons may be as consistent as possible, for we must admit that there are many possibilities of variation in personal judgment, it is wise to use

TABLE 24.2

Yarn Grades and Indexes [a]

Grade	Designation	Index
A & above	Excellent	130
B+	Very Good	120
B	Good	110
C+	Average	100
C	Fair	90
D+	Poor	80
D	Very Poor	70
BG	Below Grade	60

[a] Reprinted from "Summary of Fiber and Spinning Test Results . . ., Crop of 1954" by the U.S. Dept. of Agriculture.

consistent conditions, particularly of lighting, in making the comparisons. The sample and the master boards should be viewed from different angles and with different lighting conditions, but the lighting conditions should be controlled so that it is possible to have the same type of lighting from week to week and month to month. In this respect, a grading cabinet has been developed by the United States Department of Agriculture, which is quite useful in giving this control of conditions under which the grading is made. This cabinet has a rack to hold a standard board and side lighting which can be standardized or altered to suit the needs. It also has a cabinet for the storage of the standard boards, and this in itself is very valuable, for these photographic standards are subject to deterioration if left exposed and sitting around the laboratory. (Detailed drawings of the cabinet recommended are available both from the A.S.T.M. and the U.S.D.A.)

Continuous Filament Yarns

For continuous filament yarns the grade designations are generally as follows: First quality, second quality, and inferior. The difference in the grade designations would be due to such quality characteristics as the number of broken filaments, loops, knots, dirty yarn, poor package formation, shade, and the like. Inferior yarn would usually include other quality defects which are not classed as visible defects such as poor dye affinity, improper denier, improper filament count, weak yarn, subnormal strength or elongation, and other similar defective characteristics. These designations as to quality or grade are generally assigned by the manufacturer of the material and only in unusual cases where their quality inspection has failed would the mill find reason to question the grading.

Seriplane Evaluation

Spun Yarns

The Seriplane is a machine developed by Cheney Brothers of South Manchester, Connecticut, and sold by Alfred Suter of New York City for the semi-automatic winding of yarn onto black board panels. This machine was developed originally for the silk industry for the visual inspection of raw silk and yarn, but since then it has found wide application in other branches of the textile industry for the same purposes. This machine will wind simultaneously ten bobbins of yarn onto a black board. Accurate means of controlling the tension of individual ends is provided and the traverse of the yarn can be controlled very closely to a predetermined number of threads per inch. The machine must be started manually when all the bobbins have been placed on the holders and threaded, but the machine stops automatically on the completion of the winding. The boards themselves containing the ten individual panels of yarn can be removed from the machine merely by lifting them out of the holders and then these black boards can be placed in a special rack built to hold four of these, or eight if desired. The racks are built with casters so that the entire rack can be moved in order to take advantage of light conditions, and individual black boards can be rotated in their bearings on the rack. Thus, the entire system allows for the accurate winding of yarn quickly and easily onto well designed black boards and provides an additional means for inspecting these boards. While the Seriplane is not usually used for comparison of yarn with the photographic standards for appearance, it does provide a means for comparing many samples of yarn simultaneously. For example, as many as ten different variables can be compared on any one board. Generally, however, two, three, or four different yarns are compared at one time by winding one type of yarn on the first individual panel, the second type of yarn on the next panel, and so forth, over the entire black board to give the required samples for comparison. This method of winding, together with the means for visual examination, allows for the examination of spun yarn with the primary idea of examination for such defects as slubs, foreign matter, uneven twist distribution, neppiness, and general appearance. The Seriplane is a very valuable piece of equipment for any laboratory interested in the evaluation of spun yarn.

Continuous Filament Yarns

Inasmuch as the Seriplane was originally developed for the silk industry and used very widely by the silk industry both in this country and in Japan for evaluation of the character and quality of silk yarn, it is only natural that it should be of a similar value to the manufacturers of continuous filament yarn. One of the greatest uses by the producers of the filament yarns is for the determination of the number of defects in a given yardage of material. Several boards of the yarn are wound on the Seriplane, then the number of defects are counted and tabulated. The results are expressed as the number of broken filaments, loops, spots, and other defects based on some unit of length or weight, such as per 1,000 meters or per gram of yarn. Thus it becomes apparent that the use of the Seriplane for continuous filament yarns is primarily for quality control measures rather than for the determination of grade and appearance based on some previously established standard, as is done for cotton yarn.

In addition to examination of filament yarns on the Seriplane, they are also inspected by forming the yarn into a sheet similar to that formed when making up a warp; the number of defects are counted as this sheet of yarn is passed through a specifically lighted section in the inspection area.

Seriplane examination is for the purpose of controlling quality, but it should be emphasized that it is not used for grade classification. Generally speaking, grade classification of continuous filament yarn, which infers determination of whether the particular yarn is first, second, or inferior quality, is determined in the industry by inspection of the finished cone or cake immediately prior to packing for shipment.

Photographic Reproductions of Yarn Boards

Quite frequently in everyday quality control work, it is found desirable to compare the actual appearance of yarn being produced with that produced in the past by means other than comparison with the photographic standards of the A.S.T.M. At the same time it is inconvenient to keep yarn boards filed on hand from year to year because of the difficulty of preserving these boards, for they deteriorate in handling. The answer to this problem is to make either a yarn shadowgraph or a photograph. Such photographic records are also very valuable for experimental and research work, for it is possible to make these for different experimental samples and then these photographs can be handed around to the various interested parties without distortion of the sample itself.

There are two approaches to the problem. One of these is to make a shadowgraph, which is in effect a silhouette of the yarn, and the second is to make a direct photograph. The shadowgraph system is a little simpler; it does not require any special photographic equipment outside of the photographic paper, the light source and diffusing screen, and a dark room for development of the shadowgraph. The disadvantage of its use is that only one copy is available at a time unless several separate individual exposures are made. Where a multiple number of pictures are required, and where a proper type of camera is available, then the photographic system is preferable. The means for reproducing yarn appearance for both of these systems is described herewith.

The yarn shadowgraph method was described by de la Rama in the July 1, 1946, issue of the *Textile Bulletin* and was based on a technique developed originally at Cleveland Cloth Mills in Shelby, N. C. This system has been modified by de la Rama's associates at the School of Textiles, N. C. State College, and is as follows:

The yarn is wound on a Seriplane board to the desired width

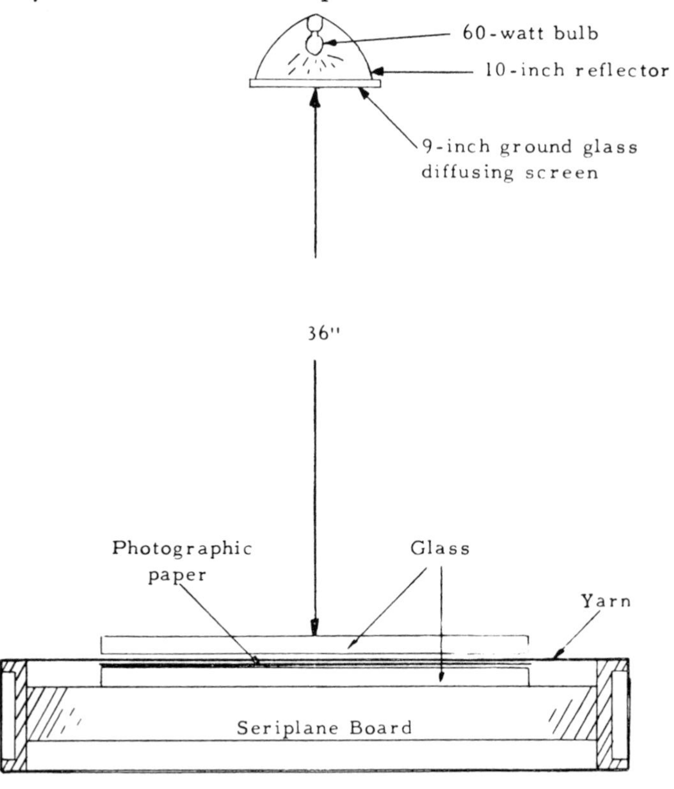

Fig. 24.2. Shadowgraph layout.

and with the proper spacing of the yarn following the A.S.T.M. specifications described previously. The Seriplane is taken to the dark room for the actual shadowgraph work. A piece of window glass is placed between the Seriplane black background and the yarn, and it will be found that there is sufficient room on the standard type of Seriplane board to do this without distorting the yarn itself. A sheet of AZO E5 photographic paper is placed between the piece of window glass and the yarn. Another technique involves using two pieces of window glass with the photographic paper between the two to form a sandwich, but it has been found that detail is lost by the use of the second piece of glass in this sandwich. Finally, a second piece of glass is placed over the yarn. This arrangement is shown in Fig. 24.2. A 60-watt bulb in a 10-inch reflector supplies the light. It will be found that a 9-inch ground glass diffusing screen placed intimately with a reflector and 36 inches above the glass on top of the Seriplane board will give a better light distribution and prevent differences in the light received by the photographic paper from center to edge. Using the 60-watt bulb, the exposure time will be found to be somewhere between 10 and 15 seconds duration. Using a photoflood bulb in place of the ordinary 60-watt bulb will cut the exposure time down considerably, but this procedure is not recommended because it makes the exposure time too critical for accurate control. After the necessary exposure, the print can be developed in the recommended type of developer (Dektol developer solution: 1 part developer, 2 parts water will be found to work very satisfactorily), then the print should be fixed in a hypo solution and washed in accordance with the recommendations of the photographic paper manufacturer. For those mills who do not have a Seriplane available for winding the yarn, it is possible to use the small yarn boards described earlier in this chapter.

When winding the yarn on a small card of the approximate dimensions of $5\frac{1}{2} \times 9\frac{1}{2}$ inches, it is suggested that the yarn be wound sufficiently tight to cause a slight curvature of the board, for this will facilitate the placing of the photographic paper under the yarn without disturbing the position of the strands of yarn on the card. It will also be found advisable to omit the lower piece of glass, provided a good piece of cardboard has been used on which to wind the yarn. In any case, it will be necessary to use a top piece of glass over the yarn. Experience will indicate the amount of tension to use in winding the yarn originally in order to provide sufficient bow in the card to allow insertion of the photographic paper and at the same time not be excessive so as to prevent the glass from holding the board flat on the table. It is also possible to make a frame or easel which can be placed over the glass which will help to keep the yarn board flat.

The second method of reproducing the yarn appearance is by photography. The yarn can be wound on either a Seriplane or a regular yarn board, but the following procedure is given for a Seriplane and can be modified for use with the regular yarn board. The yarn is wound on the Seriplane board in the desired width, usually not over 12 inches, and with the proper spacing according to A.S.T.M. specifications.

The yarn board can be mounted in either a vertical or horizontal position, whichever is more convenient. No matter which position is selected, the plane of the yarn must be at a precise right angle to the camera in order to prevent any distortion. A 4×5 view camera should be used, and this can be located 20 inches from the yarn board. Lighting is supplied by two fluorescent light fixtures each having four 20-watt bulbs. These fixtures should be 3 feet from the yarn board and at an angle of 45° to the plane of the board. The film recommended for this type of work can be either the Ansco Isopan or its equivalent, such as Eastman Panatomic X. Focusing should be done very carefully using a magnifying glass. It will be found that the exposure time will be very close to one second at f 22. The film should be developed in Ardol or its equivalent at 68 degrees Fahrenheit for $3\frac{1}{2}$ minutes. It will be noted that a normal exposure together with underdevelopment of the film has been used to keep the contrast as low as possible. Following the development, the film should be neutralized in a quick stop bath and fixed for 5 minutes in a commercial rapid fix bath; then the film is washed and dried in the usual manner. If the laboratory is interested in the long-range program of reproducing yarn boards, then the Weston Densotometer should be used to check all negatives to see that they fall within the normal limits for black and white contrast. To identify the yarn boards, a white label is generally used. The area on the negative covered by this label when tested on the Weston

Densotometer should give a density of 0.8. In enlarging, the image is blown up to actual yarn board size, which is generally $8\frac{1}{2} \times 11$ inches. The f stop to be used on the enlarger lens can be determined by the Densotometer to give a reading of 2.1 at the paper surface. Of course, this would have to be varied to suit the particular type of enlarger used, the intensity and type of lighting in the enlarger, the quality of the lens, and other factors. Likewise, the exposure time will vary under these different conditions. With an Omega enlarger, it has been found that an exposure of 5 seconds gives the best results. The enlarging paper used can be Cykora GL2 or Eastman Kodabromide F2. Whichever paper is used, the manufacturer should be consulted in order that he can supply paper which will be consistent as to either shrinkage or stretching. The paper can be developed in Vividol or Dektol, using a solution of two parts of water to one of developer. The development time should be $1\frac{1}{2}$ minutes exactly, and this time should not be varied. A fresh developing solution should be used at the rate of one gallon for each 100 sheets, and fresh batches should be used every day or more often, for even after a few hours of exposure to the atmosphere, it will be found that the black areas tend to lighten out. After development, the paper is run through the stop bath, the hypo fixing bath, and then washed. The drying should be to give a matte finish in order to make it easier to examine the photograph.

CHAPTER XXV

Twist Testing

The measurement of the number of turns per inch in yarn, whether it is single, ply, cable, or any other construction of staple or filament yarns, is of interest not only to technical personnel, but also to the production and management group. The technical group is interested in the amount of twist in yarn from the viewpoint of its effect on the physical properties, performance, and appearance of the finished product. The production supervisor is interested in knowing how much twist in being inserted as a check on machine setup, and if the yarn being produced meets the established requirements. Management is interested in twist not only for the reasons previously mentioned, but also from a production and cost standpoint. In most textile operations, a change in twist is made by either increasing or decreasing front roll speeds, which in turn means an increase or decrease in production. This change in production changes the cost of producing each pound of finished product. Thus a substantial change in the amount of twist being inserted in a yarn will affect not only the physical properties, but also the production rate and cost of the yarn.

Twist Direction

Twist direction in the yarn is designated as either Z or S twist. Whether the yarn has Z or S twist can be determined by comparing the twist convolutions with the center portion of the letters Z and S, as shown in Fig. 25.1.

Another way to identify the twist direction is as follows: Hold a short sample of yarn between the thumb and forefinger of each hand, twist the yarn with the right hand to the right (clockwise). If the twist is removed so that the yarn becomes weak or so that the individual ends of the ply become parallel, then the twist is

Z. If the yarn becomes kinky, indicating an increase in twist, then it is S twist.

The use of the letters Z and S to designate twist direction was adopted by the industry to avoid the confusion resulting from the

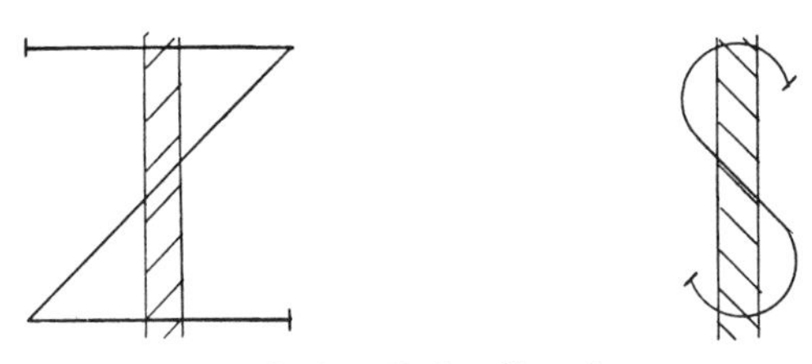

Fig. 25.1. Twist direction.

use of the terms regular and reverse or right and left twist. Curiously enough, the spinner considered regular twist of single spun yarn as having what is now called Z twist, whereas the twister considered his regular twist in a ply yarn as having what is now called S twist. It is no wonder that confusion existed and that the standard terminology of Z and S was adopted. Incidentally, most single yarn is spun in this country with Z twist and ply yarns twisted with S twist to give a balanced yarn.

Twist Measurement

Theory of Twist Tester. A twist tester is a rather simple instrument composed primarily of two jaws, one of which does not rotate and one which does rotate. The sample of yarn is placed between the two jaws and the twist removed by rotating one jaw. The nonrotating jaw can be moved horizontally along the axis of the yarn so that the distance between the jaws can be adjusted for different lengths of yarn, usually from a zero gauge length to 20 inches maximum. The rotating jaw is attached to a counter so that the number of revolutions necessary to remove twist from the test specimen can be accurately determined. In addition, the counter can be set to register for S or for Z twist.

On the newer type twist testers, the nonrotating jaw is attached to a weighting system so that a predetermined amount of tension may be conveniently applied to the test specimen. In addition, frequently this jaw is arranged so as to be free to move horizontally (usually in an arc) for the untwist-twist test.

Untwist-Twist Method

The untwist-twist method is based on the assumption that the contraction of a given length of single yarn is the same for any given amount of twist per inch no matter whether Z or S twist is used. The test itself involves the use of a known length of yarn (usually 10 inches), the untwisting of this yarn under a very slight tension, and the retwisting of the sample in the opposite twist direction. The retwisting is continued until the contracted length matches exactly the original length. The total twist in the original sample is equal to the twist removed plus the twist added, which in turn is equal to the total revolutions of the twist head divided by 2.

It is obvious that errors can arise in determining twist by this method. For example, errors can arise if the pretensioning is excessive or too slight, or if the tensioning is not removed at the zero twist length of the yarn. Yarn that has been twist-set would give results of questionable accuracy. Yarn that has been slashed or treated in any way with gums, starches, resins, or other chemicals fails to exhibit normal results, and the untwist-twist method is not used for yarns of this nature. However, with proper care exercised in setting up the tester and with experience, it is possible to get reliable test results quickly and easily for single spun yarn. It is suggested that test results from this method be checked occasionally with the short length untwist method in order to serve as a check test.

This technique is used for single yarn only, not for ply yarn, and when test lengths at least 5 inches long are available. The instrument used for this test should have some sensitive method of indicating the change in length due to the loss or gain in contraction. One method of indicating a change in length is to have a light-weight arm pivoted at one end in a delicate bearing and with the other end resting on the yarn midpoint of its length. Changes in length of the yarn are indicated by the movement of this arm through a small arc.

Another method such as is used on the U.S. Testing Company twist tester uses a pendulum arrangement. The non-rotating jaw is fastened to the end of a pendulum arm, and a crosspiece at the top of the pendulum allows for controlling the weight or reaction of the pendulum to the yarn. Another type of pendulum developed at the School of Textiles, N. C. State College, uses a longer pendulum (10 inches effective length) to provide a large radius arc with little vertical dislocation of the yarn, together with a larger scale for accuracy of reading changes in length. In addition, the pendulum is counterbalanced so that the tension on the yarn is the same at any angle of the pendulum, and the load can be preset to any desired amount within the range generally used.

Procedure for Measuring Twist

1. Adjust the distance between the clamps. The distance between the clamps is usually ten inches.
2. Set the counter to zero, taking care that it is also set for the proper twist direction.
3. Select the sample. For yarn or cord on cones, tubes, or similar packages, remove the test specimen by rotating the package and drawing the sample from the side of the package at a tension or pull not greater than that to be used on the yarn during testing. By removing the yarn from the side of the package rather than over the end, any possibility of disturbing the twist is eliminated. In removing yarn from the fabric, extreme care should be taken not to lose any twist.
4. Place the end of the yarn in the non-rotating jaw and tighten the clamp.
5. Place the other end of the yarn in position in the opened rotatable jaw, release the other jaw so that the predetermined amount of tension is on the yarn. Pull the yarn through the open clamp until the exact gauge length is between the two jaws. Then tighten the clamp on the rotatable jaw. The proper tension is shown in Table 25.1.
6. Revolve the rotatable clamp in the direction that untwists the yarn.

As the twist is removed from the singles yarn, it results in a loss of contraction or a reduction in the tension on the specimen. If the pendulum arrangement is used, the pendulum will begin to move in its arc. This movement should be stopped after it has traveled not more than 0.125 inch to prevent drafting of the test specimen. If the light arm is used, it should have an automatic stop with the arm having a traverse of not greater than 0.125 inch. This also is to prevent drafting.

Continue to rotate the jaw at a slow rate of speed. (The slow speed is used to prevent ballooning and drafting of the test specimen as the twist approaches zero.) By continuing to rotate the jaw, the contraction will eventually return and draw the test specimen back to the original gauge length.

When this point has been reached, theoretically the yarn has the same amount of twist as it had at the start of the test, but in the opposite direction. Record the number of turns registered on the dial or counter of the instrument. This test method is used not only for cotton, but also for man-made staple base spun yarns.

TABLE 25.1

Tensions Recommended by A.S.T.M. for Yarn to be Twist-Tested

Yarn	Tensions, grams per Tex
Single, Filament	0.25 ± 0.05
Single, Spun	0.25 ± 0.05
Plied	0.25 ± 0.05 [a]
Cabled	0.25 ± 0.05 [a]

[a] For plied and cabled yarns the tension must be adjusted to the equivalent of 0.25 ± 0.05 gm/Tex for the individual strand or ply before cutting away the extra strands or plies.

Calculations. The twist per inch using the untwist-twist method is equal to the total turns registered by the twist tester counter divided by 2 times the length of the yarn sample gauge length.

$$\text{T.P.I.} = \frac{\text{Dial reading}}{(2)(\text{gauge length})}.$$

EXAMPLE: The dial reading on the twist tester reads 315 for

a 10 inch gauge length. The twist per inch is:

$$\text{T.P.I.} = \frac{315}{(2)(10)} = 15.8.$$

The dial reading is divided by 2 since theoretically one-half the number of turns of the clamp was removing the twist, and the other half of the revolution was adding an equal amount in the opposite direction.

TWIST ANALYSIS DATA SHEET

Yarn No. 21/1 Date 6/15/–

Stock & Code CP34 Dept. #1 Spin.

Std. Twist, T. P. I. 14.9 Z Shift 1st

Std. Twist, T. M. 3.25 By RJM

Mach. No.	Sample No.	Twist Per Inch						Z or S
		1	2	3	4	5	Ave.	
16	1	15.8	14.6	16.7	13.6	16.4	15.4	Z
16	2	15.0	14.2	15.6	14.8	14.1	14.7	Z
7	1	15.7	15.0	16.4	14.9	16.0	15.6	Z
7	2	16.6	16.2	16.8	16.0	15.7	16.3	Z
9	1	15.5						

Fig. 25.2. Typical control form.

A typical control form for recording singles yarn twist is shown in Fig. 25.2.

Untwist Method

The untwist method of measuring the twist in a yarn is used

for both singles and ply yarns. The method involves placing the sample between the two jaws of the twist tester and removing all the twist in the sample. The amount of twist per inch is readily calculated from the total turns of the rotated jaw and specimen length.

Single Spun Yarn Test. For single yarn, the test specimen is usually 1 inch in length, although any length that is less than the staple length of the fibers can be used. This short length is necessary, for if longer lengths are used, the sample is usually destroyed before zero twist can be ascertained. This method is more accurate than the untwist-twist method, for the exact number of turns in the 1-inch length can be counted. However, it usually requires more testing to arrive at a reliable average since the variation in twist from inch to inch is much greater than it is for 10-inch lengths. The exact point of removal of all of the twist is difficult at times to locate, especially if the yarn is spun from a single end of roving in the creel of the spinning frame. If the yarn is spun from two ends of roving, the zero twist point is easier to locate due to the ease with which the two ends of roving can be separated.

Zero twist condition is determined physically by tearing the fibers of the sample apart with a sharp pick needle. Caution should be used in selecting and handling the samples so as to prevent distortion of twist.

The only calculation involved when this method is used is to obtain the average of the test data. Other statistical analysis of twist data such as range, standard deviation, standard error, and coefficient of variation may be made if deemed desirable.

Filament and Ply Yarn Test. To determine the twist in either continuous filament yarn or ply (or cable) yarns, lengths of 10 inches are generally used, although it is possible to use any available length up to the limit of the instrument. For filament and ply yarn, the jaw is rotated until practically all of the twist has been removed. At this point a sharp instrument such as a pick needle is inserted between the filaments or plies and the rotatable clamp turned until all twist is removed.

Cable Yarn Test. For cable type yarns the twist is removed from the cable and the number of turns necessary to remove the twist is recorded. Cut all but one strand of the ply yarn from be-

CABLE OR PLY TWIST ANALYSIS

Yarn Code #560 Date 6/15/-

Yarn No. 12.7/3/3 Dept. Twist

Std. Twist, T. P. I. Shift 2

Cable 10.6 S Ply 19.2 Z Single 20 Z By E

No.	Cable			Ply			Single	
	T. P. I.	L_c	L_p	T	T. P. I.	L_s	T	T. P. I.
12/1	10.5 (S)	10"	9.3"	179	19.3 (Z)	12.1"	480	19.8 (Z)
	10.7	10	9.5	182	19.2	11.9	497	20.9
	10.7	10	9.1	173	19.1	12.0	478	19.9
	10.5	10	9.2	175	19.1	12.3	490	19.9
	10.4	10	9.4	176	18.7	12.2	484	19.8
$\bar{X}$	10.6 (S)	10	9.3		19.1 (Z)	12.1		20.1 (Z)

Code 42 Example #2

	10"		14.8 (S)	10.9	366	16.8 (Z)
	10		14.6	10.8	356	16.5
	10		13.9	10.8	364	16.9
	10		14.8	10.6	371	17.5
	10		15.1	10.5	366	17.4
$\bar{X}$	10		14.6 (S)	10.7		17.0 (Z)

Fig. 25.3. Typical control form. L_c = length of original cable sample; L_p = length of ply; L_s = length of single yarn; T = number of turns.

tween the clamps and record the length of the remaining strand.

This length will differ from the original length due to contraction or elongation. Set the twist head to zero and remove all of the twist from the ply yarn. Record the twist direction and the number of turns necessary to remove the ply twist. Cut all but one end of the singles from between the clamps. Adjust the tension to the proper amount for the singles yarn, record the length of the test specimen, set the twist head to the zero point, and remove the twist. Use the method described above for filament yarn twist.

Ply or Cable Yarn Calculations. To calculate the twist in the different components of a ply and cable yarn, the data shown in Fig. 25.3 are needed.

In the first example of cabled yarn, the twist per inch is entered directly from the counter reading divided by the sample length of 10 inches (L_s). The length of the untwisted cable is entered under L_p (length of ply). Following the procedure previously given, a single strand of the ply yarn is untwisted to zero twist, and the total turns of twist counter entered under Ply, T (for example, 179 turns). From this value and the length, the T.P.I. is calculated and the length of the untwisted ply yarn entered under L_s. One end of the single is then tested by the untwist-twist method, the total turns entered, and the T.P.I. calculated by dividing by two times the length used. The direction of twist is entered at each stage.

In each case, the twist per inch is that in yarn as it is delivered from the respective machine.

The second example of a ply yarn in the same figure is included to show that the same form can be used for a ply yarn as well as for a cable yarn.

Twist Characteristics

Balanced Twist in Ply Yarns. Balanced twist is a term used to describe a ply yarn when the forces due to twist of the ply are equal and opposite to those of the component single yarns. The most common method used to test a ply yarn for balance is to take approximately a 1-yard length, hold one end in each hand, apply a slight tension to the yarn, and then bring the two ends together.

If the yarn hangs freely in a loop, the twist is considered balanced; if the loop tends to twist into a kink, the twist is not balanced. For a Z twist singles and S twist ply construction, if the loop turns in a clockwise direction, an excess of S twist is present; if the loop turns in a counterclockwise direction, there is an excess of Z twist present.

Corkscrew Twist. Corkscrew twist is a condition in which one end of a ply yarn forms a core, and the other end or ends twist around the center or core yarn. The core yarn supports most of the load until the yarn ruptures, and only then do the other yarns support any load. This results in a chain type of reaction with respect to the individual yarn breakage, and also in a considerable decrease in the total load the yarn is capable of supporting.

The test for this type of condition is either by visual examination or by use of the twist tester. To use the twist tester, a sample of yarn is placed between the jaws and all twist removed from the specimen. If all ends of the ply or cable yarn are under the same tension after the twist has been removed, there is no evidence of a corkscrew condition. On the other hand, if in a two-ply yarn one end is slack and the other is tight, the slack end will wrap around the tight end to form the corkscrew effect. The same condition can occur with multiple ply yarn.

Corkscrew yarns are generally caused by having uneven feed or tension on some of the component yarns being fed into the twisting zone of the twister. In fabric, corkscrew yarn can be the cause of second quality.

Effect of Contraction Due to Twist. Basically the contraction of a yarn due to twist is the reduction in length of the yarn caused by a change of the helical angle. The amount of contraction is largely dependent on the amount of twist and yarn diameter. Contraction is expressed as a per cent based on the change in length to the original length of the yarn before reduction in length occurs. This is represented by the following equation:

$$\%\ C = \frac{L_o - L_f}{L_o}\ (100)$$

where $\% C$ = per cent contraction; L_o = original length in inches; L_f = final contracted length in inches.

EXAMPLE. In Fig. 25.3, the ply yarn in the example at the bottom of the form (Code 42) has the following values:

$$L_p = L_f = 10''$$

$$L_s = L_o = 10.7''$$

$$\% C = \frac{10.7 - 10}{10.7}(100) = 6.5$$

For the cable yarn shown at the top of the same analysis form, the calculations are the same, although it will be noted that the ply length is shorter than the cable length. This condition in which the cable yarn elongates rather than contracts is not unusual in tire or industrial cord yarns. It occurs when single yarns are plied with both the single and ply yarn having the same direction of twist. The "twist on twist," as it is called, causes a high contraction in the ply yarn. Then when this ply yarn is used to make a cable construction, which would have in this case an S direction of twist, the S twist releases the high contraction forces in the ply, resulting in an elongation. This change is designated as a minus contraction.

In this portion of the example, for cable contraction,

$$L_c = L_f = 10.0$$

$$L_p = L_o = 9.3$$

$$\text{Cable contraction} = \frac{9.3 - 10}{9.3}(100) = -7.5\%$$

For the ply yarn contraction,

$$L_p = L_f = 9.3$$

$$L_s = L_o = 12.1$$

$$\text{Ply contraction} = \frac{12.1 - 9.3}{12.1}(100) = 23.1\%$$

Estimate of Contraction. Prediction curves for the contraction of combed cotton yarns have been drawn from a formula based on work by R. Soedibjo Hardjopertomo (1). Although the experimental work was based on only four combed yarns (20/1, 30/1, 40/, and 50/1) of high commercial quality, it is believed that results would be quite reliable if used beyond the range studied. Furthermore, the contraction for single carded yarns could be predicted from these same curves with the expectation of reasonable accuracy. The laboratory procedures used in obtaining the data followed a new approach: The spun yarns were analyzed to determine the contraction existing in yarns "as is" by untwisting to zero twist using short lengths and determining length changes with a cathatometer. Twist was both removed and added to existing yarn to get contraction changes over the range from zero to maximum twist used in the industry. The formulas developed from this study are as follows:

$$
\begin{aligned}
C &= 10^{(-1.305-0.022N)}\, T^{(1.93)} \\
&= 10^{(-1.305-0.022N+0.965 \log N)}\, (TM)^{1.93}
\end{aligned}
$$

where C = per cent contraction; N = cotton yarn number; T = twist per inch; TM = twist multiplier.

The corrected data produce straight lines on log-log paper. The contraction curves based on the work are shown in Fig. 25.4.

EXAMPLE. The solution of the equation for a twist of 25 turns per inch in 35/1 yarn is as follows:

$$
\begin{aligned}
C &= 10^{(-1.305-0.022N)}\, T^{(1.93)} \\
&= 10^{(-1.305-0.022\times 35)}\, 25^{(1.93)} \\
&= 10^{(-2.075)}\, 25^{(1.93)} \\
\log C &= -2.075 \log 10 + 1.93 \log 25 \\
&= -2.075 + 2.698 = 0.623 \\
C &= 4.2
\end{aligned}
$$

Yarn Number Correction (Indirect System). It is sometimes necessary to get a corrected yarn number from the uncontracted yarn number and contraction. The formula for this is developed as follows. Let N_o = original uncontracted yarn number; N_f = final or corrected yarn number; W = yarn weight; $\%\, C$ = per cent contraction; L_o = original length; L_f = final contracted length.

From the basic relationship of yarn number to length and weight,

$$N_o = \frac{L_o}{W} \tag{1}$$

and

$$N_f = \frac{L_f}{W} \tag{2}$$

As W remains constant,

$$\frac{L_o}{N_o} = \frac{L_f}{N_f} \tag{3}$$

The formula previously developed for contraction is

$$\%C = \frac{L_o - L_f}{L_o} (100) \tag{4}$$

or,

$$L_f = L_o\left(1 - \frac{\%C}{100}\right) \tag{5}$$

Substituting L_f from (5) in equation (3),

$$\frac{L_o}{N_o} = \frac{L_o\left(1 - \frac{\%C}{100}\right)}{N_f}$$

$$N_f = N_o\left(1 - \frac{\%C}{100}\right) \tag{6}$$

For a ply yarn, the equivalent yarn number of the ply would be calculated as above, with the correction for the plies made by dividing by the number of plies.

For ply (or cable) yarn,

$$N_f = \frac{N_o}{\text{Plies}}\left(1 - \frac{\%C}{100}\right) \tag{7}$$

EXAMPLE. Two ends of 20/1 yarn are plied. A sample of the single yarn used in the ply contracted during the twisting process from 18 inches to 17 inches. Determine the yarn number of the 20/1 after twisting.

$$\%C = \frac{18 - 17}{18}(100) = 5.5\ \%$$

$$N_f = 20\left(1 - \frac{5.5}{100}\right) = 18.9.$$

Thus, each end of the two-ply yarn has contracted from 20/1 to 18.9/1, and the ply yarn size would be 9.45 equivalent, or 18.9/2.

Example. Refer to Fig. 25.3, Yarn Code 560. Assume the cable yarn construction is 12.7/3/3. Determine the final cable yarn number, correcting for contraction at each process.

For the 12.7/3 yarn,

$$N_f = \frac{12.7}{3}\left(1 - \frac{23.1}{100}\right)$$

$$= \frac{12.7}{3}(0.769)$$

$$= 3.25 \text{ equivalent yarn number.}$$

For the cable (made of three ends of 3.25),

$$N_f = \frac{3.25}{3}\left(1 - \frac{-7.5}{100}\right)$$

$$= \frac{3.25}{3}(1.075)$$

$$= 1.16 \text{ equivalent single yarn number.}$$

If the correction for contraction were omitted, the final yarn size would have been simply

$$12.7 \div 9 = 1.41 \text{ equivalent single yarn number.}$$

This failure to consider contraction introduces a large error into the calculation: By the correct method, the 1.16 yarn has 974.4 yards to a pound, whereas the latter incorrect method shows 1184.4 yards to a pound, a difference of 210 yards in each pound of yarn.

Yarn Number Correction (Direct System). In the direct systems of yarn numbering, such as the denier system, the yarn number is always equal to the ratio of weight to length.

By following the same pattern of reasoning as was used for the cotton numbering system, it can be shown that

$$N_f = N_o \div \left(1 - \frac{\%C}{100}\right)$$

Therefore, for the denier, Tex, and other direct systems, use the above relationship.

Effect of Contraction on Turns Per Inch. In calculating the twist and twist gear to be used in a spinning or twisting process, due regard must be paid to the effect of contraction on twist. This holds true for all ring-type machines, where twist constants are based on delivery roll sizes in relation to spindle speeds. For upstroke twisters the effect of contraction is not a part of the twist calculation, for the take-up drum winds the contracted length.

As an example of this effect of contraction in a cotton-type ring twister, suppose that the machine is geared to insert 16 turns per inch, based on the twist constant. During twisting, the yarn contracts 20 per cent. Thus, the 16 turns are inserted in the 1-inch length, which contracts to a length of (1 — 0.20) or 0.8 inch. The twist per inch has become

$$16 \div 0.80 = 20 \text{ turns}$$

Control Program. The first important point in a quality control program covering twist is to check the twist per inch in any yarn whenever a frame is first started after a changeover. In fact, after a changeover, no spinning frame or twister should be allowed to produce any more yarn than is needed for test purposes without written approval from the laboratory. Once the frame is in production, the amount of routine testing depends to some extent on the end product. For yarns going into mechanical fabrics and for industrial use, a weekly check on each frame should cover most needs. For yarns for non-mechanical or industrial use, if the mill has a sound policy for controlling changeovers, then a monthly check should be ample. The number of tests for routine work vary from four to six bobbins for each frame, and six to eight on the start-up after a changeover. The number of tests per bobbin should be five when 10 inch test specimens are used, or at least ten if 1-inch samples are used.

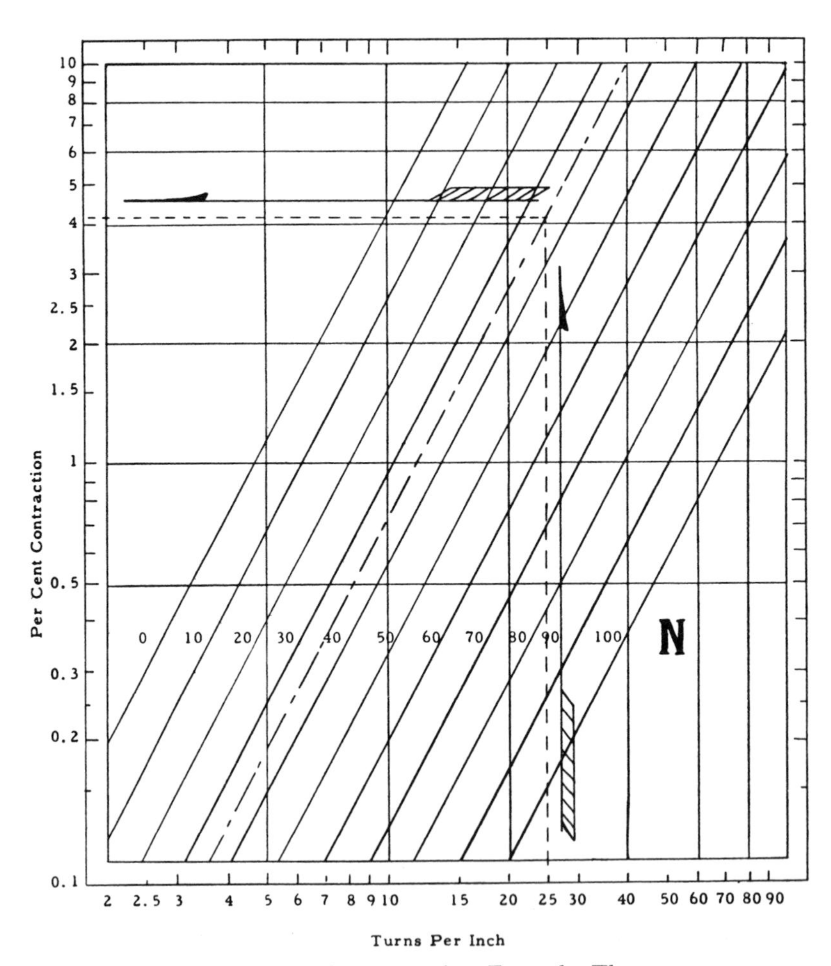

Fig. 25.4. Prediction curve for contraction. Example: The per cent contraction for a 35s yarn with a T.P.I. of 25 will be 4.2 %.

The coefficient of variation for twist test on staple synthetic yarns should range from approximately 4 to 12 per cent. For filament yarns, the value should range from 2 to 8 per cent, depending upon the type of yarn and the amount of twist.

A.S.T.M. recommends a coefficient of variation of 20 per cent for cotton yarns. The coefficient of variation is influenced by yarn evenness; nevertheless, average to high-quality mills should seldom approach this 20 per cent value.

References

1. Hardjopertomo, R. S., *Thesis*, School of Textiles, N. C. State College, 1953.

CHAPTER XXVI

Additional Tests for Fibers and Yarns

Weight per Unit Length of Cut Staple

Weight Method. The linear density of cut staple is determined by weighing known lengths of fibers, counting, and calculating the denier from these data. Tests should be carried out under standard conditions, which gives the fiber number at standard regain rather than at commercial regain. If the staple is cut irregularly or in random lengths, and if the fibers cannot be measured for length without distortion resulting from crimp, then this test is not recommended.

The procedure recommended by Committee D-13, A.S.T.M., is as follows: A small tuft of 500 to 1,000 fibers is selected, made parallel and free of short fibers by hand combing. The group of fibers is then held under tension to remove crimp and cut to a length preferably longer than 22.5 mm. Groups of fibers of 0.5 mg. are weighed simultaneously to the nearest 0.002 mg.

The average fiber number in denier units is calculated from the formula

$$\text{Denier} = \frac{(\text{Mgms. weight})(9000)}{(\text{Mms. length})\ (\text{No. of fibers})}$$

Vibroscope Method. The Vibroscope method of determining the weight per unit length is useful where single fiber measurements are desired, and in those cases where the direct weighing or direct measurement of length is difficult. The method is based on the principle that a known length of material when held under a known tension has a natural frequency of transverse vibration. This frequency is also a function of the weight per unit length. Therefore, if a fiber is held at one end and subjected to a tension by a hanging weight on the other end with a predetermined length between the two, and if the fiber can be subjected to a source of alternating energy so as to cause transverse vibration in the fiber, then the weight per unit length (i.e., denier) can be calculated.

The basic formula (1) for the linear density is

$$d = \frac{T}{4L^2f^2}$$

where d = density in grams per centimeter; T = tension in dynes; L = length in centimeters; f = the frequency in cycles per second. To convert d to units of denier, multiply by 0.9×10^6; if T is measured in grams, multiply the result by 980 to convert.

Instruments have been developed and are available from commercial sources (Insco Corporation, Groton, Mass., and G. F. Brush Associates, Princeton, N.J.). Vibration of the fiber can be attained by attaching one end of the fiber to a transducer to give mechanical vibration, or by supporting the fiber between two electrodes to give electrostatic vibration. The vibration of the fiber is detected by viewing it through a low-powered microscope, or by projection of the enlarged image onto a screen. The vibration should be fundamental, i.e., the fiber forms a single balloon shape over its entire length, rather than the second or third harmonic, i.e., forming double or triple balloons of this shape:

◇◇, ◇◇◇

Air Flow Method. The average linear density of a discrete sample can be determined by using the air flow method. The procedure follows that described for cotton using the Micronaire. However, the scale used must be one calibrated for the class of fiber being tested, with the class being based on the specific gravity of the fiber. Mixed deniers of the same class of fibers can be determined using this instrument, but mixed deniers of fibers of radically different specific gravities will give erroneous results.

Wet Strength and Elongation of Filament Yarn

To measure the change in physical properties of filament yarn when it is completely saturated with water, the strength and elongation of yarns are measured after the yarn has been immersed in distilled water containing a wetting agent. This test gives a measure of the reaction of the yarn to wearing and laundering conditions. It also provides a means of getting inter-laboratory tests which have been performed under identical conditions. Many fiber producers use this method of testing for their routine quality control, for it saves time and eliminates the need for controlled atmospheric conditions.

Procedure for Test

1. Prepare a skein on a standard reel and cut the skein. The skein may be of any given length, depending upon the number of individual lengths needed.
2. Place the group of yarns on a convenient rack and make the desired number of breaks from the conditioned sample. (Note: If more than one yarn is to be tested, all dry or conditioned tests should be completed before any wet yarns are placed in the test instrument. This is to prevent wetting the clamps before all dry tests are completed. One suggestion is to make all dry tests during the first part of the day and all wet tests during the latter part of the day.)
3. Take a sufficient number of the remaining yarns and immerse in a water bath containing a suitable wetting agent.
4. Remove one piece of yarn from the water after it has become thoroughly wet and place the specimen in the testing machine. (Care should be taken that the sample is not stretched.)
5. Continue testing until the desired number of tests are completed.

Calculations. The test results should be reported as the per cent loss in strength based on the dry yarn strength, or as wet strength in grams per denier. For example, if the dry strength is 4.20 g.p.d. and the wet strength is 2.73, the per cent loss in strength would be:

$$\%\ \text{loss} = \frac{4.20 - 2.73}{4.20}(100) = 35\ \%$$

The elongation should be calculated in the same manner as for dry yarn and expressed as a per cent.

Knot Strength of Filament Yarn

The knot strength test is a measure of the strength of a test specimen with an overhand knot tied in the yarn. It gives an indication of the brittleness of the yarn. The elongation of the yarn in this test is not of interest.

Procedure for Test

1. Select a sample of the yarn to be tested and tie a single overhand knot at approximately the middle of the specimen.
2. Place the sample in the testing machine with the knot at approximately equal distance from each jaw.
3. Break the specimen and record the strength in grams per denier.
4. Make a sufficient number of standard tests from the same package and record the results in units of grams per denier.

Calculations. The results should be expressed in units of per cent loss in strength. For example, if the regular strength of a yarn is 3.15 g.p.d., and the knot strength is 2.86, the per cent loss in strength would be:

$$\% \text{ loss} = \frac{3.15 - 2.86}{3.15} (100) = 9.2\,\%$$

To eliminate calculating the grams per denier for each break, the total grams strength may be used; for example, if the yarn in the previous example had been 100 denier, then the regular and knot break would have been 315 grams and 286 grams respectively, and the answer would have been the same.

The strength is also expressed as knot-breaking strength in grams, or in grams per denier.

Elongation resulting from the knot test is of no significance and is generally not calculated.

Loop Strength of Filament Yarn

The loop test consists of determining the strength of a compound strand formed when one strand of yarn is looped through another and then broken. The results of this test when compared with the standard strength of the yarn are an indication of the brittleness of a yarn. However, this test is not generally considered as sensitive as the knot test.

Procedure for Test

1. Select two strands of yarn.
2. Form one strand into a loop by placing the ends together.
3. Place the loop in one jaw of the tester so that the loop extends approximately half the distance between the jaws.
4. Place the other strand through the loop, bring the two ends together, and fasten in the other jaw.
5. Break the sample and record the results in grams.
6. Repeat until a sufficient number of tests have been made.
7. Make a sufficient number of standard tests from the same package and record the results in grams per denier.

Calculations. The results of the loop test should be reported as the per cent loss in strength.

$$\% \text{ loss} = \frac{\text{g.p.d. of standard break} - \frac{1}{2}(\text{g.p.d. of loop break})}{\text{g.p.d. of standard break}} (100)$$

To eliminate calculating the grams per denier for each test, the formula can be reduced to

$$\% \text{ loss} = \frac{\text{gms. of standard break} - \frac{1}{2}(\text{gms. of loop break})}{\text{gms. of standard break}} (100)$$

Additional Tests for Man-Made Yarns

In addition to the tests outlined above, other measurements used primarily for filament yarns are filament count, commercial weight, examination for visual defects, dye affinity, and shrinkage.

A description of the procedure for measuring the number or count of yarn is given in the chapter on yarn numbering systems. The commercial weight is discussed in both the chapter on moisture and the chapter on staple synthetic fibers.

The counting of filaments and the measuring of dye affinity and shrinkage are routine control tests by the fiber manufacturer. These tests should also be made occasionally at the spinning or weaving mill. The shrinkage and dye affinity tests are more important than the filament count and should be made on a routine basis. The filament count is generally used only when a case or shipment of mixed yarn of the same denier is received.

To determine the dye affinity, it is customary to select some direct dye which is easy to apply and dye knitted samples of the yarn and either compare the color with a standard sample, or compare the shades within the same sample.

The shrinkage test usually consists of measuring a length of yarn under a given tension and then scouring the yarn for thirty minutes at 80 to 90° C. in a mild soap solution. The yarn is rinsed, dried, conditioned, and the length again measured under the same tension. The change in length is shrinkage and is expressed as a per cent of the original length.

The procedure suggested for yarns spun of Orlon uses a 120-yard length skein, reeled in the conventional manner, and laced loosely at four places. The skeins are hung over a rod, and loaded by a dead weight of $3\frac{1}{2}$ pounds for yarns to 10/1 cotton count, $2\frac{1}{2}$ pounds for 10/1 to 20/1; $1\frac{1}{2}$ pounds for 20/1 to 30/1, and $\frac{3}{4}$ pound for 40 to 60. After a delay of 30 seconds, the skein length under the load is measured. The skeins are then removed, twisted loosely, wrapped in cheesecloth, boiled for 30 minutes in water with mild soap, extracted, removed from the cheesecloth, and air dried. The final length is determined in the same manner as was used prior to the treatment. Shrinkage is then calculated. For complete test details, refer to Du Pont Textile Fibers Technical Information reports.

The filament count is normally made by one of two methods. One technique is to separate the filaments with a pick needle and count the filaments as they are separated. When using this method, the yarn should be placed on a black velvet board or on some contrasting color if the yarn has been dyed. The other procedure is to make cross sections of the yarn and count the filament with the aid of the microscope. By using this method, the cross section of the filaments can be examined for inconsistencies. However, this technique is more time-consuming and requires more expensive equipment. A third method possible with certain synthetics consists of inducing a static charge in zero twist yarn. By this charge, the filaments can be made to repel one another and form a balloon-like structure, enabling the technician to count the filaments.

The Manra yarn filament counter made by Newmark (London) Distributors Ltd. under license from the British Rayon Research Association provides another method recently developed for this work. The instrument is designed to "count rapidly and automatically the number of filaments in a continuous filament yarn. The measurement technique consists of loading the yarn into the instrument clamps, when the number of filaments is displayed on a decatron type counter. The time for a test is approximately eight seconds."

Reference

1. "Proposed Method for Test for Linear Density Fibers by the Vibroscope," ASTM Committee D-13.

CHAPTER XXVII

Approach to Evenness Measurements

The most significant advance in the field of textile quality control that has occurred in the past few years has been in the measurement of evenness or uniformity. One impetus to this development in the United States was the introduction by the Saco-Lowell Shops of their evenness tester, designed to measure mechanically the variations that exist in sliver and coarse rovings. A second unit, the Saco-Lowell lap meter, added a means of measuring the irregularities that exist in picker laps. The information supplied by these testers gave a clear indication of the need to be able to analyze more rapidly and accurately the uniformity of all textile slivers, rovings, and yarns. The Saco-Lowell evenness tester was a valuable tool, but its use, confined to sliver and coarse rovings, failed to meet all the needs of quality control men.

As a result of this need for new methods of measuring uniformity, three instruments were placed on the market in this country in quite rapid succession. The first of these, the Uster evenness tester, because of its early entry into the field, is perhaps the one in widest use. The Brush uniformity analyzer, which was introduced at about the same time, following its design at the Institute of Textile Technology, has also received wide acceptance. The third machine, the Pacific evenness tester, depending on mechanical measurement of the textile material, has found its widest acceptance in the woolen and worsted industry, and where heavier materials are involved. All these instruments are reliable and have been major contributions to this growing branch of quality control.

Prior to the development and introduction of these newer testers, a considerable amount of basic research work was performed in England by such researchers as Gregory, Martindale,

Foster, and Cox. Most of the results of this research work have been published in either the *Journal of the Textile Institute* or the *Shirley Institute Memoirs*. A partial bibliography of these works is at the end of this chapter. Resulting from some of this work was the now famous Martindale theory, which is the basis for modern concepts of yarn uniformity, and which will be discussed later. Other major contributions to the science of evenness measurements have been made by Hans Locher of the Zellweger Corporation at Uster, Switzerland.

Classification of Variation. It has been suggested by Locher that the types of variation be classed in three categories: short-term, medium-term, and long-term variations. The short-term variation would be confined to those irregularities occurring over distances not greater than ten times the fiber length, medium-term variation to irregularities occurring over distances from ten to one-hundred times the fiber length, and long-term variation would be irregularities occurring with a frequency equivalent to or greater than one-hundred times the fiber length. Other suggestions concerning the classification of variation include only short-term and long-term where the short-term is variation equivalent to approximately the draft of the drafting element, and long-term variation is any variation greater than the immediate preceding draft. In any event, the long-term variation includes such non-uniformity as is found from inside to outside of a package or between packages from the same producing agent. Where drafts are high, short term variation in the materials fed tend to become long-term variations in the material delivered. This also means that the short term variation in the preparatory operations such as picking and carding become long-term variation in the yarn. In this connection, a considerable amount of discussion has centered around the importance of picker lap uniformity and how this may influence the uniformity and appearance of the finished yarn. Most quality control engineers agree that the importance of picker lap evenness is a function of the number of doublings in the entire production line. Since the present trend in the industry is to reduce the number of doublings, in some cases to as low as six, it is obvious that the control of picker lap evenness is a critical point in any quality control program. One statement often heard is "it is impossible to make a uniform yarn unless the lap and slivers used in manufacturing this yarn are uniform."

Factors Influencing Uniformity. There is almost an unlimited source of variables which may contribute to the unevenness of a textile yarn. Among these are the fiber properties of length and fineness. Fiber length and length distribution have a direct bearing on fiber control during the drafting operations, and influence directly the evenness of the finished yarn. This particular aspect of evenness has been studied very extensively, and the behavior of these variables has been entitled "drafting waves."

Fiber fineness influences yarn evenness, for fineness determines the number of fibers in the average cross section for any given size of yarn. Thus, a yarn made from cotton of a 4.0 microgram per inch fineness will average 25 per cent more fibers in a cross section than the same size yarn spun from cotton of 5.0 micrograms per inch fineness. According to the Martindale theory to be discussed later, uniformity is a function of the number of fibers in the cross section of sliver, roving, or yarn.

In addition to fiber properties, there are many machinery defects that may contribute to the variation in laps, sliver, roving, and yarn. In fact, improper adjustment and poor maintenance are the chief causes of abnormal unevenness. Listed below is a partial compilation of the causes of such defects:

Pickers. 1. Extreme variation in level of stock in the hopper. *2.* Worn belts on evener motion. *3.* Waste in evener motion or sluggish action of evener motion. *4.* Improper setting of blending reserve. *5.* Waste in air ducts or on condenser screens. *6.* Uneven pressure on lap during formation.

Cards. 1. Eccentric doffer, cylinder, or licker-in. *2.* Improper setting of the card. (Improper setting could cause defects at different periods, depending upon which setting was causing the non-uniformity in the sliver.) *3.* Wrong tension draft between doffer comb and condenser trumpet. *4.* Uneven or jerky operation of coiler.

NOTE: A period varying from 14 to 20 inches frequently occurs in card sliver which is due to the fold or twist of the sliver while being fed into a 12-inch can. The source of this defect is very frequently overlooked and is very difficult to locate.

Drawing frames. 1. Rolls out of round; concave or convex rolls. *2.* Worn and eccentric top rolls. *3.* Improper roll weights. *4.* Improper roll setting. *5.* Dirty rolls. *6.* Chipped or worn gears in drafting element. *7.* Incorrect tension draft. *8.* Improper trumpets.

Roving frames. 1. Bent or eccentric bottom or top rolls. *2.* Improper roll settings. *3.* Dirty bottom rolls. *4.* Improper weighting of top rolls. *5.* Stretched or grooved aprons, dirty aprons, or apron cradles. *6.* Bent, rusty, or out-of-balance flyers. *7.* Incorrect tension gear, causing stretch. *8.* Worn spindles. *9.* Mixed diameter roving bobbins.

NOTE. Long-term variation at the roving process may be traced and found to be short-term variation in the previous process.

Spinning frames: *1.* Eccentric or worn top roll coverings. *2.* Bent or eccentric bottom rolls. *3.* Improper weighting of top rolls. *4.* Improper roll settings. *5.* Trash or waste in the aprons of the drafting system. *6.* Improper delivery of roving due to faulty skewers or skewer holders. *7.* Improper twist in roving. *8.* Badly worn rings or travelers. *9.* Out-of-set spindles, guides, or rings.

NOTE. Long-term variation in yarn very frequently is due to short-term variation in the roving.

Fibers Per Cross Section, Cotton Yarn. The fibers per cross section, N, are calculated by the following relationship for cotton yarns:

$$N = (\text{micrograms/gram})\ (\text{grams/pound}) \div (\text{hanks/pound})\ (\text{yard/hank})\ (\text{inches/yard})\ (\text{micrograms/fiber inch})$$

$$= (1{,}000{,}000)(453.59) \div (\text{yarn no.})\ (840)(36)\ (\text{fineness})$$

$$N = 15{,}000 \div (\text{yarn no.})\ (\text{fineness})$$

For a fineness of 4 micrograms and a yarn count of 40/1, the number of fibers per cross section would be:

$$N = 15{,}000 \div (40)(4) = 93.7$$

For a 40/1 made from a 3-microgram cotton, the fibers per

cross section would be

$$N = 15{,}000 \div (40)(3) = 125.$$

Expected Coefficient of Variation, Cotton Yarns. The simplified version of the Martindale theory is based on the assumption that the random distribution of fibers in a yarn follows the Poisson type. Since the standard deviation for this type of distribution is estimated by the $\sqrt{N}$, and the coefficient of variation is

$$\frac{\sigma}{\bar{X}}(100)$$

the coefficient of variation for the fiber distribution in a yarn becomes:

$$V = \frac{\sqrt{N}}{N}(100) = \frac{1}{\sqrt{N}}(100)$$

where N = number of fibers per cross section.

For the 40/1 spun from 4-microgram cotton, the expected coefficient of variation would be:

$$V = \frac{1}{\sqrt{93.7}}(100) = 10.4\ \%$$

For the 40/1 spun from 3-microgram cotton, the expected coefficient of variation would be:

$$V = \frac{1}{\sqrt{125}}(100) = 9.0\ \%$$

By comparing the expected variation for the 40/1 made from the two cottons, the influence of fiber fineness on yarn variation becomes apparent.

In combining the relationships of

$$N = 15{,}000 \div (\text{yarn no.})(\text{fineness})$$

(where the fineness is in micrograms per inch), and

$$V = \frac{1}{\sqrt{N}}(100)$$

by substituting for N, the relationship becomes

$$V = \frac{1}{\sqrt{15{,}000 \div (\text{yarn no.})(\text{fineness})}}(100)$$
$$= 0.82\sqrt{(\text{yarn no.})(\text{fineness})}$$

Using the previous example of a 40/1 with a cotton having a fineness of 4 micrograms, the expected coefficient of variation is

$$V = 0.82\sqrt{(40)(4)} = 10.4\ \%$$

Fibers Per Cross Section, Staple Synthetic Yarns. The number of fibers per cross section, N, of a yarn manufactured from staple synthetic fibers is calculated by dividing 5315 by the yarn number times the average denier of the staple, i.e.,

$$N = 5315 \div (\text{yarn no.})\ (\text{ave. denier})$$

The yarn number is on the English System. The value 5315 is the conversion constant to go from denier to cotton number. Thus, 35/1 cotton yarn is equivalent to 5315 ÷ 35 = 152 denier. By dividing the total denier of a yarn by the average denier per fiber, the result is the number of fibers in the cross section.

Expected Coefficient of Variation, Staple Synthetic Yarns. Make a similar substitution to the one used for cotton to calculate the coefficient of variation.

$$V = \frac{1}{\sqrt{N}}(100)$$

$$V = \frac{1}{\sqrt{5315 \div (\text{yarn no.})\ (\text{ave. denier})}}(100)$$

Simplified, the relationship becomes:

$$V = 1.37\sqrt{(\text{yarn no.})\ (\text{ave. denier})}$$

For a 20/1 yarn spun from 2.0 denier staple, the coefficient of variation is:

$$V = 1.37\sqrt{(20)(2)} = 8.6\ \%$$

Irregularity Index. In connection with the expected and actual coefficient of variation of a yarn, it has been suggested by Locher and others that the ratio between the two variations be used. This is to eliminate primarily the influence of material diameter, and is known as the irregularity index, *I*, which is the actual coefficient of variation divided by the expected coefficient of variation, or

$$I = \frac{\text{C.V. a}}{\text{C.V. t}}$$

Obviously, for a perfect yarn this value would be one; however, in actual practice it varies according to the type of material being tested. The standards for *I* which have been established are shown in Table 29.12.

Partial Bibliography on Evenness

Gregory, J., *J. Textile Inst.*, **41**, Tl–T52 (1950).
Martindale, J. G., *J. Textile Inst.*, **41**, P340–P355 (1950).
Keggin, J. F., Morris, G. and Yuill, A. M., *J. Textile Inst.* **40**, T702 (1949).
Priestman, H., *Bull. Natl. Assoc. Manufacturers, U.S.A.*, p. 129 (1910).
Martindale, J. G., *W.I.R.A. Publication* **177**, 9 (1945).
Balls, W. L., *Studies of Quality in Cotton*, Macmillan, London, 1928, Ch. 8.
Foster, G. A. R., *Shirley Inst. Memoirs*, **A2**, 313 (1929).
Foster, G. A. R., *J. Textile Inst.*, **41**, P357–P375 (1950).
Goodings, A. C., *J. Textile Inst.*, **22**, T1 (1931).
Foster, G. A. R., and Martindale, J. G. *Shirley Inst. Memoirs*, **A5**, p. 125, 1941.
Cox, D. R., Proc. Royal Soc. (London), **A197**, 28 (1949).
Martindale, J. G., *J. Textile Inst.*, **36**, T35 (1945).
Martindale, J. G., *W.I.R.A. Publication* **180** (1947).
Barella, A. T., *Textile Inst.*, **43**, A929–A930 (1952).
Cox, D. R., and Townsend, M. W. T., *Textile Inst.*, **42**, P152–P161 (1951).
Enrick, N. L., *Textile World*, **103**, 130–131 (1953).
Hasler, A., and Honegger, E., *Textile Research J.*, **24**, 73–85, (1954).
Honegger, E., *Textile Research J.*, **26**, 351–354 (1956).
Keyser, W. R., Middleton, J. O., and Dougherty, J. E., *Textile Research J.*, **27**, 466 (1957).

CHAPTER XXVIII

Evenness Testers

Nearly all evenness testers currently in use in the textile industry may be classed as either capacitance or mechanical types, or a combination of these two. However, Beta gauge and photoelectric types are being used to a much lesser extent. As mentioned previously, the testers most commonly used by American textile mills are the Uster, Brush, and Pacific testers. Other testers which have appeared on the market are the Fielden Walker, Serc, and Turl-Boyd. The Uster and Brush instruments have proven very popular with mills processing cotton and spun synthetic fibers and yarns. The Pacific tester has become more popular with woolen and worsted mills that process heavy slivers, rovings, and yarns.

Uster Evenness Tester

The Uster evenness tester (Fig. 28.1), one of the first capacitance-type testers developed, is manufactured by the Zellweger Company of Uster, Switzerland, and sold in this country by the Uster Corporation of Charlotte, North Carolina. The Uster tester operates on the principle that a change in the mass of the dielectric (non-conducting material) in the condenser will change the capacitance. The measuring comb of the tester is actually a set of condensers with an air dielectric. When some material other than air is placed between the plates of one of the condensers, the capacitance is changed in proportion to the mass of that material. As the mass of the material changes, the capacity of the condenser changes in proportion. The shape of the material must also be considered. As long as the cross section of the material is approximately round, the theory outlined above is true. If the material is ribbon-shaped, as for instance low twist, continuous filament rayon yarn, considerable error will result unless auxiliary equipment such as a "Rotofil" is used. The Rotofil simply inserts a false twist in the yarn so that the ribbon-effect is eliminated.

The method used to determine the amount of change of capacitance is to measure the change in frequency of an oscillator in

Fig. 28.1. Uster evenness tester with recorder and integrator.

which the condenser is part of the circuit. In the Uster tester, two oscillators are balanced against each other with an adjustment knob provided on the panel to manually equalize the frequencies. Inserting material in a slot changes the frequency of one oscillator. This oscillator is compared then with the other base oscillator by a mixer which determines the mass of material in the slot at the instant. The result is then amplified, sent through the range dividers, gain control, amplified again, and sent to the integrator and recording device. This is shown in Fig. 28.2.

The Uster tester has proven to be a versatile instrument for measuring the uniformity of materials ranging from extremely heavy sliver to very fine yarns. In addition to this, the instrument can be equipped with such additional attachments as an integrator, a lap Vari-Meter, the sliver lap unwinding device, and the Rotofil. The integrators which will be discussed in greater detail later in this chapter are supplied in two models. One model measures the mean linear deviation, and the other model measures the coefficient of variation. The Vari-Meter is an attachment which is mounted in the calender section of the picker, and,

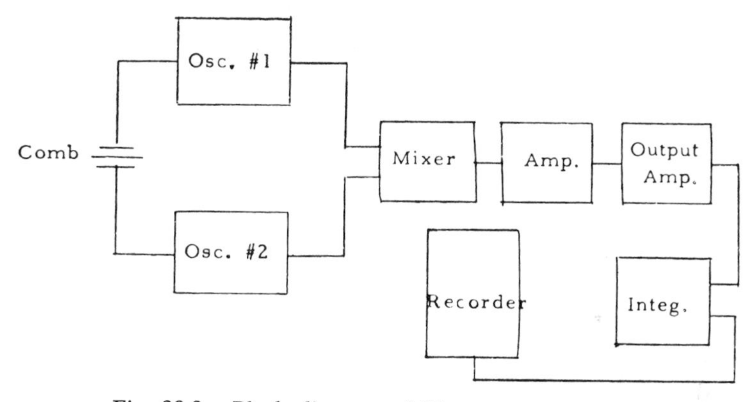

Fig. 28.2. Block diagram of Uster evenness tester.

through the circuit of the evenness tester, measures and records the irregularity of the picker lap. The sliver lap unwinding device is for testing sliver laps on the Model B evenness tester. The Rotofil inserts false twist into low twist filament yarn to eliminate the shape effect of the ribbon-like yarn.

It is not always advantageous to test all materials under identical conditions. Differences in the strand thickness and magnitude of thickness variation, spacing of variations, ability to withstand the stresses encountered at high speeds, as well as the purpose of the analysis, demand consideration. The several pairs of measuring combs provide a means of accommodating strands of various thickness; the speed at which the material passes through the Model B machine can be set at either 2, 4,

8, 25, 50, or 100 yards per minute; chart speeds can be controlled at either 1, 2, ,4 10, 20, or 40 inches per minute; and full scale sensitivity can be pre-selected at either ± 100 %, ± 50 %, ± 25 %, or ± $12\frac{1}{2}$% variation. A combination of these variables may be selected to exaggerate certain types of defects for more careful study.

The dimensions of the electrodes used in the scanning system are 8 millimeters for yarn, 16 millimeters for roving, and 32 millimeters for sliver. The length of the scanning distance determines whether or not the short-term variations existing, especially in yarn, can be recorded. This scanning is not done ideally by cross section of infinitely short length but by finite length, which is called the measuring field length. For example, for yarn this is 8 millimeters. The finitely limited length of the scanning distance results in the determination of an average value of the cross section of the material between the electrodes. The cross sections of slivers, rovings, and yarns represent a multitude of superimposed wave lengths, ranging from a fraction of an inch to several yards in length.

Fundamental investigations have revealed that the irregularity of spun yarns, i.e. the variations in fibers per cross section, follow a wave length that corresponds to two to three times fiber length, and furthermore that these wave lengths are of the highest amplitude. For example, the irregularity of a cotton yarn spun from 1-inch staple occurs in waves of approximately 2 to 3 inches in length; the amplitude within these waves is 95 per cent of actual using an 8-millimeter scanning field. On the other hand, if the field were increased to 16 millimeters length, the wave amplitude would be considerably reduced, for the sensitivity to very short length irregularities is lowered by the averaging effect of the longer measuring field. With a 32-millimeter field, the reduction would be more pronounced, so that an inert type of test condition becomes evident.

In addition to testing materials under the "normal" setting, there is also an inert test which may be used when variation over extremely long lengths is to be studied. As a result of setting the instrument on the inert test, the electrodes of the measuring field

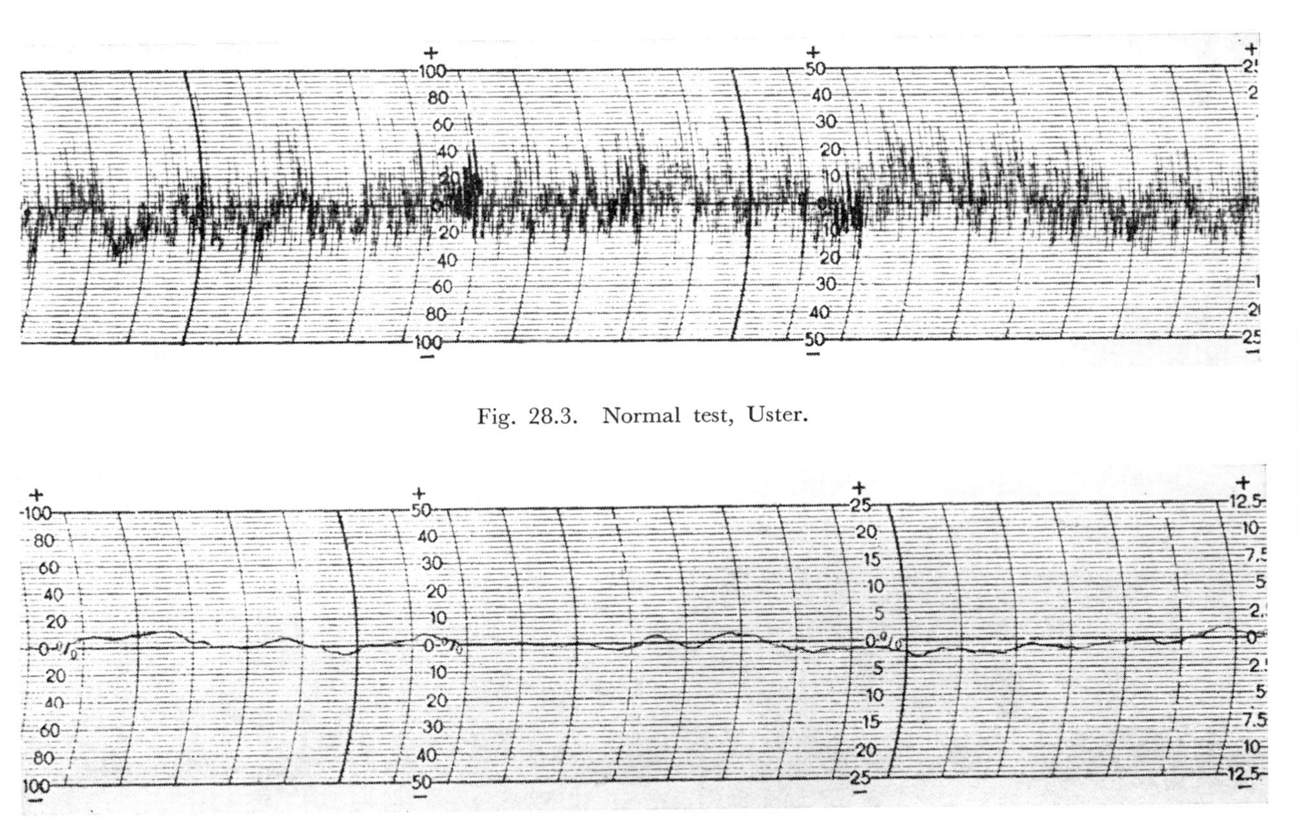

Fig. 28.3. Normal test, Uster.

Fig. 28.4. Inert test, Uster.

are in effect lengthened. The effect resulting from this is shown in Fig. 28.4. The yards per minute speed of the material and the apparent or effective measuring length of the electrode, are shown in Table 28.1. This apparent length is equivalent to a cut length in inches and is comparable with the results of weighing of similar lengths of materials cut to the specific lengths shown.

TABLE 28.1

Effect of Material Speed Using the Inert Test on the Uster Tester [a]

Material speed, yards per min.	Apparent electrode length, ins.
1	2.5
2	5
4	10
8	20
25	60
50	114
100	225

[a] Source: Uster Corporation.

It is beyond the scope of this book to present all the technical information pertaining to the Uster and other capacitance type evenness testers. Mr. Hans Locher and his staff at the Zellweger Corporation and Uster Corporation in Charlotte, North Carolina, have issued a series of technical bulletins and books pertaining to this type of tester. They have done an excellent job in supplying both theoretical proof and also practical applications to the science of evenness testing. The authors are indebted to both the Zellweger and Uster Corporations for supplying technical information and suggestions pertaining to the presentation of this material on evenness testing.

Brush Uniformity Analyzer

The Brush uniformity analyzer is an electronic instrument which will semiautomatically measure the variations in weight per unit length of yarn, roving, or sliver. These variations, which are usually recorded as per cent non-uniformity, encompass both

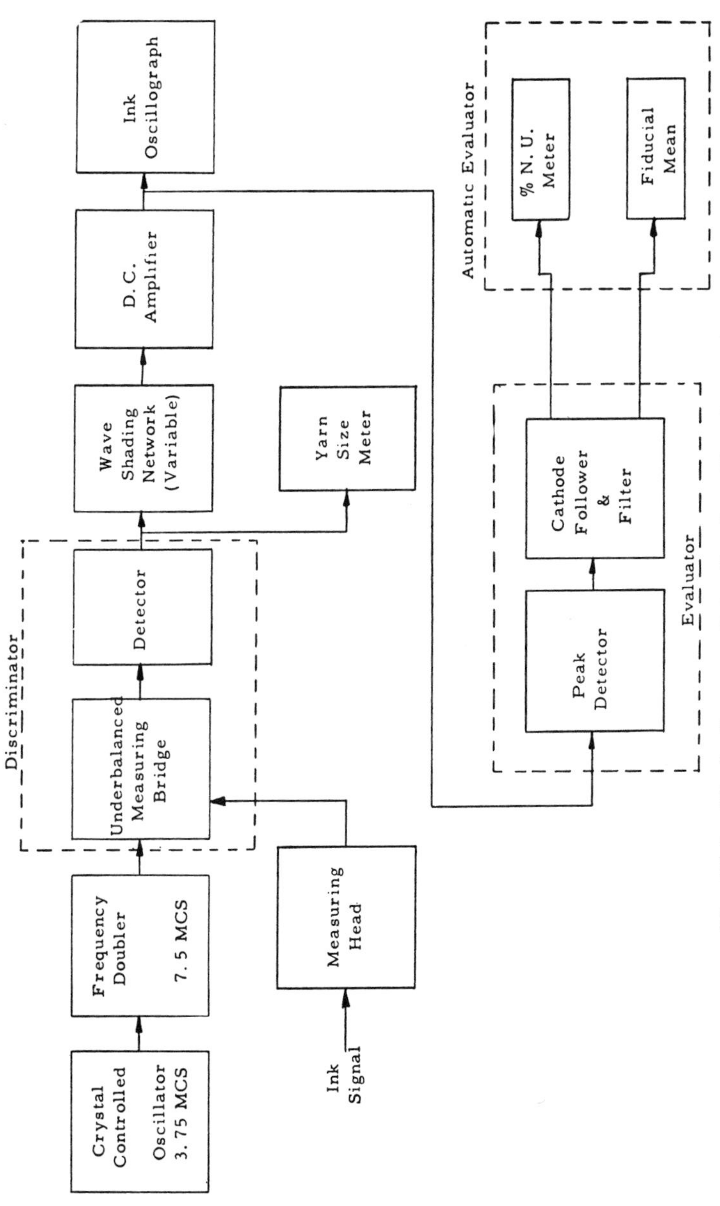

Fig. 28.5. Block diagram of the Brush uniformity analyzer.

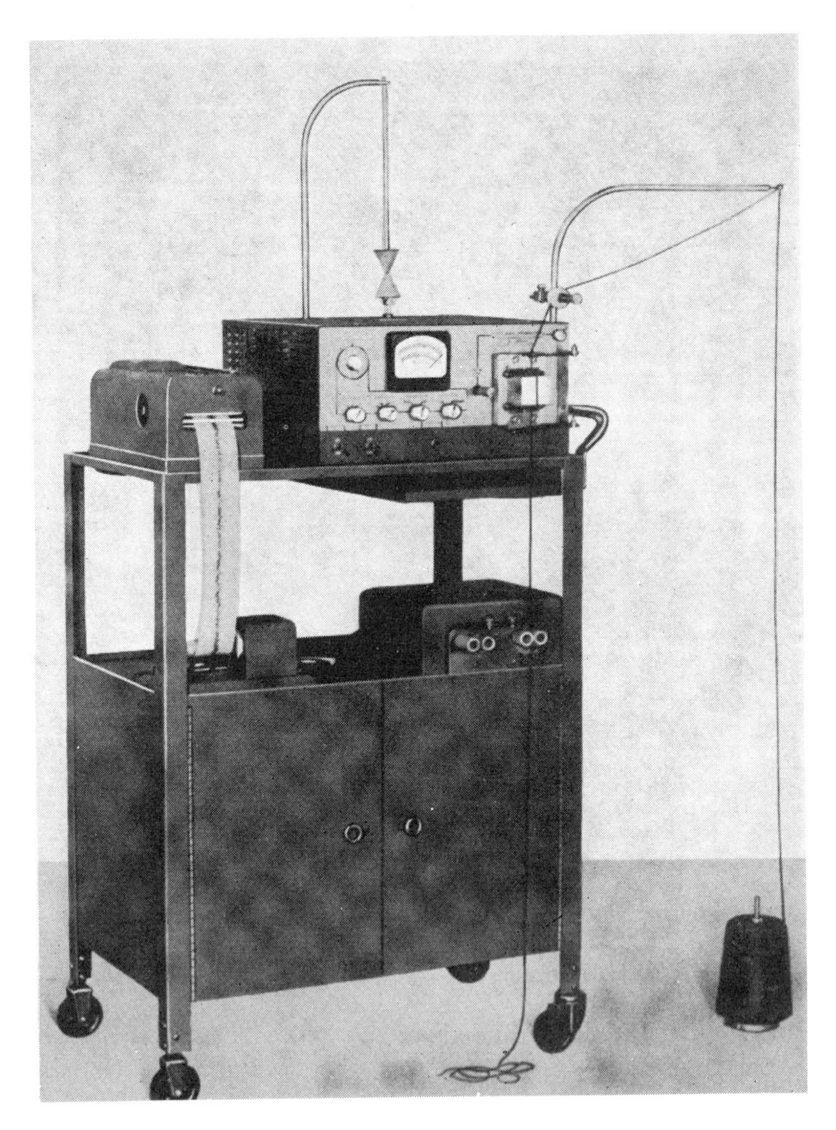

Fig. 28.6. Brush uniformity analyzer.

peak variations and average mass density of the material being measured.

The material to be measured is passed between the two plates of a measuring capacitor, thereby changing the effective dielectric constant for this capacitor. As the weight per unit length varies, so does the capacitance of the measuring head. These minute capacitance variations are injected across two arms of an unbalanced bridge, so that the output voltage of the bridge is a direct function of these variations.

This output voltage is then applied to a detector system which rejects high frequency components and a signal proportional to the input variations is amplified and applied to a direct-writing oscillograph. This oscillograph records a pattern proportional to the capacitance variations of the measuring head on a chart. (See Figs. 28.5 and 28.6.)

A second method available for recording nominal per cent non-uniformity is provided by an evaluator circuit and appropriate meters. The evaluator consists of an electronic detector which creates a DC voltage proportional to the peak average of the yarn signal. This signal is applied to an appropriately calibrated meter which will enable direct reading of nominal per cent non-uniformity of the material being tested. An additional voltage is supplied to a fiducial meter to correct for the variations in the average mass density of the material under test. This unit is known as the automatic evaluator, which in effect indicates results in terms of average peak-to-peak variation rather than area reading. The non-uniformity ratings obtained from both the oscillograph and

TABLE 28.2

Sensitive Length Measurements on External Evaluator

Material speed, yards per min.	Setting of sensitive length switch, ft.	Meter reading is ave. % N.U. per
100	1	1 foot
100	10	10 feet
10	1	0.1 foot
10	10	1 foot

evaluator agree, thereby facilitating direct comparison between the charts from the oscillograph and the data from the evaluator.

The range or amplitude of the chart is controlled by the "pen sensitivity." Three positions are provided: position *1* is used for yarn, position *2* is used for roving, and maximum is used for sliver. These positions correspond to the 100 per cent, 50 per cent, and 25 per cent ranges respectively on the Uster tester.

For normal testing, yarns are run at 100 yards per minute with the *evaluator sensitive length* set at 10 feet. Roving and sliver are run at 10 yards per minute with the evaluator sensitive length set at 1 foot. These relationships are summarized in Table 28.2. together with the length to use in translating the reading from the meter to average per cent non-uniformity.

In running a test, readings are taken at five-second intervals (as shown by the flashing of a light) of (first) the per cent N.U. meter and (second) the fiducial meter until a total of twenty readings from each meter have been recorded. The per cent non-uniformity is calculated from the following equation

$$\%\ \text{N.U.} = \frac{\text{Sum of \% N.U. readings}}{\text{Sum of fiducial readings}}(100)$$

Table 28.3 will help to evaluate the charts.

TABLE 28.3

Chart Evaluation

Material speeds, yards per min.	Chart speeds, blocks per sec.	Ratio of material to chart	
		blocks	ft. material
100	1 (low speed)	1	5
100	5 (medium speed)	1	1
100	25 (high speed)	5	1
10	1 (low speed)	1	½
10	5 (medium speed)	1	0.1
10	25 (high speed)	5	0.1

High and lows for each yard or length desired are read from the base line (zero mm.) of chart. Substitute into equation:

$$\frac{2\,(H - L)}{H + L}(100) = \%\ \text{N.U. for each yard}$$

or

$$\frac{2(\Sigma H - \Sigma L)}{\Sigma H + \Sigma L}(100) = \text{Ave. \% N.U. per yard}$$

Influence of Moisture on Capacitance-Type Testers. There has been considerable interest in the effect that moisture variations in a yarn would have on the apparent variation in uniformity determined by capacitance-type testers. This problem becomes more acute if tests are made on the textile material in a non-conditioned atmosphere where variations can occur from bobbin to bobbin, lot to lot, day to day, or between tests. In one such study it was found that for cotton yarns tested with approximately zero per cent moisture regain, the per cent unevenness was rather high, in the range of 75 per cent maximum variation. The same sample of yarn when conditioned to 6 per cent regain had a maximum variation of between 60 and 50 per cent. In an experiment performed on the same sample of yarn with a regain ranging from zero to 16 per cent, it was found that there was a gradual decrease from the 75 per cent maximum variation to approximately 45 per cent for the 16 per cent regain. The greatest change in unevenness values was associated with the range of from zero to 5 per cent regain for the cotton yarns tested. The results of this study indicate that small changes in moisture regain in the order of 3 or 4 per cent are relatively unimportant. Larger changes will influence the tester to some extent.

As a further result of this study, some technicians have expressed the belief that blends of spun synthetic yarns would likely be influenced by the amount of moisture present in the component fibers in the blend as much or more than some processing defects. For this reason, some companies limit the use of the tester to locate trends in processing rather than to attempt to analyze every section of the curve for maximum variation values.

One point that should be remembered is that moisture will have an extremely pronounced effect on the apparent evenness of the yarn if the moisture is not homogeneously distributed throughout the sample. This can be demonstrated very easily by dampening one side of a bobbin and passing the yarn through the evenness tester. Generally, the influence of moisture can be disregarded, provided that the moisture is homogeneously distributed throughout the yarn and that the range of variation in moisture is not large.

Pacific Evenness Tester

The Pacific tester (Fig. 28.7) is a combination mechanical, electromagnetic, and electronic tester, differing in theory and operation from capacitance-type testers. The three component units of the tester may be described as follows:

1. Mechanical. The purpose of the mechanical unit is to compress the yarn, roving, or sliver in slots or grooves of certain dimensions with weights proportioned to the groove width.

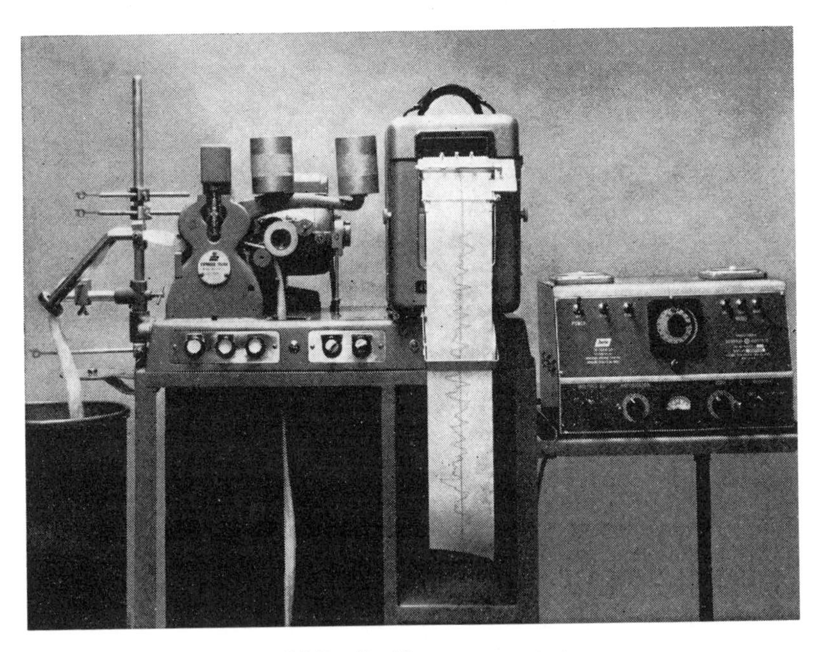

Fig. 28.7. Pacific evenness tester.

Each pair of rolls consists of a top tongued roll and a bottom grooved roll. The bottom grooved roll is positively driven; the top tongued roll is friction driven. When the sliver or yarn is introduced between the top and bottom rolls in the desired groove, variation in thickness of the compressed mass is translated into a vertical movement of the top roll.

2. Electromagnetic. The movement of the top or tongued roll is converted into an electrical impulse by means of an electromagnetic displacement gauge, the operation of which is as follows:

A rod with a micrometer attached transmits the vertical movement of the upper roll to a movable armature which is positioned in an air gap approximately midway between two coils of an impedance comparison circuit. As the air gap in the displacement gauge head changes, the coil impedance changes. Because the air gap of the displacement gauge head is adjusted so that the circuit is initially balanced, any changes in the coil impedance of the displacement gauge head causes a change in the output of the impedance comparison circuit. The change in output is proportional to the change in length of the displacement gauge air gap. A power unit with a voltage regulating type of transformer is used in conjunction with the displacement gauge forming an integral unit. This unit is quite simple, containing a dry disc rectifier rather than a vacuum tube rectifier and furnishes a current upon which a varying line voltage has minimum effect.

3. Photoelectric recorder. This unit is a recording milliammeter with high-speed response, temperature and drift compensation, and a switch with seven ranges of current used to denote the amount of magnification. The recorder is linear from center zero to the edges of the chart, and there are adjustments for setting both the mechanical and electrical zero.

CHAPTER XXIX

Evaluation and Interpretation of Evenness Measurements

Interpretation of Evenness Charts

The accurate and proper interpretation of evenness charts is of utmost importance if the fullest benefit is to be derived from this type of testing. To interpret the charts properly, a thorough knowledge of the different factors likely to influence such test results is necessary. Some of these are the capacity of the tester being used, the strong and weak points of the apparatus, the proper method of operating the tester, the process being measured, and the way the test results are to be used.

The job procedure for the Uster tester stipulates that the instrument be set as nearly as possibly on zero before starting the test. This is of utmost importance, especially if the direct reading method is to be used in analyzing the charts.

The influence of material speed is shown on the charts Figs. 29.1 and 29.2. Examination of these charts reveals the type of autograph that may be expected at a few of the speeds available on the Uster tester. By using a relatively slow material speed and fast chart speed, the variation within and between individual yards of material can be studied. However, with the combination of high material speed and slow paper speed, it becomes increasingly difficult to study the variation from yard to yard since the chart tends to give a very close pattern. Such combinations of high material speed with low chart speed are used to study long-term variation, whereas slow material speed with high chart speed may be used to study short-term variation.

Three methods of calculating per cent maximum unevenness are shown in Tables 29.1, 29.2, and 29.3. The method shown in Table 29.1 requires less time than the other two methods, but it

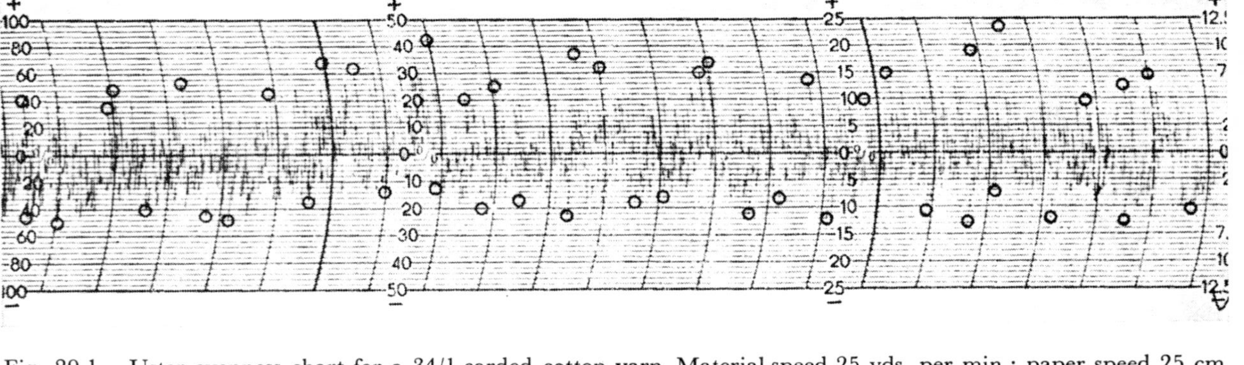

Fig. 29.1. Uster evenness chart for a 34/1 carded cotton yarn. Material speed 25 yds. per min.; paper speed 25 cm. per min.; 100 % scale — normal test.

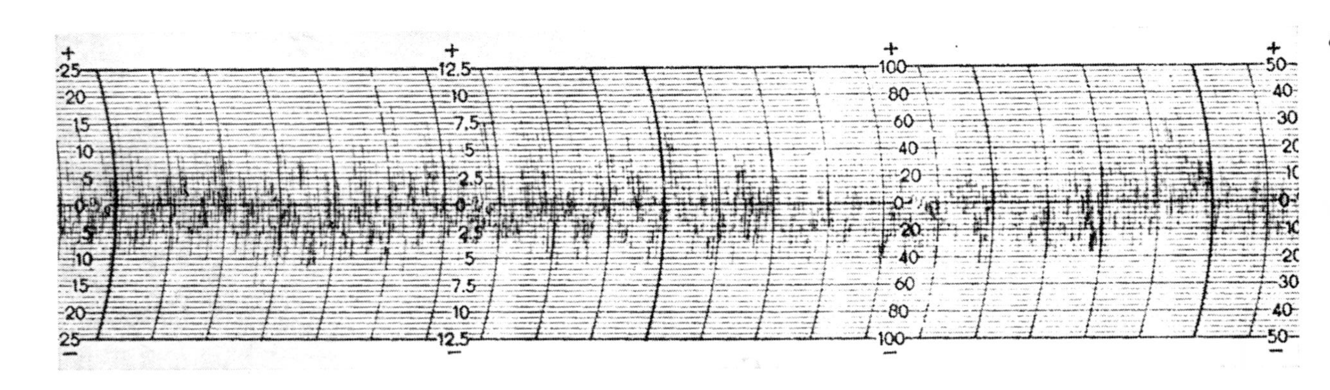

Fig. 29.2. Uster evenness chart for a 34/1 carded cotton yarn. Material speed 50 yds. per min.; paper speed 25 cm. per min.; 100 % scale — normal test.

is likely to introduce some error in the results if the machine is not perfectly zeroed throughout the test. To use this method, the maximum per cent unevenness values per yard are read

TABLE 29.1

Tabulation of Per Cent Maximum Unevenness Per Yard from Uster Chart Shown in Fig. 29.1[a]

Yard number	Plus side of zero line	Minus side of zero line	Sum
1	58	42	100
2	50	50	100
3	38	48	86
4	94	28	122
5	76	52	128
6	60	44	104
7	40	48	88
8	56	32	88
9	68	44	112
10	60	32	92
11	64	36	100
12	74	44	118
13	48	36	84
14	40	40	80
15	84	24	108
16	64	28	92
17	68	36	104
18	44	48	92
19	52	44	96
20	48	40	88
21	34	50	84
22	40	44	84
	$\Sigma = 1260$	890	2150

$$\% \text{ unevenness} = \left(\frac{1260}{22} + \frac{890}{22}\right) = 97.7$$

$$\text{or } \frac{2150}{22} = 97.7$$

[a] The method used for this table was the "Direct Reading Method." The per cent unevenness was read from right to left directly from the chart using the 100 per cent scale.

directly from the chart using the same scale on the chart as was used on the tester during the time of the test. In this example, the 100 per cent scale was used. The maximum deflection on each

TABLE 29.2

Tabulation of Maximum Deflection Per Yard in Millimeters from Uster Chart [a] Shown in Fig. 29.1

Yard number	Above center line	Below center line	Sum
1	14.5	10.5	25
2	12.5	12.5	25
3	9.5	12	21.5
4	23.5	7	30.5
5	19	13	32
6	15	11	26
7	10	12	22
8	14	8	22
9	17	11	28
10	15	8	23
11	16	9	25
12	18.5	11	29.5
13	12	8.5	20.5
14	10	10	20
15	21	6	27
16	16	7	23
17	17	9	26
18	11	12	23
19	13	11	24
20	12	10	22
21	8.5	12.5	21
22	10	11	21
Σ	315.0	222.0	537.0

$$\% \text{ unevenness} = \left(\frac{315.0}{22} + \frac{222.0}{22}\right)(4) = 97.6$$

[a] The millimeter deflections in the table are measured from the center line of the chart.

side of the center line per yard of material is noted and recorded. These values may be added together as they are read from the chart, or they may be tabulated as shown in the table and the sum calculated after all the points have been read. Using this method, it would not be necessary to calculate the maximum variation per yard for each yard of material but simply to summarize the data as shown by the values 1260 and 890 in the table. The average of each column is calculated, and these two averages added together give the per cent maximum unevenness. When summarizing the values as they are read directly from the chart, take the sum which is represented in the table by 2150 and divide by the number of yards to get the average per cent unevenness.

The method illustrated in Table 29.2 is very similar to the above, except that the scales on the chart paper are disregarded and the deflection is recorded in millimeters. Horizontal lines on the chart are spaced in millimeters; this facilitates the recording of the date in terms of millimeters. When using this method the operator has the same alternatives as with the first method regarding the recording and summarizing of the data. The values can be added together as recorded or they can be summarized as shown by the values 315.0 and 222.0. Another variation of this method is to count the number of lines or millimeters between the high and low points. This value would be the same as the sum of the two separate readings, thus reducing the possible errors in reading the chart. Regardless of which method is used to calculate the average deflection, this value is multiplied by the appropriate constant k. When using the 100 per cent scale, the k value is 4, due to the fact that each horizontal line has a value of 4. When using the 50 per cent scale, the k value is 2 since each horizontal line has a value of 2 on this particular scale. For the same reasons, 1 is used for the 25 per cent scale, and 0.5 for the $12\frac{1}{2}$ per cent scale. This method is also sensitive to the proper centering of the instrument at the start of the test, and so care should be made to center the tester before the test is started.

The third method for evaluating the Uster charts is presented in Table 29.3. When using this method, the maximum and minimum deflection per yard is measured from the bottom or base line of the chart, as shown in the table. After the chart has been read and the data tabulated, either the per cent unevenness per yard or the average per cent unevenness for the entire test may be calculated.

TABLE 29.3

Tabulation of Maximum Deflection Per Yard in Millimeters from Uster Chart[a] Shown in Fig. 29.1

Yard number	Maximum	Minimum
1	39.5	14.5
2	37.5	12.5
3	34.5	13
4	48.5	18
5	44	12
6	40	14
7	35	13
8	39	17
9	42	14
10	40	17
11	41	16
12	43.5	14
13	37	16.5
14	35	15
15	46	19
16	41	18
17	42	16
18	36	13
19	38	14
20	37	15
21	33.5	12.5
22	35	14
	$\Sigma = 865.0$	328.0

$$\% \text{ unevenness} = \frac{2(865 - 328)}{(865 + 328)}(100) = 90$$

[a] These millimeter deflections are measured from the bottom line of the chart.

In the two methods described previously, the total deflection was measured by adding the deflection above the center line to that below the center line. This total deflection, with an allowance to correct for the scale being used, measures the per cent unevenness for each station sampled. In this third method, the per cent unevenness is calculated by dividing the total deflection at each selected length by the average distance expressed as a per cent of this total deflection above the base line. This tends to relocate the center line for each length tested, thereby reducing the error induced by failure to obtain perfect centering of the test.

The total deflection, as illustrated in Table 29.3, is designated by $H - L$, in which H is the distance above the base line for maximum deflection, and L is for the minimum deflection, both in millimeters. The center line at each station is located a distance of $L + \frac{1}{2}(H - L)$ mm. above the base line. The unevenness, expressed as per cent for the full 100 per cent scale is:

$$\frac{H - L}{L + \frac{1}{2}(H - L)}(100) = \frac{2(H - L)}{H + L}(100)$$

The per cent unevenness for different scales is as follows:

$$\text{for } 50\% \text{ scale, } \frac{H - L}{H + L}(100)$$

$$\text{for } 25\% \text{ scale, } \frac{H - L}{2(H + L)}(100)$$

$$\text{for } 12.5\% \text{ scale, } \frac{H - L}{4(H + L)}(100)$$

Although this method of analysis tends to compensate for errors in centering the autograph on the chart, it would be inconsistent to neglect to try for optimum centering. The evidence of the results from the three methods of analysis indicates that the chart in Fig. 29.1 was not centered perfectly. Confirm this by comparing the results of the first and second analysis methods (97.6 per cent unevenness) with the results of the third method (90.0 per cent unevenness).

Periodicities and their Analysis

In addition to the routine analysis of evenness charts, an inspection of the chart for repeating patterns or periodicities has proven to be helpful. Periodicities are frequently caused by mechanical troubles within the processing equipment, with the drafting elements the worst offenders. Unevenness patterns can be caused by eccentric top or bottom rolls, worn roll stands, worn aprons, and other defects within the drafting elements. The location and corrcction of such defective machine parts has proven to be one of the most valuable uses of the evenness testers.

Example of Periodicity. An example of a drafting defect is shown on the roving chart, Fig. 29.3. The roving that was tested was found to have a repeat pattern spaced at approximately 6 cm. lengths on the chart. As the material was fed at the rate of 8 yards per minute with a chart speed of 25 cms. per minute, the ratio of material to chart is 0.32 yards or 11.52 inches of material per centimeter of paper. With the periodicity occurring at intervals of 6 cms., there are 11.52 times 6, or 69.12 inches of roving per pattern or per defect. If this value 69.12 is divided by the draft on the roving frame of 12.2, the result of 5.7 inches indicates that the defect is not occurring at the roving frame but rather in the sliver. The draft of the drawing frame producing the sliver is 6.0. By dividing a frequency of defect of 5.7 by this draft of 6.0, the answer is found to be 0.95. Inasmuch as this ratio is less than 1, this is a clear indication that the defect is the result of some malfunctioning of drafting at the drawing frame.

Summary of Calculations

$$\frac{8(36)}{25} = 11.52 \text{ in./cm}$$

$$11.52(6) = 69.12 \text{ in./defect in roving}$$

$$\frac{69.12}{12.2} = 5.7 \text{ in./defect in sliver}$$

The roll sizes and draft distribution of the drawing frame used are given in Table 29.4.

With the above drafts and roll sizes, the periodicities in Table 29.5 that might result from eccentric rolls or the like can be calculated. Knowing these possible periodicities, the results from the chart analysis can be compared to them, and the cause of the trouble located.

The calculated periodicity closest to the chart value of 5.7

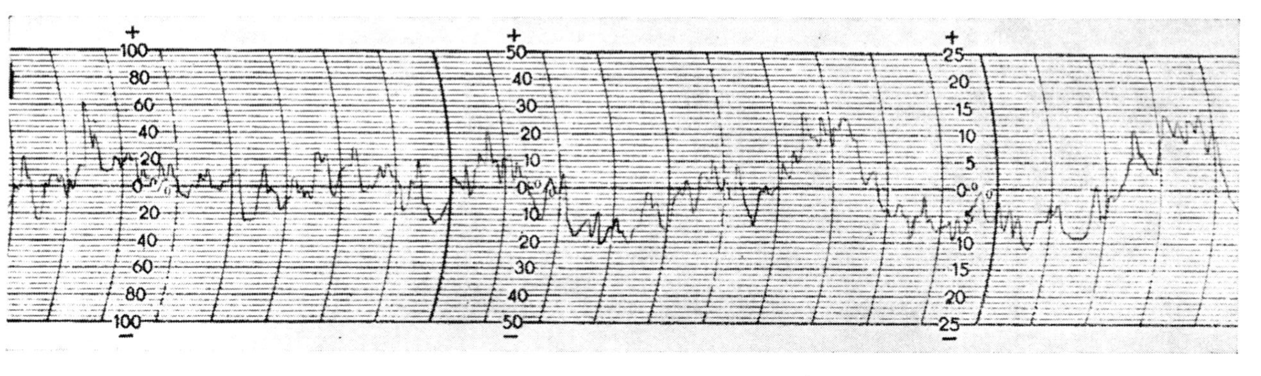

Fig. 29.3. Uster evenness chart for a carded cotton roving. Material speed 8 yds. per min.; chart speed 25 cm. per min. — normal test.

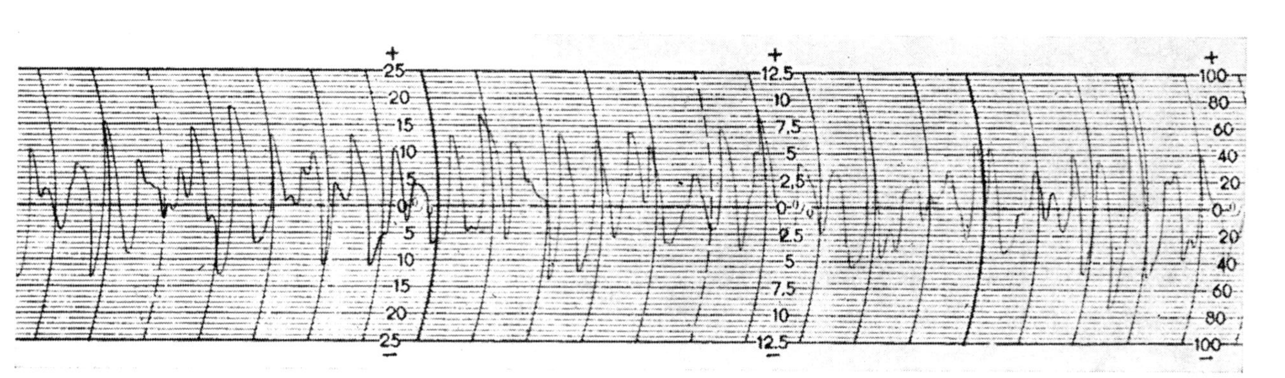

Fig. 29.4. Uster evenness chart for a carded cotton sliver. (This is a sample of the sliver used to make the roving in Fig. 29.3.) Material speed 8 yds. per min.; chart speed 25 cm. per min. — normal test.

is 6.28; the difference between the two values can be due to a reading error of the chart. It can be assumed with little doubt that the cause of the trouble lies in the calender roll, and the action

TABLE 29.4

Drawing Frame Roll Diameters and Draft Distribution

Roll		Draft
Back roll (1.125″)	to 3rd roll	1.15
3rd roll (1.0″)	to 2nd roll	1.80
2nd roll (1.0″)	to front roll	2.84
Front roll (1.25″)	to calender roll (2″)	1.02
Total draft, back roll	to calender roll	6.00

TABLE 29.5

Periodicity Resulting from Eccentric Rolls

Possible cause of trouble	Periodicity calculation	Periodicity
Back roll	1.125(π) (6.0)	21.21
Third roll	1.0(π)(1.8) (2.84) (1.02)	16.38
Second roll	1.0(π) (2.84) (1.02)	9.10
Front roll	1.25(π) (1.02)	4.01
Calender roll	2.0(π)	6.28

taken should be directed to a close inspection of the calender rolls in question.

To follow this example further, the sliver from which the roving was made was tested and the results of this are shown in the chart, Fig. 29.4. Examination of this chart shows that the difficulty is occurring at intervals of approximately 0.5 cm. With a material speed of 8 yards per minute and a chart speed of 25 cms. per minute, the ratio is 11.52 inches of material per centimeter of chart. With the defect occurring at 0.5 cm. intervals, the spacing of the defect in the sliver would be 11.52 times 0.5, which equals 5.76 inch intervals. The calculations made previously showed that the interval was 5.6 inches and thus the two charts match very closely in the results indicated.

Other examples of calculations used in locating periodicities can be worked out very easily. For instance, at carding, a 27-inch doffer will deliver approximately 86.0 inches of web per revolution; with a tension draft of 1.12, this means that 96.3 inches of sliver would be delivered into the can. Repeat patterns in card sliver that appear with this frequency could be traced to trouble at the doffer.

If a repeating defect is located in the yarn, and through calculations it is reasoned that the defect is occurring at carding, then sample the product at the card. In other words, it is better to sample the product from the defective equipment in order to make the final analysis. There is doubtful value in the inspection of a yarn sample chart to locate a defective machine part where this defective part is located at the early stages of processing. It is easier to isolate defective equipment and test the product delivered by that piece of equipment than to carry out the study after subsequent processings have confounded the pattern, especially where a doubling operation has been interposed. The irregularity patterns introduced by low twist in continuous filament yarns and the means of correction by use of false twist devices have been previously mentioned.

Tests of sliver coiled in cans may also reflect to a very limited extend the effect of twist. Each coil of sliver in a can has one turn of twist over its length. Study of these coils shows that the twist is uniformly distributed. Removal of the sliver from the can in the conventional manner results in removal of this twist. Therefore, only those lengths of sliver that have concentrated twist convolutions pressed in tightly would cause irregularity patterns. It is more likely that apparent thin places will show up on the evenness chart as a result of the disfiguration of short sections of the sliver. Cross-overs of the sliver in the coil or sharp kinks in the coiled sliver will be reflected as non-uniformity by the instrument. These cross-overs frequently show on the Spectrograph chart as a repeat pattern at a frequency determined by the size of the coil. If this condition is suspected, it is suggested that tests be made of the sliver before it has been passed through the coiler mechanism as well as after it has been passed through the coiler but not into a can. Evidence of different irregularity patterns in the sliver between the two test samples would indicate malfunctioning of the coiler rolls.

Spectrograph. To facilitate the accurate and quick location of periodicities in textile sliver, roving, and yarn, the Zellweger Corporation has developed an instrument known as the Spectrograph. This instrument, when used in conjunction with the Uster evenness tester, will record the wave length for waves or irregularities that repeat in the material. It has been pointed out that these waves occur with frequencies equivalent to approximately two to three times the staple length of the fiber being processed. The Spectrograph substantiates this. In addition to recording these waves or frequencies, it also records any others that occur in the material. It is these other waves that are generally of importance to the quality control engineer. The recording paper for the Spectrograph is calibrated in such a way that the wave length is read directly from the chart in inches or yards. This method of analysis increases the speed and the accuracy with which the charts may be interpreted. Very frequently several wave lengths are superimposed on each other, and it is extremely difficult, if not impossible, to distinguish between them. However, with the Spectrograph, these wave lengths are recorded separately, so that they can be isolated and corrected. Some illustrations of these principles follow.

The actual spectrograms for a cotton and a worsted yarn are shown in Fig. 29.5 and Fig. 29.6 respectively. These two charts represent good quality yarns; despite this fact, notice the drafting waves.

An example of a good quality cotton card sliver is shown by the spectrogram in Fig. 29.7.

Examples of yarns that have definite repeat patterns are shown in the spectrograms, Fig. 29.8, through Fig. 29.12. Note in Fig. 29.9, the ideal, the normally expected, and the actual spectrum. The latter, shown by the hatched-in area, shows repeat patterns that occur at approximately two to three inches and five inches, as well as medium term variation from 24 inches to $1\frac{1}{2}$ yards.

In Fig. 29.13 and 29.14, the effect of roll run-out or eccentricity

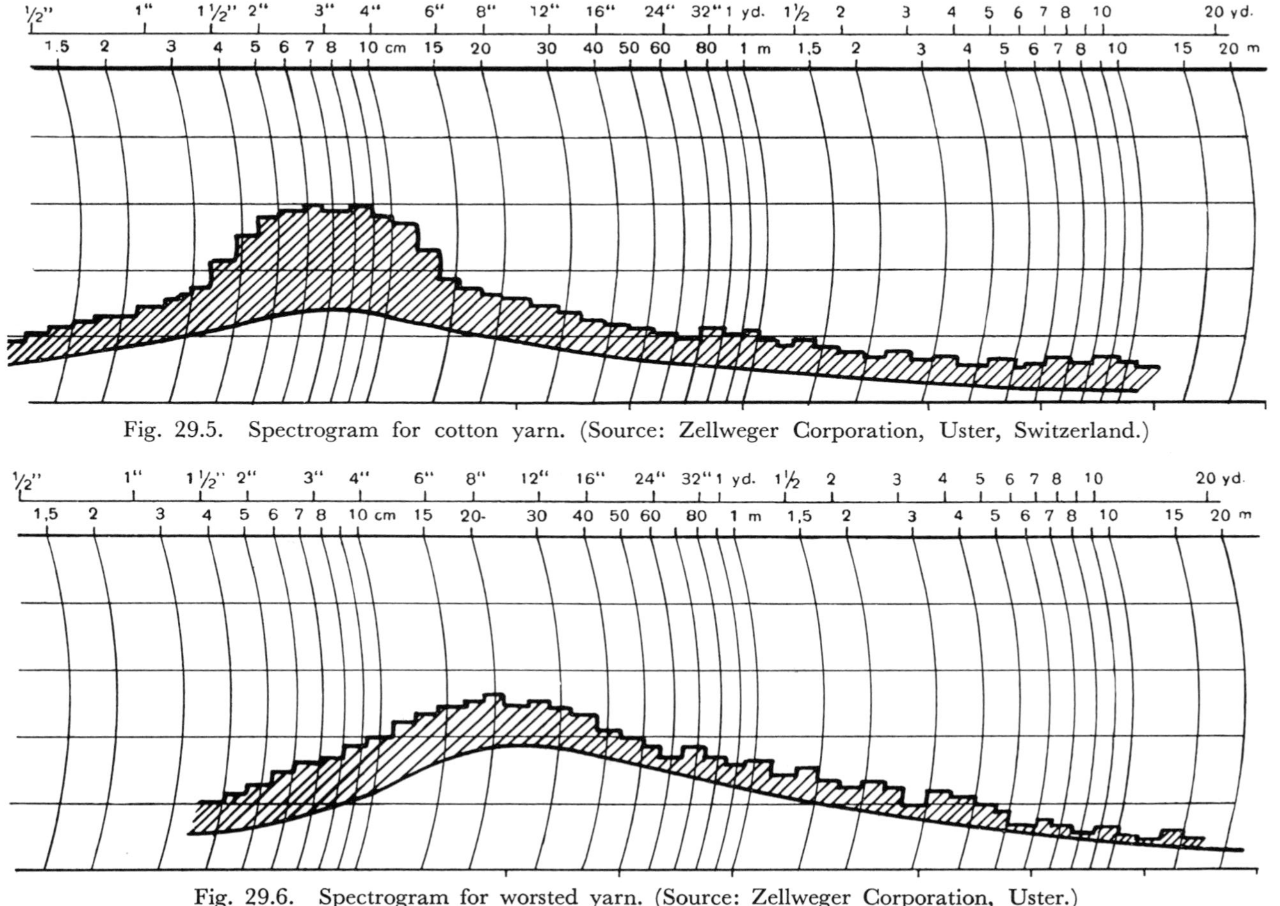

Fig. 29.5. Spectrogram for cotton yarn. (Source: Zellweger Corporation, Uster, Switzerland.)

Fig. 29.6. Spectrogram for worsted yarn. (Source: Zellweger Corporation, Uster.)

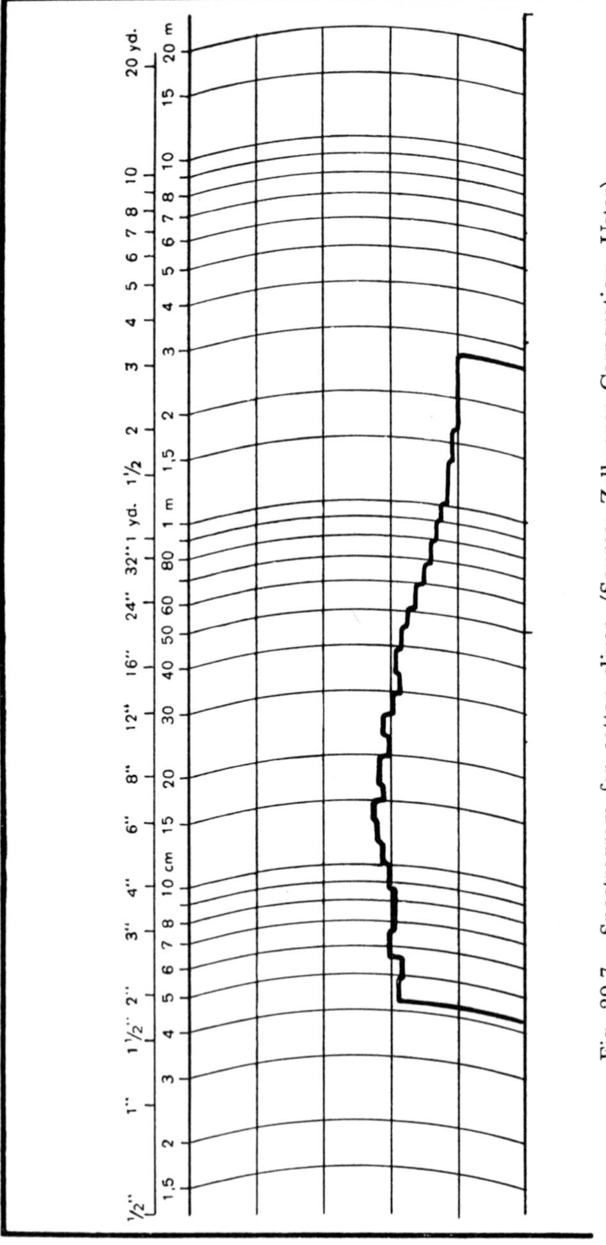

Fig. 29.7. Spectrogram for cotton sliver. (Source: Zellweger Corporation, Uster).

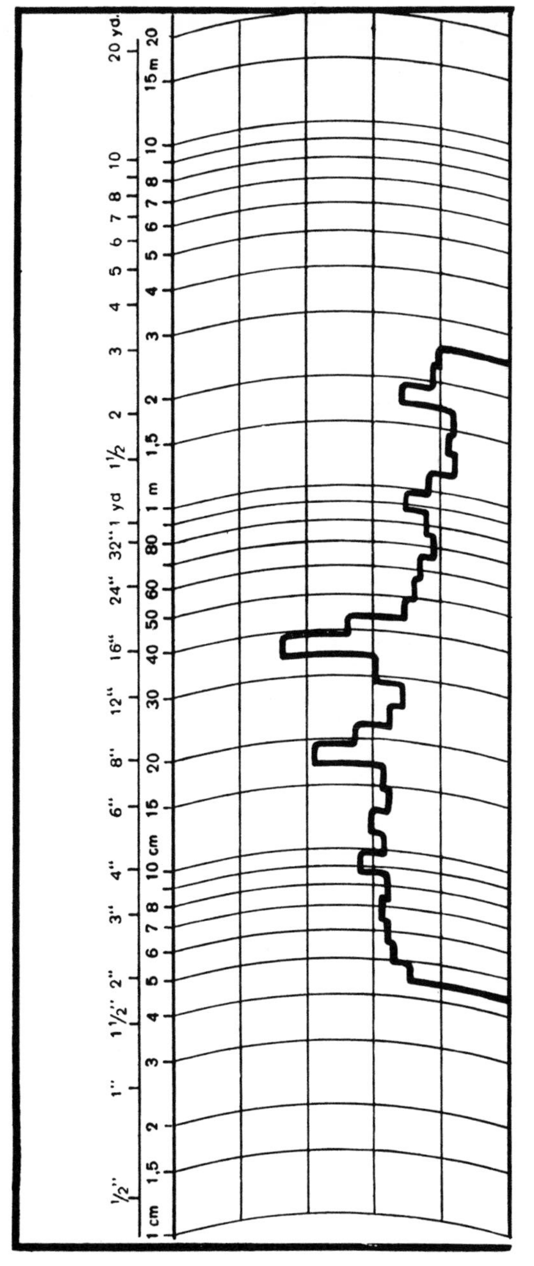

Fig. 29.8. Spectrogram for yarn showing repeat. (Source: Zellweger Corporation, Uster.)

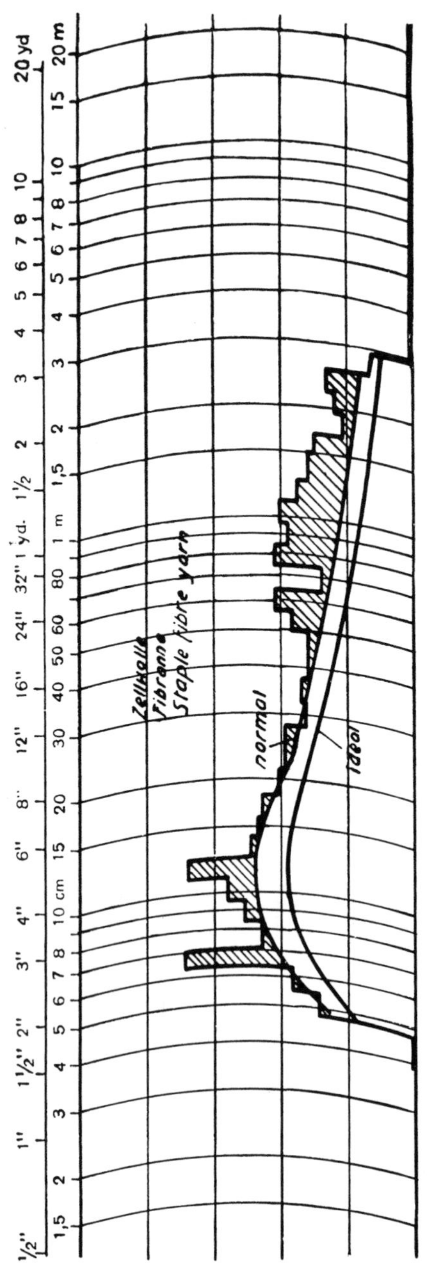

Fig. 29.9. Spectrogram for staple fiber yarn. (Source: Zellweger Corporation, Uster.)

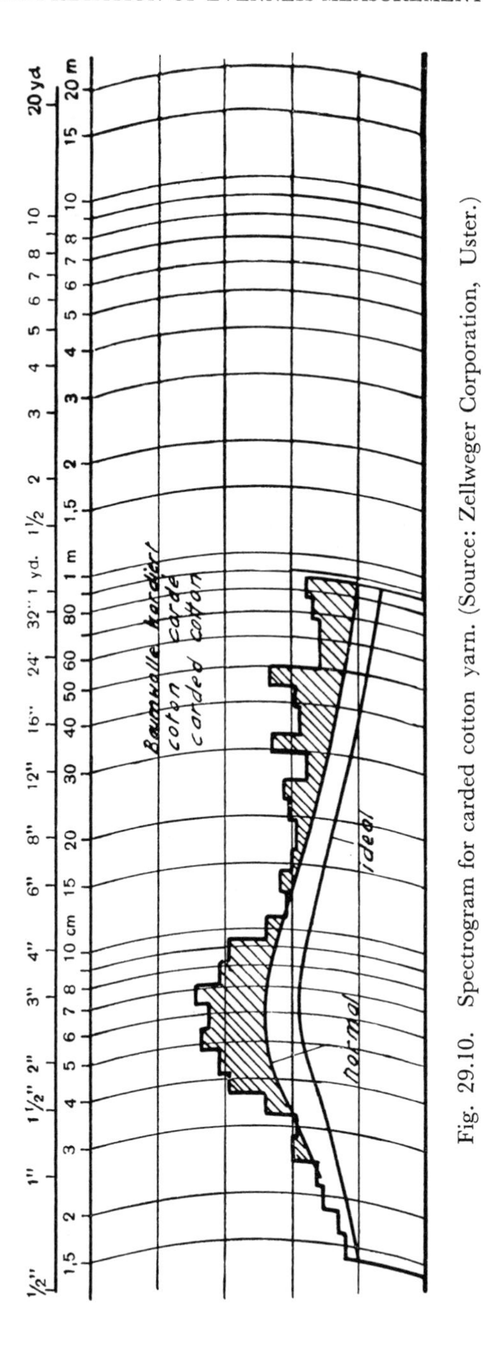

Fig. 29.10. Spectrogram for carded cotton yarn. (Source: Zellweger Corporation, Uster.)

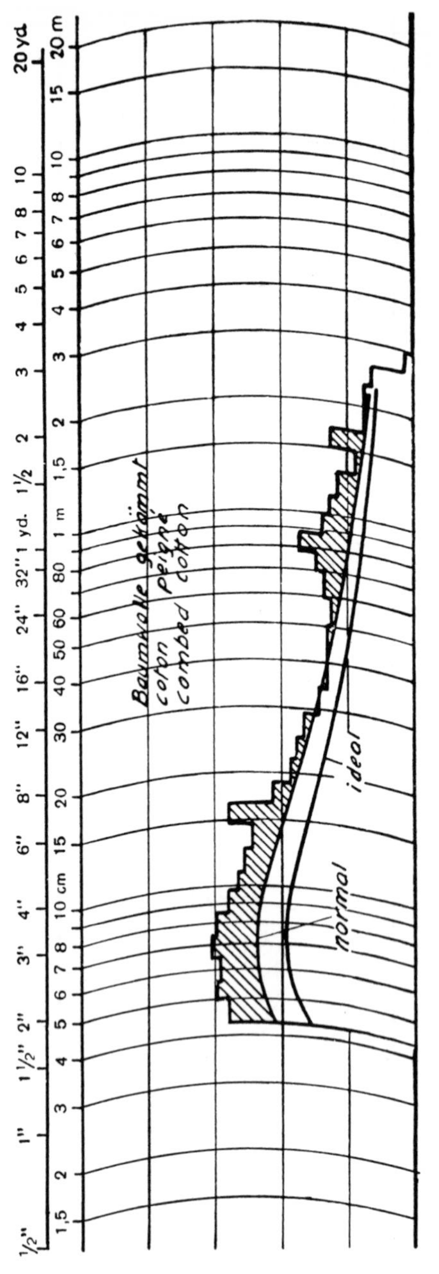

Fig. 29.11. Spectrogram for combed cotton yarn. (Source: Zellweger Corporation, Uster.)

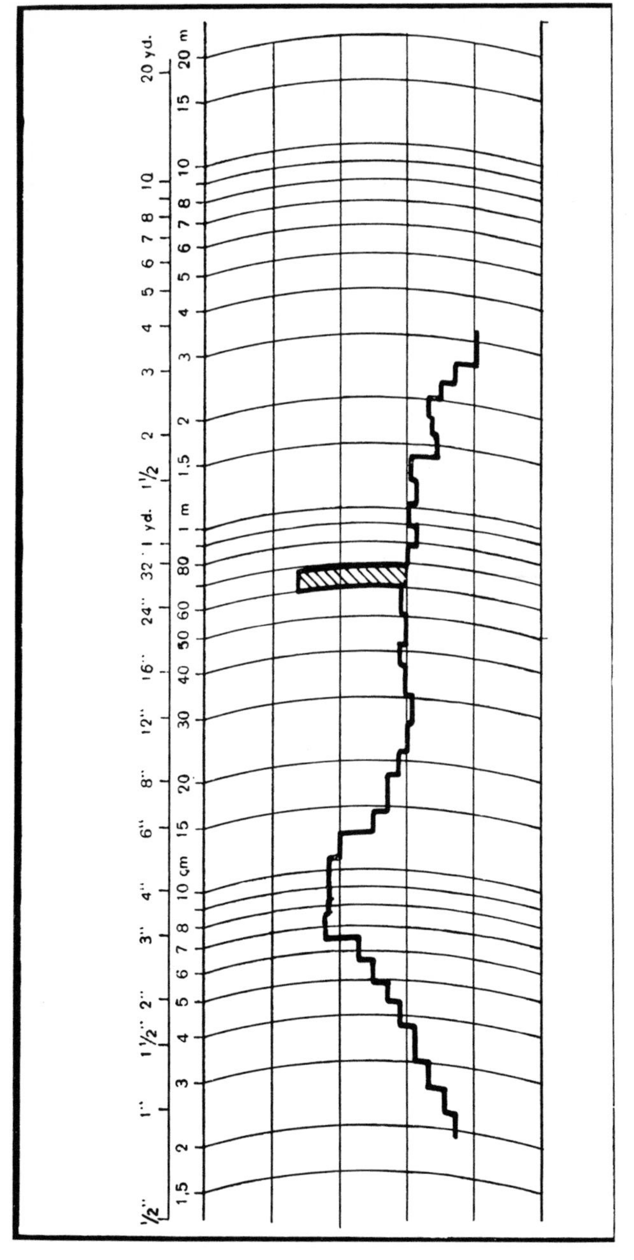

Fig. 29.12. Spectrogram showing repeat. (Source: Zellweger Corporation, Uster.)

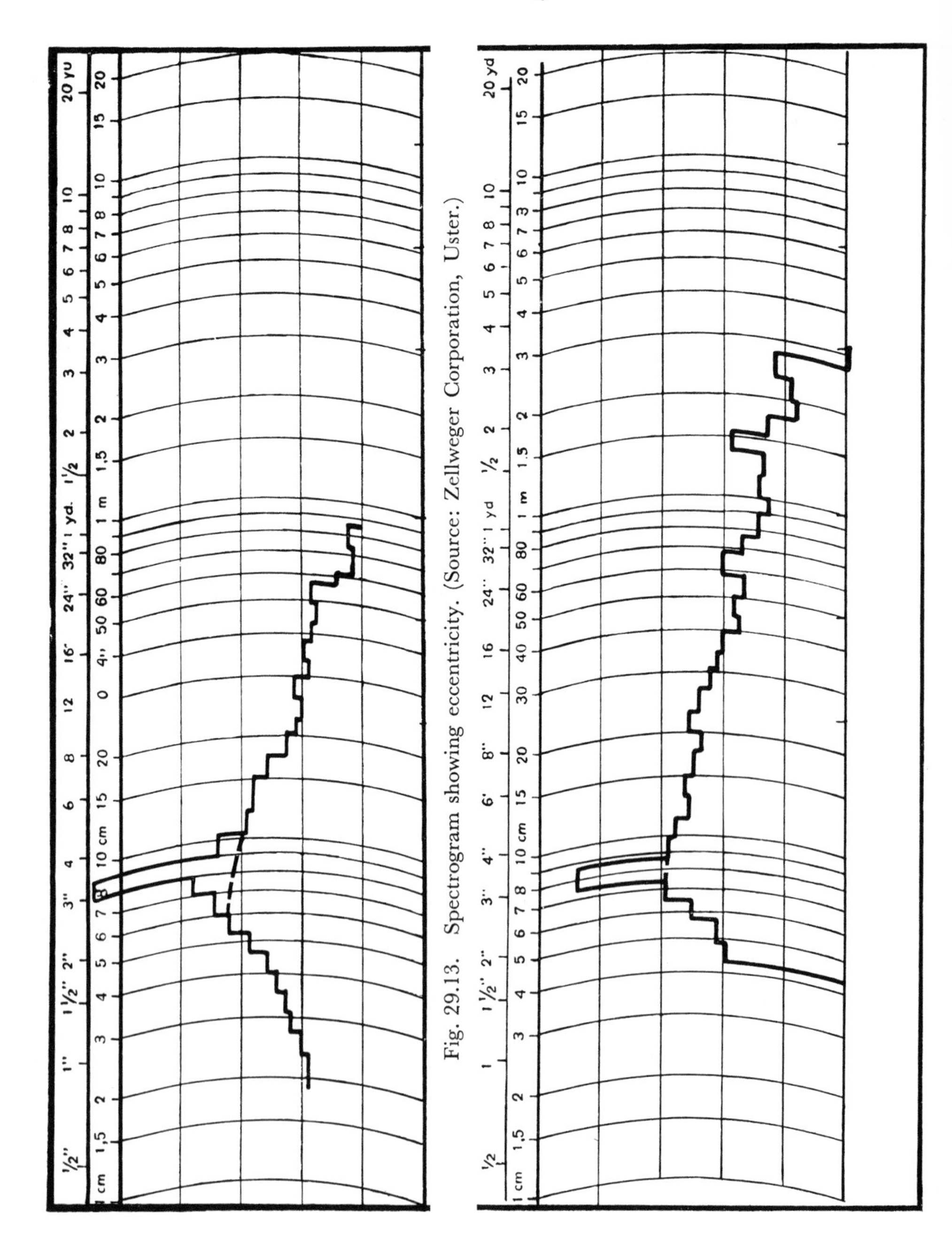

Fig. 29.13. Spectrogram showing eccentricity. (Source: Zellweger Corporation, Uster.)

is shown. The spectrogram of Fig. 29.13 was made from 45/1 yarn spun with a draft of 20; the front roll run-out in spinning was 0.008 inch. Fig. 29.14 is for a 1.6 hank roving also produced with a roll run-out of 0.008 inch.

Integrators. The Uster evenness testers may be equipped with one of two types of integrators. One type is the linear integrator which measures the mean deviation in yarn from the average. The other type is the quadratic integrator which measures the coefficient of variation. The two integrators are operated in the same manner, the only difference being that the quadratic integrator squares the areas being measured; this results in a measurement of the coefficient of variation rather than the mean linear unevenness.

The type of integration performed by the linear and quadratic integrators is illustrated by Uster Corporation by Figs. 29.15 and 29.16 together with the mathematical calculations. An example of the calculations for the theoretical chart of Fig. 29.17 amplifies the basic theory.

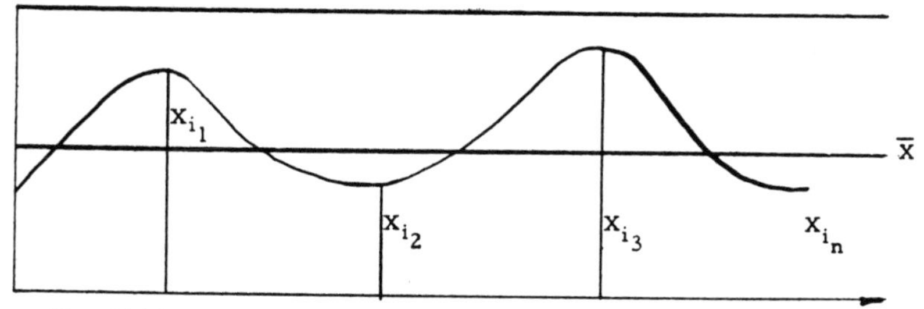

Fig. 29.15. Chart illustrating integration principles. See Fig. 29.16.

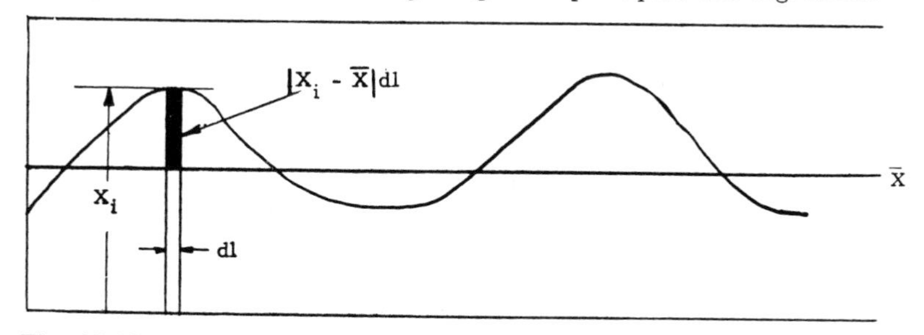

Fig. 29.16. Chart illustrating integration method of obtaining mean and standard deviations.

The mean deviation is a linear measure expressed as a percentage. As shown in Fig. 29.15, it is the mean of the individual deviations of the curve from the average material diameter $\bar{X}$, expressed as a per cent of $\bar{X}$. For example, it is the difference between the individual readings X_i and $\bar{X}$, summed irrespective of sign, i.e., irrespective of whether they are above (+) the average or below (−) the average $\bar{X}$, and divided by the number of points, n, taken. This mean deviation is then multiplied by $100/\bar{X}$ to give the result as a percentage of the average value. The same results can be obtained as shown in Fig. 29.16 by integrating the areas $(X_i - \bar{X})$ dl from O to L, dividing by L to get the average deviation (geometrically a rectangle having a base of length L and a height which equals the average deviation) and then multiplying by $100/\bar{X}$ to give results as a percentage. This is in effect what is performed by the Uster linear integrator. The result is also known as the mean linear irregularity $U(L)$.

The coefficient of variation CV is measured by the Uster quadratic integrator, which gives a power value expressed as a percentage. In Fig. 29.15, it is the square root of the sum of the squares of the individual deviations from the average diameter $\bar{X}$ divided by $n - 1$, where the minus one corrects for degrees of freedom. This is in reality the root-mean-square value or standard deviation. This value multiplied by $100/\bar{X}$ gives the coefficient of variation. The same results would be obtained as illustrated for Fig. 29.16 by first integrating the squared areas $(X_i - \bar{X})^2$ dl from O to L, then getting the average by dividing by L, taking the square root, and multiplying by $100/\bar{X}$.

$$\text{Mean deviation} = \frac{100}{\bar{X}}\left(\frac{1}{n}\sum_{i=1}^{n}|X_i - \bar{X}|\right)$$

or

$$= \frac{100}{\bar{X}}\left(\frac{1}{L}\int_0^L |X_i - \bar{X}|\, dl\right)$$

$$\text{Coefficient of variation} = \frac{100}{\bar{X}}\sqrt{\frac{1}{n-1}\Sigma\,(X_i - \bar{X})^2}$$

$$= \frac{100}{\bar{X}}\sqrt{\frac{1}{L}\int_0^L (X_i - \bar{X})^2\, dl}$$

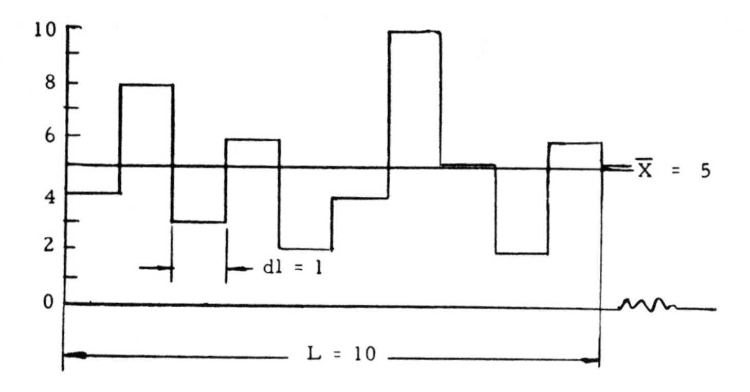

Fig. 29.17. Idealized chart or histogram.

X_i	$\lvert X_i - \bar{X} \rvert$	$(X_i - \bar{X})^2$
4	1	1
8	3	9
3	2	4
6	1	1
2	3	9
4	1	1
10	5	25
5	0	0
2	3	9
6	1	1
	$\Sigma = 20$	$\Sigma = 60$

$\bar{X} = 5,\ n = 10$

$$\text{Mean deviation} = \frac{100}{\bar{X}} \left(\frac{1}{n} \sum_{i=1}^{n} \lvert X_i - \bar{X} \rvert \right)$$

$$= \frac{100}{5} \left[\frac{1}{10} (20) \right] = 40\ \%$$

$$\text{Coefficient of variation} = \frac{100}{\bar{X}} \sqrt{\frac{1}{n-1} \sum_{i=1}^{n} (X_i - \bar{X})^2}$$

$$= \frac{100}{5} \sqrt{\frac{1}{9} (60)} = 51.5\ \%$$

The coefficient of variation may be estimated from the mean deviation by multiplying the mean deviation by 1.25. It is not generally recommended that this conversion be made; however, if it is done, the results should be used with caution.

Evenness Test Frequency. The evaluation of short-term variation or unevenness has become a prime criterion of the effectiveness of quality control for the carding, drawing, combing, roving, and spinning processes. There is increasing experience upon which to base the frequency of these evenness tests, but also some difference in opinion as to the classification of evenness. Obviously, the test frequency depends in a large measure upon the laboratory and personnel facilities of the mill and the ability of the personnel to interpret the results so as to produce a higher grade of work. One of the weaknesses that can occur with any quality control program is the failure to analyze the source of trouble, for the instruments used produce either tabulated or graphic information and do not in themselves indicate the source of trouble.

On three-shift operation, it is suggested that each delivery of drawing be tested once a month. The sampling should be done in much the same manner as described for sliver weight except that samples should be taken from the drawing frame in the proper length. Incidentally, it is wise to check the sliver from the can occasionally to see the effect of the coiling on the sliver, and this can be done by sampling at one time from the bottom of the can and other times from the middle and then from the top of the can.

Every roving frame should be tested at least once a month as a spot check measure for evenness. Other studies should be made at more frequent intervals in order to build up experience as to the quality of the production.

Some mills have found it advantageous to have evenness testers in the operating areas of the mill so that spot checks can be made every day on certain of the machines. Claims have been made that this is a very effective means of spotting mechanical troubles in a hurry, and certainly there must be advantages to such a system. In any event, it is strongly recommended that every mill have an evenness tester of one type or another and that a system be set up to keep this tester in active operation, evaluating the evenness of material from sliver through yarn. The establishment of the routine and the frequency of testing would evolve through experience, but every operating machine should be evaluated once each month.

Evenness Standards. The standards established by Uster for per

TABLE 29.6
Uster Standards: Range of Variation in Per Cent [a]

Stock	Combed			Carded		
	very even	Ave.	Uneven	very even	Ave.	Uneven
Slivers						
card	8	11	15			
comber	20	30	40			
breaker drawing	16	22	32			
finisher drawing	12	18	25	13	20	30
Rovings (H.R.)						
0.6 to 1.5				23	32	45
1.0 to 2.0	20	27	40			
1.5 to 3.0				23	35	50
2.0 to 3.5	18	30	45			
6.0 to 9.0	23	35	50			
Yarns (Count)						
16				71	88	110
20	52	69	88	74	91	115
24	55	71	91	77	93	121
30	58	74	96	80	99	126
32				82	102	129
35 to 36	60	80	102			
40	64	82	104	88	110	136
50	66	88	110			
60	71	93	115			
72 to 76	77	99	124			
80	80	102	126			
85	82	104	129			
95 to 105	91	110	135			
140	110	132	154			

[a] Source: Uster Corporation, Charlotte, N. C.

TABLE 29.7

Brush Standards for Uniformity Analyzer [a]

Material	Good, %	Fair, %	Poor, %
Slivers [b]			
card	7–10	9–15	16 up
breaker drawing	10–12	11–15	17 up
finisher drawing	12–14	13–18	19 up
Rovings [b] (H.R.)			
1.00	22–26	23–27	28 up
1.25	24–27	25–29	30 up
1.50	26–29	27–33	34 up
1.75	28–32	29–37	38 up
2.00	30–35	30–40	41 up
Yarns [c] (Count)			
10–14	75–83	80–87	88 up
15–18	78–86	83–90	91 up
19–24	80–92	85–97	98 up
25–30	85–98	91–106	107 up
31–35	90–105	97–112	113 up
36–40	94–110	101–120	121 up
41–50	96–115	108–132	133 up
	Combed Cottons		
Slivers [b]			
combed sliver	10–15	13–19	21 up
breaker drawing	9–13	11–16	17 up
finisher drawing	11–15	13–20	21 up
Combed Rovings [b] (H.R.)			
1.00	18–21	20–24	27 up
1.25	18–23	20–25	27 up
1.50	19–23	22–27	29 up
1.75	21–25	24–32	34 up
2.00	22–27	25–38	40 up
2.25	22–29	25–40	42 up
2.50	23–32	27–41	43 up
Combed Yarns [c] (Count)			
22–26	79–85	83–90	92 up
31–35	83–89	86–95	97 up
36–40	85–93	90–98	100 up
41–50	90–100	95–106	108 up
51–60	96–107	101–114	116 up
61–70	105–116	110–121	123 up
71–85	113–121	116–130	132 up

[a] Source: Brush Development Company. [b] 3 ft. sens. length. [c] 10 ft. sens. length.

cent variation are shown in Table 29.6. The standards established by Brush for variation are shown in Table 29.7. Uster standards for mean linear irregularity (U %) and mean deviation for carded cottons are listed in Table 29.8, and the Uster standards for mean linear irregularity (U %) or mean deviation for combed cottons are listed in Table 29.9.

The Uster standard values for the coefficient of variation (CV %) for carded cottons are in Table 29.10, and for combed cottons, Table 29.11.

The Uster standard values of the irregularity index (I %) for carded cotton are shown in Table 29.12, and for combed cottons, Table 29.13.

The correction factors for the Uster integrator are shown in Table 29.14.

TABLE 29.8

Uster Standard Values for Mean Linear Irregularity (U %) for Carded Cottons [a]

Material	Even, U %	Ave., U %	Uneven, U %
Sliver [b]			
card, 46–60	2.5–2.8	4.0–4.5	5.0–5.6
breaker drawing, 46–76	2.6–3.4	4.0–5.0	6.1–7.8
finisher drawing, 30–46	2.8–3.5	4.5–5.6	6.7–8.2
Rovings [c]			
slubbing, 0.6–1.2	4.1–5.8	5.6–8.0	7.6–10.8
single process, 1.2–2.4	4.4–6.1	5.8–8.2	8.0–11.2
Yarn [c]			
6.0	9.0	12.0	15.5
8.0	10.2	13.1	17.0
12.0	11.4	15.2	19.7
16.0	12.2	15.3	20.0
20.0	13.0	16.5	20.0
24.0	13.5	17.3	20.6
30.0	14.4	18.1	21.8
36.0	15.0	18.7	22.3
40.0	15.0	18.8	22.5

[a] Source: Zellweger Ltd.

[b] Grains per yard.

[c] English system, 840-yard hank.

TABLE 29.9

Uster Standard Values for Mean Linear Irregularity (U %) for Combed Cottons [a]

Material	Fiber Fineness					
	Denier 1.3 Arealometer 362 Micronaire 3.7			Denier 1.6 Arealometer 329 Micronaire 4.5		
	Even U %	Ave. U %	Uneven U %	Even U %	Ave. U %	Uneven U %
Sliver [b]						
card, 35–60	2.5	4.0	5.0	2.8	4.5	5.6
comber, 46–60	2.5	5.0	10.0	3.0	6.0	11.3
breaker drawing, 44–69	2.4	3.8	5.7	2.6	4.2	6.3
finisher drawing, 24–52	2.2	3.4	4.9	2.5	3.8	5.5
Rovings [c]						
slubbing						
0.8	2.8	3.8	6.0	3.1	4.3	6.7
1.2	3.4	4.7	7.4	3.7	5.2	8.2
single process						
1.5	2.7	3.9	5.7	3.0	4.4	6.4
1.8	3.0	4.3	6.2	3.3	4.8	7.0
2.0	3.1	4.5	6.6	3.5	5.0	7.3
3.0	3.8	5.5	8.1	4.2	6.1	9.0
4.0	4.4	6.4	9.3	4.9	7.0	10.3
5.0	5.0	7.2	10.5	5.5	7.9	11.6
roving						
6.0	4.8	6.0	8.4	5.3	6.6	9.4
7.0	5.2	6.5	9.1	5.7	7.2	10.0
9.0	5.9	7.4	10.3	6.5	8.2	11.4
12.0	6.8	8.5	12.0	7.6	9.4	13.2
Yarn [c]						
20.0	9.4	12.2	15.5	10.3	13.4	17.1
24.0	9.9	12.7	16.3	11.0	14.0	17.6
30.0	10.8	13.5	17.0	12.0	15.0	18.7
36.0	11.5	14.5	17.8	12.7	16.0	19.7
40.0	11.7	14.8	18.0	12.9	16.4	19.8
46.0	12.5	15.1	18.1	13.8	17.0	20.3
50.0	12.8	15.4	18.3	14.2	17.1	20.3
60.0	13.8	16.2	19.0	15.3	18.0	21.1

[a] Source: Zellweger Ltd. [b] Grains per yard. [c] English system, 840-yard hank.

TABLE 29.9 (*Continued*)

Material	Fiber Fineness: Denier 1.3, Arealometer 362, Micronaire 3.7			Fiber Fineness: Denier 1.6, Arealometer 329, Micronaire 4.5		
	Even U %	Ave. U %	Uneven U %	Even U %	Ave. U %	Uneven U %
Yarn [c]						
70.0	14.4	17.0	20.1	16.0	18.8	22.2
75.0	14.7	17.4	21.6	16.2	19.2	22.8
80.0	14.9	17.7	21.0	16.5	19.7	23.3
85.0	15.2	18.0	20.9	–	–	–
90.0	15.2	18.1	20.5	–	–	–
100.0	15.3	18.5	21.0	–	–	–
120.0	16.2	18.9	21.6	–	–	–

TABLE 29.10

Uster Standard Values for Coefficient of Variation (CV %) for Carded Cottons [a]

Material	Even, CV %	Ave., CV %	Uneven, CV %
Sliver [b]			
card, 46–60	3.1–3.5	5.0–5.6	6.2–7.0
breaker drawing, 46–76	3.3–4.2	5.0–6.3	7.6–9.8
finisher drawing, 30–46	3.5–4.4	5.6–7.0	8.4–10.4
Rovings [c]			
slubbing, 0.6–1.2	5.1–7.2	7.7–10.0	9.6–13.6
single process, 1.2–2.4	5.5–7.6	7.2–10.2	10.0–14.0
Yarn [c]			
6.0	11.3	15.0	19.5
8.0	12.7	16.3	21.0
12.0	14.3	19.0	24.6
16.0	15.2	19.0	25.0
20.0	16.2	20.0	25.0
24.0	17.0	21.7	25.8
30.0	18.0	22.6	27.2
36.0	18.8	23.3	27.8
40.0	18.8	23.5	28.2

[a] Source: Zellweger Ltd. [b] Grains per yard. [c] English system, 840-yard hank.

TABLE 29.11

Uster Standard Values for Coefficient of Variation (CV %) for Combed Cottons [a]

Material	Fiber Fineness: Denier 1.3, Arealometer 362, Micronaire 3.7			Fiber Fineness: Denier 1.6, Arealometer 329, Micronaire 4.5		
	Even CV %	Ave. CV %	Uneven CV %	Even CV %	Ave. CV %	Uneven CV %
Sliver [b]						
card, 35–60	3.0	5.0	6.2	3.4	5.6	7.0
comber, 46–60	3.1	6.3	12.5	3.7	7.5	14.1
breaker drawing, 44–69	3.0	4.8	7.2	3.3	5.3	7.9
finisher drawing 24–52	2.8	4.2	6.1	3.1	4.7	6.8
Rovings [c]						
slubbing						
0.8	3.4	4.8	7.6	3.8	5.3	8.4
1.2	4.2	5.9	9.3	4.7	6.5	10.2
single process						
1.5	3.4	4.9	7.1	3.8	5.5	8.0
1.8	3.7	5.4	7.8	4.1	5.9	8.7
2.0	3.9	5.7	8.3	4.3	6.2	9.2
3.0	4.8	6.9	10.2	5.3	7.6	11.2
4.0	5.5	8.0	11.7	6.1	8.8	12.9
5.0	6.2	9.0	13.0	6.8	9.9	14.5
roving						
6.0	6.0	7.5	10.5	6.7	8.3	11.7
7.0	6.5	8.1	11.4	7.2	9.0	12.6
9.0	7.4	9.2	12.9	8.2	10.2	14.3
12.0	8.5	10.7	15.0	9.4	11.8	16.5
Yarn [c]						
20.0	11.7	15.2	19.3	13.0	16.8	21.4
24.0	12.4	15.8	20.4	13.8	17.5	22.0
30.0	13.5	16.9	21.2	15.0	18.7	23.4
36.0	14.3	18.0	22.2	15.8	20.0	24.6
40.0	14.6	18.5	22.4	16.2	20.5	24.8
46.0	15.7	19.3	23.0	17.3	21.3	25.4

[a] Source: Zellweger Ltd.

[b] Grains per yard.

[c] English system, 840-yard hank.

TABLE 29.11 (*Continued*)

Material	Fiber Fineness					
	Denier 1.3 Arealometer 362 Micronaire 3.7			Denier 1.6 Arealometer 329 Micronaire 4.5		
	Even *CV* %	Ave. *CV* %	Uneven *CV* %	Even *CV* %	Ave. *CV* %	Uneven *CV* %
Yarn [c]						
50.0	16.0	19.3	23.3	17.8	21.3	25.4
60.0	17.3	20.3	23.8	19.1	22.5	26.4
70.0	18.0	21.3	25.1	20.0	23.5	27.8
75.0	18.3	21.7	25.8	20.3	24.0	28.5
80.0	18.6	22.1	26.2	20.7	24.5	29.1
85.0	18.9	22.4	26.3	–	–	–
90.0	19.0	22.6	26.4	–	–	–
100.0	19.1	23.0	26.4	–	–	–
120.0	20.2	23.6	27.0	–	–	–

TABLE 29.12

Uster Standard Values of the Irregularity Index (I %) for Carded Cottons [a]

Material	Even, I %	Ave., I %	Uneven, I %
Sliver [b]			
card, 46–60	5.0	8.0	10.0
breaker drawing, 46–76	6.0	9.0	14.0
finisher drawing, 30–46	5.0	8.0	12.0
Rovings [c]			
slubbing, 0.6–1.2	4.0	5.5	7.5
single process, 1.2–2.4	3.0	4.0	5.5
Yarn [c]			
6.0	2.8	3.7	4.8
8.0	2.7	3.5	4.5
12.0	2.5	3.3	4.3
16.0	2.3	2.9	3.8
20.0	2.2	2.8	3.4
24.0	2.1	2.7	3.2
30.0	2.0	2.5	3.0
36.0	1.9	2.4	2.8
40.0	1.8	2.3	2.7

[a] Source: Zellweger Ltd. [b] Grains per yard. [c] English system, 840-yard hank.

TABLE 29.13

Uster Standard Values of the Irregularity Index (I %) for Combed Cottons [a]

Material	Even, I %	Ave., I %	Uneven, I %
Sliver [b]			
card, 35–60	5.0	8.0	10.0
comber, 46–60	5.0	10.0	20.0
breaker drawing, 44–69	5.0	8.0	12.0
finisher drawing, 25–52	3.6	5.5	8.0
Rovings [c]			
slubbing, 0.8–1.2	2.5	3.5	5.5
single-process, 1.5–5.0	1.8	2.6	3.8
roving, 6–12	1.6	2.0	2.8
Yarn [c]			
20.0	1.7	2.2	2.8
30.0	1.6	2.0	2.5
40.0	1.5	1.9	2.3
50.0	1.5	1.8	2.1
60.0	1.5	1.7	2.0
80.0	1.4	1.6	1.9
100.0	1.3	1.5	1.7
120.0	1.2	1.4	1.6

[a] Source: Zellweger Ltd. [b] Grains per yard. [c] English system, 840-yard hank.

TABLE 29.14

Correction Factors for Integrator [a]

Average value (reading)	Range of scale			
	± 100 %	± 50 %	± 25 %	± 12.5 %
−10	1.67	1.25	1.11	1.05
− 9	1.56	1.22	1.10	
− 8	1.47	1.19	1.09	1.04
− 7	1.39	1.16	1.08	
− 6	1.32	1.14	1.06	1.03
− 5	1.25	1.11	1.05	
− 4	1.19	1.09	1.04	1.02
− 3	1.14	1.06	1.03	
− 2	1.09	1.04	1.02	
− 1	1.04	1.02	1.01	1.01

TABLE 29.14 (Continued)

Average value (reading)	Range of scale			
	± 100 %	± 50 %	% 25 %	± 12.5 %
0	1.00	1.00	1.00	1.00
+ 1	0.96	0.98	0.99	
+ 2	0.92	0.96	0.98	0.99
+ 3	0.89	0.94	0.97	
+ 4	0.86	0.92	0.96	0.98
+ 5	0.83	0.91	0.95	
+ 6	0.81	0.89	0.94	0.97
+ 7	0.78	0.88	0.93	
+ 8	0.76	0.86	0.93	0.96
+ 9	0.74	0.85	0.92	
+10	0.72	0.83	0.91	0.95

[a] Source: Zellweger Ltd.

CHAPTER XXX

Control in Winding, Warping, and Slashing

Yarn Tests: Cones, Tubes, and Cheeses. It is not general practice to use the production of the winders as a quality control center except for package characteristics, inasmuch as the yarn spun on bobbins provides a more suitable place for control. However, particularly in yarn mills, the cones are used for checking on the quality of the yarn and package. The purchaser of yarn on cones, tubes, or cheeses would check these packages as these are his raw material. The test methods used are the same as for spun yarn insofar as yarn size, twist, and strength are concerned. Committee D-13 of the A.S.T.M. has set up under tests and tolerances for cotton yarns a tabulation (Table 30.1) covering

TABLE 30.1

Number of Tests (A.S.T.M. Designation D–180–54T)

Test	No. of tests	Coeff. of variation	Precision, %	Probability
Yarn number	25	8.0	3.0	0.95
Single end break	40	15.0	4.65	0.95
Skein break	25	10.0	3.9	0.95
Ply twist	10	8.0	5.0	0.95
Single twist	15	20.0	10.0	0.95

the number of samples for the different tests listed for determining the quality of sales yarn.

The number of tests given in the table is based on lots of sales yarn of 3000 pounds, but fewer tests can be made if agreeable to the parties concerned.

Yarn Number: Warper Beams. The yarn number calculated from warper beams gives an excellent opportunity for determining the effectiveness of yarn number control. The yarn number so calculated does not give short range or individual yarn number variations, but it does give the yarn number on which fabric weight is based and represents the true average of the population.

It is almost universal practice in mills to keep a record of the specifications of each beam made, that is, the number of ends on the beam, the length in yards, the gross weight, and the tare weight. Knowing these, the yarn number can be calculated very simply and quickly. In doing this, it is important to be sure that all empty beams are properly tare-weighted and that beam trucks or other means of conveyance are weighed carefully. Some mills have found it advantageous to use scales with automatic printing devices in order to get accurate and recorded results. It is recommended that the yarn numbers calculated from beam weights be made a part of the quality control program and that the yarn sizes taken at the spinning frame be compared with these beam yarn number averages.

Dye Package Control. Probably the most critical winding operation performed in the yarn mill is in the production of packages for dyeing.

The success of any package dyeing process depends first on the package formation, second on the dye house procedures and materials. A well-formed package is one that has the optimum diameter, winds, weight, and density. The number of winds per inch or the angle of wind and the arrangement of the wind at the end of each traverse has much to do with the uniformity of dye penetration over the length of the package. Hard ends on a dye package will result in poor dye penetration at the ends. The cure may not be wholly affected by cam design on the winder; in many instances it is necessary to "break" the package ends, i.e. to distort them to form a rounded shape before dyeing. Too soft a density will result in poor unwinding of the dyed package, but more important is the unnecessarily high cost involved. Excessive density of the dye package resulting from excessive yarn tension, winding speed, or pressure of the package against the winding drum will result in uneven dye penetration and excessive variation in shade within the package. Thus, it becomes clear that the construction of a dye package must be correct if the dye house is to be successful. But one more step must be taken beyond this: The individual dye packages in a batch must match one another as nearly as possible in order to avoid color variation from package to package.

The quality control steps to meet the above requirements involve the standardization of mechanical settings on the winders and the evaluation of the results of such standardization. Assuming that the well-organized winding department can set up the winders properly, then the other quality control features can be followed more easily.

The first step of control is to establish package standards: traverse length (dictated by machine design), package diameter (limited to some extent by the dye house equipment design), and package weight and density. Weight and density are linked together, for if package dimensions are constant, then weight is proportionate to density.

Package diameter on the winder is controlled at the individual winding position by feelers or other mechanical devices. These should be checked periodically by the department mechanic. It is recommended that occasional checks be made of each winder by the quality control department in order to verify the consistency of size and to keep the department alert.

Package density is a function of winding tension. Therefore, a program of checking and regulating winding tensions should be in force. The simple types of tension measuring devices can be used by winding department personnel for day to day control. Such devices are the Sipp-Eastwood Tensometer and the Uster thread tension gauge. It is suggested that each end be checked at least twice a week of three-shift operations. More critical results can be obtained by using the Tension Analyzer, marketed by Universal Winding Company, Providence, Rhode Island. This instrument has a high sensitivity of 0.1 gram per millimeter of chart width, or 4 grams for full chart width. The highest range gives a 0 to 400-gram sensitivity, and with an adapter this can be extended to 2,000 grams. The frequency response of 100 cycles or tension changes per second is very good, so that with high winding speeds the tension in materials can be recorded accurately on an oscillograph. Tests of each end will reveal errors and inconsistencies in equipment and setting. Once these have been corrected an occasional, possibly bi-weekly or monthly, check of winders will provide the means of maintaining quality.

The final check of dye packages is that of weighing individual packages, or of checking them for density. The weight method would seem to be the simplest way to judge the package, but many mills prefer to check the actual density or dye permeability.

One instrument in limited use designed for this purpose is the Suter-Manville density tester. This instrument measures the penetration of four finger-like feelers under load into the surface of a supported package. The fingers are mounted so as to move individually but all are carried on a single small rod connected directly with the shaft of a thickness measuring gauge. The average movement of the fingers is read directly from the thickness gauge dial in thousandths. In operation, the package to be tested is placed on a mandrel beneath the feeler fingers, which are temporarily moved by a lever so as to clear the package. The fingers are then lowered to the package surface, at which time the gauge reading is noted. Next, by pressing a release lever, the fingers are forced into the yarn package by a relatively heavy weight. The penetration reading is then taken; the difference between the two readings gives the finger penetration in inches. The free dropping of the weight from the surface of the package to its final penetration eliminates the creeping of the fingers deeper and deeper as is found if the weight is allowed to act slowly; once released, the penetration is final and no tapping or jarring of the instrument will change the effective reading. Two readings at 90° from each other on each package are taken. Results are penetration in thousandths of an inch, and these are a measure of package density. The frequency of test is a matter of judgement and depends on the existing conditions in any particular mill. It has been found that a bi-weekly check of all packages (identified as to their position) will correct bad conditions and allow maintenance of very high quality standards. A spot check of ten per cent of all packages is an alternate method found to be effective.

Another method of testing is to measure the permeability of the dye package by using air flow. The principle involved is much like that used in measuring fiber fineness with the Micronaire, in that there is a pressure change in a tube feeding air through an orifice to a dye package. The principles involved are described in an article by D. L. Worth in *Textile Industries* for September, 1953. Some of the findings of tests performed using this permeability tester are interesting. For example, these tests verify those of the authors in showing that high density packages and the outside of all packages tend to lighter shades.

Yarn Inspection. One of the continuing problems in the production of ply yarn materials from spun blends is that of mixed yarn. The use of fugitive tints has helped to minimize the trouble, but unfortunately sometimes there are an insufficient number of different tints to meet the mill's needs. Also some tints tend to look alike with fluorescent lighting and this can lead to mixed yarn. One place of checking for mixed yarn or odd ply is at the warper; this is not a cure for the odd ply or mixed yarn, but the elimination of such bad yarn does prevent spoilage of cloth.

It is the duty of warper tenders to watch for bad yarn during warping. This can be supplemented by a test program consisting of an inspection of the yarn at the beginning of each warp; this means in effect the inspection of yarn at the completion or end of each warp. When a warp has been completed, before removing the beam and while the ends are in their orderly position and taut, transparent tape is used over and under the ends to bind them in position. A second binding of the ends is made at about $\frac{1}{2}$ inch spacing from the first binding. The sample is identified and cut from the warper at $\frac{1}{2}$ inch from the tapes, thus giving a sample with yarn between the tapes and extending a little beyond the tapes. The sample can then be examined under a large magnifying glass. A trained technician can spot off-shade yarn, odd ply, and other defects readily. Mixed blends will not be apparent in most cases and can be caught only by a head-end dye test from woven samples. The reason for allowing the $\frac{1}{2}$ inch of yarn to extend beyond the tape becomes apparent when the inspection is made, for wrong twist and off-size yarns will tend to curl differently from the correct yarn in this short length. Inspection can be made in the department, and this is preferable because of the time factor.

For continuous filament synthetic yarn inspection at the warping operation, a suitable instrument is the Lindly automatic yarn inspector (sold by Forster Machine Co., Westfield, Mass. The description of this instrument by Lindly & Company, Inc., is as follows:

> A variable sensitivity optical electronic device capable of detecting variations in yarn diameter such as slubs, broken filaments, and puff balls — at high speeds. Generally used in conjunction with the warping operation for 100 % inspection and repair, or for statistical and quality control. For all types of continuous filament yarn, whether raw or thrown. Users report up to 90 % improvement in quality, and increased production.

Sizing Content. The amount of starch and other ingredients added at slashing is vital in cost and quality work. Some mills are content to compare the total weights of unsized yarn with the sized yarn and calculate the per cent of sizing added from these figures. However, differences in moisture regain, waste losses, and stretch factors tend to invalidate such results. The desizing of samples is strictly speaking a chemical test; however, many small mills unequipped for chemical testing can set up a desizing unit with little expense.

The test for determining the total sizing in a material is described in detail under the Qualitative Analysis of Textiles (D629-54) in A.S.T.M. Standards on Textile Materials, Committee D-13. Reference should be made to this for details. In general terms, the test involves extraction with carbon tetrachloride, drying and washing, immersion in an aqueous solution of a starch and protein solubilizing enzyme preparation, boiling in water, and drying. In cases where the mill knows what ingredients are being used in their sizing bath, the test procedure can be simplified. This test is primarily for materials sized with starch. Other types of tests are necessary where synthetic resins or rubber and pyroxylin materials are used.

CHAPTER XXXI

Woven Fabric Testing, General

Width

To measure the width of a fabric, it is desirable to have a full length roll or cut. If a sample of such a length is available, it is passed over an inspection machine or similar rewinding mechanism and the width measured to the nearest $\frac{1}{16}$ inch at not less than five different points throughout the length. If only a short length is available, it is laid out smooth and without tension on a horizontal surface and the width measured to the nearest $\frac{1}{16}$ inch.

The critical point in measuring the width of a sample of fabric is to get the material smooth, while at the same time to prevent distortion of the width due to excessive tension. It will be found necessary to smooth the fabric out widthwise in order to eliminate any wrinkles. Some technicians will spread the fabric widthwise by rubbing their hands across the surface of the fabric from the center out and then allowing the fabric to come into its normal or relaxed position. If a perfectly smooth surface is not available for use in measuring the width, then it will be found helpful to place the measuring stick under the fabric. By smoothing out the fabric on the stick, the width can be measured with normal accuracy. This method also works satisfactorily when measuring the fabric in the fold. At all times be sure that the measuring stick is perpendicular to the selvages.

Weight

Mill Practices. Practically all fabric is bought and sold on yardage and weight. For many consumer dress goods fabrics of cotton, the units of length are cuts or pieces. Under the Worth Street Rules, a single cut is a length of 60 yards, and a double cut is a length of 120 yards. For other markets and fibers, piece lengths are set at different standards. For these same cotton dress goods fabrics, the weight is specified in units of yards per pound for definite widths. For fabrics for military use, usually the weight is specified in units of ounces per square yard, and widths are also specified. For certain other fabrics, notably ducks and heavy mechanical fabrics, the weight is specified in ounces per linear yard for the specified width.

No matter what method may be used by the textile manufacturer, there are established tolerances and standards that must be met in order to have the fabric acceptable, or to avoid rejections or price rebates. For cotton fabrics bought and sold under Worth Street Rules, a series of specifications spell out the tolerances on total yardage, width, sley, picks, and weight. For contracts with the Federal Government, a complex set of specifications spell out general and individual requirements. As a result, one of the prime objectives in the mill is that of meeting fabric weight specifications. There is scarcely any surer way of losing money in the mill than by delivering cloth that is overweight in comparison to standards. Thus, the testing and quality control program set up in the yarn mill has as one of its first objectives the control of fabric weight.

Quality control programs on fabric weight should have three facets: the checking of head ends of new samples, the checking of first cuts, and the checking of baled fabric. On new samples, or on new styles being initially put into loom production, head-end samples (a few yards cut from the loom at the start) should be taken to the laboratory for complete analysis and comparison with standards. Approval of the head end should precede any further operation of the loom.

For normal weave room operation, once head-end samples are approved for a new style, first cuts should be checked in the cloth room for weight, width (at five places), and ends per inch warp and filling.

The first cut signifies that a new warp has been started at the loom. By checking the first cut, if errors exist, they can be detected, and if the cloth is satisfactory, there is definite assurance that the loom is under control for the characteristics measured. The check-up of first cuts is particularly important for fancy mills, that is, those operating on contract for a wide range of consumer fabrics over relatively short periods of time. For mills on repetitive work such as print cloths, sheetings, or the like, the establishing of fabric quality can be less frequent. For example, the fabric from one warp from each slasher set would suffice in most cases.

The third facet involves the unit weight based on baled batches. Each bale of fabric is made up of a definite yardage which is normally entered on a baling slip. The net weight of the several pieces of fabric is obtained from individual piece weights (in some mills) or from a total net weight. From weight and length can be calculated the unit weight. With proper control in the mill, the average unit weight taken from bale lots should vary not more than one per cent either way from standard (with a maximum of 2 per cent under abnormal conditions). As a matter of interest, filling yarn standard numbers should be controlled from fabric weights. For instance, if the specified weight of a fabric is 4.25 yards per pound, and the filling is 50/1, actual fabric weights of 4.21 in the bale would call for a change to the lighter side of 1 per cent in the 50/1, or to 50.5 as standard. Where the same filling yarn is used in several fabrics, this method may well prove impractical, so other standards must be modified in order to meet the quality standards of the fabric. In any event, unit weights taken from the bale will certify to the degree of control in the mill. Daily reports on baled weights should be a part of the data made available to the quality control department.

Fabric Weight Measurement

The weight of fabric can be determined by one of three general methods. One is to weigh an entire cut, roll, or bale of fabric, where the width and yardages are known. This method is found useful where the fabric is weighed immediately prior to shipping and serves as a check against sales specifications. This gives an over-all average type measurement. One disadvantage of using this method is the fact that the material is not exposed to standard atmosphere conditions and the moisture content of hydrophilic materials will vary from shipment to shipment, except in those cases where cloth room atmospheric conditions are regulated within fairly close tolerances. A second method is to weigh 1-yard lengths of a fabric. The yard lengths may be either linear yards or square yards. The third method is to weigh small specimens which have been cut from the fabric. These specimens should not be smaller than 2 inches square (4 square inches), and if sufficient fabric is available, the desirable minimum is 20 square inches. The small samples may be cut from the master sample with scissors; however, it is advisable to use a die of the proper size so that the samples will always be exactly the same size. The samples can be prepared much faster with the use of a die.

Fabric weight is generally expressed in one of three ways: ounces per square yard, ounces per linear yard for standard widths, and yards per pound for given widths. The simplest of the three methods is to express the weight in ounces per square yard, for using this method the weight of the material is not a function of fabric width. When using the ounces per linear yard for standard widths, it is necessary to remember which widths are standard for different types of fabric. This can be very confusing due to the many different widths which have been accepted as standards. Table 31.1 is a partial list of the different widths for

TABLE 31.1

Partial List of Different Fabric Widths Currently Used as Standards for Expressing Fabric Weight

Standard width, ins.	Fabric	Oz./linear yard for standard width
22	Numbered Duck	7 to 30
28½	Army Duck	7 to 15
29	Flat Duck	7 to 15
30	Drills	3 to 8
36	Chafer Duck	8 to 14
37	Shoe Duck	6 to 9
38	Enameling Duck	7 to 10
40	Hose Duck	10 to 24
42	Belt Duck	20 to 48

duck. The third method of expressing fabric weight is to give the number of yards in 1 pound of the material for a given width.

For example, the statement "32-inch, 6.25-yard sheeting" means there are 6.25 yards in one pound of the sheeting which is 32 inches wide. This fabric has a square yard weight of 2.88 ounces. This value would naturally remain constant; however, the yards per pound would vary with the width. For example, the same fabric in a 36-inch width would have 5.56 yards per pound, and in a 40-inch width would have 5.00 yards per pound. Regardless of which method is used in the merchandising or selling of fabrics, the weight is generally reduced to ounces per square yard by laboratory and quality control personnel, for this is a common basis for most calculations involving fabrics.

Procedure for Preparation of Test Sample. If the weight of a cut, roll, or bale of fabric is to be used in calculating the average square yard weight, the net weight, the length, and the width of the fabric are recorded.

If a square yard or a linear yard is to be weighed, it should be cut to accurate dimensions and conditioned prior to weighing. When using small samples, the fabric should be laid out smooth and without tension. The fabric should reach moisture equilibrium prior to weighing.

Schiefer (1) has suggested the use of circular dies 2.4126, 3.4119, and 3.8146 inches in diameter because the weight in grams of such specimens when multiplied by 10, 5, and 4 respectively gives the weight in ounces per square yard of the fabric.

If four samples with a diameter of 3.8146 inches are weighed simultaneously, the weight in grams will be equal to the square yard weight in ounces. The same is true for five specimens 3.4119 inches in diameter and for ten specimens 2.4126 inches in diameter.

Calculations: Rolls, Cuts, or Bales. The weights used for bales would be the net weights used in shipping or receiving. For rolls or cuts, the weights would be those of the cloth department or laboratory as obtained on accurate scales suitable for the work.

The following examples illustrate the calculations followed in getting unit weights from large samples.

Basic information: 4 pieces of fabric, each 120 yards in length weigh 194.5 pounds. The average width matches the standard width of 40 inches.

1. For yards per pound:

$$\text{yds/lb} = \frac{\text{total yards weighed}}{\text{total weight in pounds}}$$

$$= \frac{120(4)}{194.5} = 2.47$$

Also,

$$\text{yds/lb} = \frac{16}{\text{oz/linear yd}}$$

2. For ounces per linear yard:

$$\text{oz/linear yd} = \frac{\text{lbs (16)}}{\text{yds}}$$

$$= \frac{194.5(16)}{480} = 6.48$$

Also,

$$\text{oz/linear yd} = (\text{oz/sq. yd}) \left(\frac{\text{width}}{36}\right)$$

3. For ounces per square yard:

$$\text{oz/sq. yd} = \frac{(\text{lbs})(16)(36)}{(\text{yds})(\text{width, ins.})}$$

$$= \frac{194.5(16)(36)}{480(40)} = 5.84$$

Calculations: One-Yard or Short Lengths. One of the problems encountered in getting unit weights from 1-yard lengths is that of obtaining accurate lengths. This is particularly true with finished fabric where bow resulting from finishing operations tends to lead to inaccuracies. One obvious method of overcoming the problem is to mark the 1-yard length on the fabric before cutting. Certain grey goods may be torn to size.

The following examples illustrate the calculations used in getting unit weights for short samples.

Basic information: A single yard of 28-inch fabric weighs 8.0 ounces.

1. For yards per pound:

$$\text{yds/lb} = \frac{\text{yds}}{\text{lbs}} = 1 \div \frac{8}{16} = 2$$

Also,

$$\text{yds/lb} = \frac{16}{\text{oz/linear yd}}$$

2. For ounces per linear yard: Obviously the weight of 8 ounces for a 1-yard length gives ounces per linear yard.

Also,

$$\text{oz/linear yd} = \text{oz/sq. yd}\left(\frac{\text{width}}{36}\right)$$

3. For ounces per square yard:

$$\text{oz/sq. yd} = \frac{\text{oz}(36)}{\text{yd (width, inches)}} = \frac{8(36)}{1(28)} = 10.3$$

The calculations for short lengths (other than for 1-yard lengths) follow the same simple calculations.

Calculations: Small Samples. For weighing small samples, a balance accurate to 0.01 grams should be used. When the sample or samples are weighed, the size and weight of each sample should be recorded. Where multiple samples can be die-cut from a master sample, this is considered the best procedure. However, frequently samples submitted for analysis are either too small for die cuts or cannot be mutilated; in these cases the actual sample with whatever trimming is needed is used for weight analysis.

The following examples illustrate the calculations used for small samples.

Basic information: Five 2-inch by 2-inch samples cut from a 38-inch width fabric weigh (a) 1.80 grams on gram scale, or (b) 27.77 grains on grain scales.

$$\text{Total area sampled is } 5(2 \times 2) = 20 \text{ sq. in.}$$

$$\text{Conversion of grams to pounds: } \frac{\text{grams}}{453.59} = \text{lb}$$

$$\text{Conversion of grains to pounds: } \frac{\text{grains}}{7000} = \text{lb}$$

1. For yards per pound:

(a)
$$\text{yds/lb} = \frac{\text{square inches weighed } (453.59)}{\text{grams (width)}(36)}$$

$$= \frac{20(453.59)}{1.80(38)(36)} = 3.68$$

(For, 1.80 grams for 20 square inches would be equivalent to $1.80\,\frac{(38)(36)}{20}$ grams for a linear yard. Also, $\frac{1.80}{453.59}\left(\frac{(38)(36)}{20}\right)$ gives the pounds for a linear yard. Therefore, the reciprocal of this gives yards per pound.)

(b) If grains are used in weighing, then

$$\text{yds/lb} = \frac{\text{square inches weighed } (7000)}{\text{grains (width)}(36)}$$

Also, as for other examples given,

$$\text{yds/lb} = \frac{16}{\text{oz/linear yard}}$$

2. For ounces per linear yard:

(a)
$$\text{oz/linear yard} = \text{grams}\left(\frac{16}{453.59}\right)\left(\frac{(\text{width})(36)}{\text{sq. inches}}\right)$$

$$= 1.8\left(\frac{16}{453.59}\right)\left(\frac{(38)(36)}{20}\right) = 4.35$$

Also, as for other examples given,

$$\text{oz/linear yard} = \text{oz/sq. yd.}\left(\frac{\text{width}}{36}\right)$$

(b) Using grains,

$$\text{oz/linear yard} = \text{grains}\left(\frac{16}{7000}\right)\left(\frac{(\text{width})(36)}{\text{square inches}}\right)$$

3. For ounces per square yard:

$$\text{(a) oz/sq. yd} = \text{grams}\left(\frac{16}{453.59}\right)\left(\frac{(36)(36)}{\text{sq. inches weighed}}\right)$$

$$= 1.8\left(\frac{16}{453.59}\right)\left(\frac{(36)(36)}{20}\right) = 4.11$$

For,

$$\frac{1.80}{20} = \text{grams/sq. in.}$$

$$1.80\left(\frac{(36)(36)}{20}\right) = \text{grams per square yard}$$

and as

$$\frac{16}{453.59} = \text{oz/gram}$$

then

$$1.80\left(\frac{16}{453.59}\right)\left(\frac{(36)(36)}{20}\right) = \text{oz/sq. yd.}$$

(b) For oz/sq. yd., when using grain scales, the calculation becomes:

$$\text{oz/sq. yd} = \text{grains}\left(\frac{16}{7000}\right)\left(\frac{(36)(36)}{(\text{sq. inches weighed})}\right)$$

$$= 27.77\left(\frac{16}{7000}\right)\left(\frac{(36)(36)}{20}\right)$$

$$= \frac{27.77(2.962)}{20} = 4.11$$

Note that the constant 2.962 equals $16(36)(36) \div 7000$.

Effect of Moisture on Fabric Weight. Some technologists recommend that the weight of samples should be determined with the specimens in an oven dry condition. Others suggest that if the samples are conditioned from the dry side and reach moisture equilibrium, the weight and subsequent calculations are sufficiently accurate. The latter method is more commonly used and is recommended for quality control testing.

As an example, if a 25-square-inch sample of viscose fabric weighs 32.85 grains when oven dry, the calculations are:

$$\text{Ounces per square yard, oven dry} = \frac{\text{weight in grains of sample}}{\text{square inches weighed}} (2.962)$$

$$= \frac{32.85}{25} (2.962) = 3.9$$

The weight of this fabric with 11-per cent regain would be calculated as follows, letting x equal the weight with 11-per cent regain.

$$\frac{x - 3.9}{3.9} = 0.11$$

$$x = 0.11(3.9) + 3.9 = 3.9(1 + 0.11)$$

$$= 4.33 \text{ ounces per square yard with } 11\,\% \text{ regain}$$

The formula reduced to one equation becomes:

$$\text{Weight at any given regain} = (\text{oven-dry weight}) \left(1 + \frac{\text{given \% regain}}{100}\right)$$

The weight of the sample of fabric with 11-per cent moisture content would be calculated as follows, letting y represent the weight with 11-per cent moisture: (Remember that moisture content is based on conditioned weight.)

$$\frac{y - 3.9}{y} = 0.11$$

$$y = \frac{3.9}{0.89}$$

$$= 4.38 \text{ ounces per square yard with } 11\,\% \text{ moisture content}$$

This formula reduced to one equation becomes:

$$\text{Weight at any given content} = \frac{(\text{oven-dry weight})}{1 - \left(\frac{\text{given \% content}}{100}\right)}$$

Summary of Calculations for Fabric Weight

For yards per pound

(a) $\dfrac{\text{total yards}}{\text{net pounds}}$

(b) $\dfrac{16}{\text{ounces per linear yard}}$

(c) $\dfrac{\text{total yards}}{\text{ounces}}(16)$

(d) $\dfrac{\text{square inches weighed }(453.59)}{\text{grams (width)}(36)} = \dfrac{\text{square inches weighed}}{\text{grams (width)}}(12.60)$

(e) $\dfrac{\text{square inches weighed }(7000)}{\text{grains (width) }(36)} = \dfrac{\text{square inches weighed}}{\text{grains (width)}}(194.4)$

For ounces per linear yard

(a) $\dfrac{\text{ounces weight}}{\text{yards weighed}}$

(b) $\dfrac{\text{pounds }(16)}{\text{yards}} = \dfrac{16}{\text{yards/pound}}$

(c) ounces/square yard $\left(\dfrac{\text{width}}{36}\right)$

(d) $\dfrac{\text{grams (width) }(16)(36)}{453.59\text{ (square inches weighed)}} = \dfrac{\text{grams (width)}}{\text{square inches weighed}}(1.270)$

(e) $\dfrac{\text{grains (width)}(16)(36)}{\text{square inches weighed }(7000)} = \dfrac{\text{grains (width)}}{\text{square inches weighed}}(0.0823)$

For ounces per square yard

(a) $\dfrac{\text{pounds }(16)(36)}{\text{yards (width)}}$

(b) $\dfrac{\text{ounces }(36)}{\text{yards (width)}} = \text{ounces per linear yard}\left(\dfrac{36}{\text{width}}\right)$

(c) $$\frac{36(16)}{\text{width (yards per pound)}} = \frac{576}{\text{width (yards per pound)}}$$

(d) $$\frac{\text{grams }(16)(36)(36)}{\text{square inches weighed }(453.59)} = \frac{\text{grams weight of sample}}{\text{square inches weighed}}(45.72)$$

(e) $$\frac{\text{grains }(16)(36)(36)}{\text{square inches }(7000)} = \frac{\text{grains weight of sample}}{\text{square inches weighed}}(2.962)$$

Sley and Pick

The term "construction" is used to denote the number of ends and picks in a woven fabric and is designated by such figures as 90 × 80. This means a sley of 90 ends per inch in the warp and 80 picks per inch in the filling. The first number should always designate the warp ends per inch.

A pick glass or some other convenient system of magnification is the method generally used to count the ends and picks in a fabric. If the fabric is of such a dense construction that it is not feasible to use a pick glass, another conventional method is to unravel the sample. Frequently it is advantageous to use a combination of the pick glass and unraveling. If a fabric has a high concentration of ends, a few picks can be unraveled, leaving an exposed fringe of ends. A pick glass and pick-out needle or a mechanical pick counter can then be used to get an accurate count of the ends per inch.

One system that is used to alleviate eye strain is to project the fabric onto a large screen. The projector used for this is a slide type with suitable holders to accommodate fabric.

Low-powered microscopes also are available on the market for counting the sley and pick and for examining surface characteristics of the fabric. One such model is mounted over a metal cabinet containing a light source and a translucent glass. The sample can be held in a small vise-like attachment (with one-way control of tension in the sample), and in turn this sample can be held over the translucent glass so that an even light passes through the sample. Another arrangement can be used where the light is directed onto the surface of the fabric.

Lighting of samples plays an important part of fabric analysis. It will be found useful to set up a permanent translucent light source in a convenient place on some laboratory table. One method is to cut out a table top to accommodate a pane of frosted glass at least 8 × 10 inches in size. Below this can be built in a light fixture carrying two or more bulbs. It will be found that many finely woven samples can be more easily counted with this type of light. In other cases, a desk light with a flexible goose neck that can be adjusted to any desired position will be found to be effective.

Another system used to determine ends per inch is through the use of glass or other transparent materials with ruled lines or gratings. For instance, a piece of glass with gratings of 10 lines per inch placed over a fabric with the same number (10) ends per inch would give no interference fringe pattern. But, if the fabric had 9 ends per inch, an interference fringe would result at the ninth grating, or interference fringes would be formed every 9/10 of an inch. For fabric with 11 ends per inch, the interference fringes would be formed at the 11 grating, or every $1\frac{1}{10}$ inches. The location of these interference bands serves to give the number of ends per inch. Used in this manner, a set of glasses ruled in different parallel gratings is necessary to cover the range of fabric constructions to be used or studied. Of incidental value is another feature of the use of gratings: Irregularity of the spacing of the ends or picks in the fabric becomes strikingly evident when viewed through such ruled glass.

For convenience, glass ruled with converging lines is commercially available. Because of the converging lines, it is possible to use a single device for a wide range of ends per inch, for the interference effect reaches peaks just above and below the position corresponding to the exact number of ends per inch at that location in the fabric.

Regardless of the method used, no counts should be made near the selvage or ends of the sample. The sample should be laid out smooth on a horizontal surface. If the fabric has more than 25 ends per inch, the number of ends in 1 inch are counted in at least 5 different points on the fabric with no two spaces counted including the same yarns. If the fabric has less than 25 ends per inch, the number of ends in 3 inches are counted. The average number of ends per inch are calculated from the measurements. At least five 3-inch counts are made with no two including the same yarns.

If the fabric is 3 inches or less in size, all the ends are counted and the results expressed as the average ends per inch.

Bow

The bow of a fabric is the curvature of the filling thread from its normal path perpendicular to the selvages. Usually there is very little bow in grey goods just taken from the loom. However, some bow may be induced in shearing or brushing operations and can become quite pronounced in certain fabrics during or after finishing operations. The amount of bow is determined by measuring the amount of deviation of the filling thread from the imaginary straight line position that the yarn should take. To measure bow, the path of a filling thread is marked by pencil across the width of the fabric. The bow is then determined by measuring in inches to the nearest sixteenth the greatest distance that this drawn line varies from the imaginary straight line perpendicular to the selvages. Bow is frequently expressed as the per cent of the width of the fabric. The calculation consists merely of dividing the amount of bow by the width in inches and multiplying by 100.

Crimp

The value of measuring crimp is overlooked by many technicians. It is only too seldom realized that crimp affects the finishing process as well as such fabric physical properties as strength, elongation, abrasion, and serviceability. Even though a fabric has a square construction, that is, the same yarn size in warp and filling and the same number of ends and picks, it will not have a balanced strength unless the crimp is the same in both the warp and filling yarns. For fabrics which are subject to a bursting type of pressure, the balancing of crimp is of primary importance. For certain other applications, a complete unbalancing of crimp is of primary importance and necessary. For example, in fabrics used for conveyor belts, the warp crimp should be as low as possible, and the filling crimp should be high. In this case, low warp crimp is desirable to prevent stretching since the warp yarns are continually under strain when in operation. The filling yarns need the crimp so that the belt would be free to "trough", i.e., the sides of the belt can be turned up to form a trough while in operation. From a manufacturing point of view, it is difficult to produce a fabric with no crimp in the warp since the warp yarn will invariably crimp more than the filling during weaving. If the fabrics are to be finished, some balancing of the crimp is possible by controlling the tension under which the fabrics are processed.

There are two methods of expressing the amount of deformation of the yarn due to the interlacing in weaving. The term used in the previous discussion, crimp, is generally used by laboratory personnel, finishers, cutters, and customers. Crimp is calculated by expressing the change in length from the straight yarn to the woven yarn as a per cent of the woven length. Since the yarn is in the woven form when it reaches the laboratory, finisher, or such, it is only natural for this group to think in terms of woven length, and thus, of crimp. On the other hand, the term take-up is used by the weaver due to the fact that take-up is based on the straight or non-woven length, and the yarn is in that form when delivered to the weave room.

There are several methods available for measuring the change in length of the yarn as a result of weaving by analysis of the woven fabric. One of these is the "hand method," which is the most common and yet the least accurate method used. This technique involves the measuring of a given length of yarn in the fabric, marking the fabric or snipping several ends at this length, unraveling the yarn, and straightening the yarn by hand while measuring the straightened length. The source of error in this method is in straightening the yarns to the extreme ends, for it is extremely difficult to straighten yarn in this fashion to its extremes without stretching the yarn. This is especially true when one technician analyzes a wide variety of fabrics.

The Brighton Crimp Tester

The Brighton crimp tester is an instrument for measuring crimp, which eliminates the straightening of the yarns by hand. Another added feature of the instrument is that crimp and take-up can be read directly from the dial. However, such instruments as this must be calibrated. The calibration consists of determining the amount of load that can be applied to different size yarns without stretching the yarn. After such a calibration, the use of such instruments will reduce the time considerably for measuring and calculating.

The Brighton crimp tester is simple to use, as shown in the sketch, Fig. 31.1. An 8-inch length is marked in the fabric. The

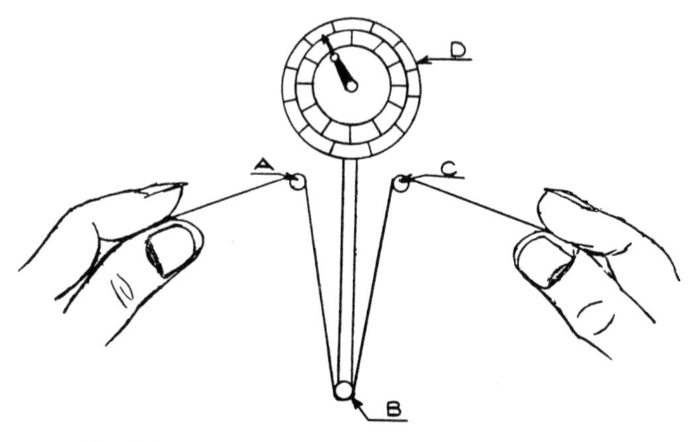

Fig. 31.1. Principle of the Brighton crimp tester.

end is removed and placed in the instrument as shown, with the loop under B. The hook B moves in a vertical plane, with movements actuating dial D to show crimp or take-up. The end is moved by hand until the marks that were 8 inches apart in the fabric are at the position shown by the arrows at A and C. The crimp is removed by the stress applied by the weight on hook B and its position gives the desired reading.

Another method which is accurate and rapid is to draw a stress-strain relationship of the yarn. This can be done by marking a given length, usually ten inches, of yarn in the fabric. The yarn is then unraveled and placed in a recording type strength tester such as an IP-2 (or a vertical pendulum tester of very low capacity). The yarn is placed in the jaws without tension and with the two marks at the face-edge of the clamps. The machine is then started and a stress-strain curve of the type shown in Fig. 31.2 is autographed by the machine. Once the relationship becomes linear, the crimp has been removed and the normal stress-strain relationship

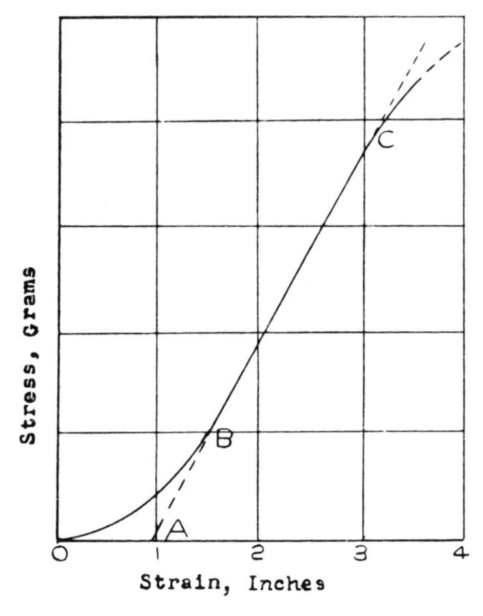

Fig. 31.2. Stress-strain curve. Line *CBA* is the stress axis.

of the material is being drawn. By extending the linear portion of the curve *CB* to the stretch axis (line *CBA*), the point *A* on the strain axis can be located. Point *A* represents the origin of the stress-strain curve without crimp. Therefore, the distance *OA* represents the change in length of the yarn. This distance when expressed as a per cent of the woven length is crimp.

Other methods have been used for obtaining crimp values. For example, one method that has been used for precision work is as follows: A 10-inch (or longer) length of yarn is pre-marked in the fabric and removed. One end of the sample is placed in a jaw, and the other end of the marked length is attached to the end of a fine chain. The chain is supported at one end of its catenary by the yarn, and the other end is on a movable element. Lowering vertically this latter end imposes a greater load on the yarn, thus removing crimp. The relationship of load to crimp removal can be plotted by getting the movement of the loaded end of the yarn sample at each stage of loading. This entire technique actually has more value in research than in everyday quality control or fabric analysis.

Another technique that can be used advantageously in the mill for setting up tables of crimp is as follows: Ten-inch (or longer) lengths are marked accurately on the yarn in the loom, with excess tension removed. After weaving, the crimped lengths are measured and the per cent crimp calculated.

For crimp in the filling, a comparison of the width in the reed with grey cloth width will give an accurate measure of the crimp. Obviously this must be true, for the reed carries the straightened filling into the fell of the cloth. From this point on, change in width is due to, or results in, the crimp in the filling.

Of course, in production it is customary to compare slashed lengths of yarn with the length of fabric woven from this yarn, thus providing the planning department, the designer, and the cost accountant with basic information on crimp or take-up. If cut marks are put on the yarn in slashing at predetermined and accurate spacings, the distance between cut marks in the woven fabric gives the complete information to calculate crimp or take-up. In fact, a history or record of the crimp in different fabrics is essential to designing, cost, and planning if these functional departments are to do their respective jobs properly. For example, suppose that a mill takes a contract for a broadcloth; the sales agreement calls for delivery in double-cut lengths of 120 yards. Records indicate that this fabric normally has 8 per cent crimp. The planning department must make allowances not only in poundages processed, but also in the lengths warped and slashed to allow for take-up (crimp) in weaving. The spacing between the cut marks stamped on the yarn in slashing must be such that when the cloth is cut from the looms in the weave room, the lengths between marks are the specified 120 yards. The calculations show that 129.6-yards lengths must be marked at the slasher.

Calculations. A comparison of the calculation of crimp and take-up is given below. A 10-inch length of yarn is marked in the cloth straightened to $10\frac{3}{4}$ inches.

Length in fabric	10″
Straightened length	$10\frac{3}{4}$″
% crimp	7.5 %
% take-up	7.0 %

$$\text{Crimp, \%} = \frac{\text{straightened length} - \text{woven length}}{\text{woven length}}(100)$$

$$= \frac{10.75 - 10}{10}(100) = 7.5\ \%$$

$$\text{Take up, \%} = \frac{\text{straightened length} - \text{woven length}}{\text{straightened length}}(100)$$

$$= \frac{10.75 - 10}{10.75}(100) = 7.0\ \%$$

For the stress-strain relationship shown in Fig. 31.2, assume that a 10-inch length was marked in the cloth and placed in the machine.

The gain in length is measured from the strain side of the curve and in Fig. 31.2 is 1 inch; therefore, the straightened length becomes the original length plus this length, which is 10 plus 1, or 11 inches.

$$\text{Crimp} = \frac{11 - 10}{10}(100) = 10\ \%$$

$$\text{Take-up} = \frac{11 - 10}{11}(100) = 9.1\ \%$$

Frequently it is necessary to convert from crimp to take-up or *vice versa* without benefit of the original measurements. This can be done by using the following relationships:

$$\% \text{ take-up} = \frac{(\% \text{ crimp})(100)}{(\% \text{ crimp}) + 100}$$

$$\% \text{ crimp} = \frac{(\% \text{ take-up})(100)}{100 - (\% \text{ take-up})}$$

These relationships may be derived from the basic definitions of crimp and take-up. If each equation is solved for a common value, such as original length, the resulting relationships may then be equated and solved for the above formulas. T = per cent take-up; C = per cent crimp; L_o = original length; L_w = woven length.

$$T = \frac{L_o - L_w}{L_o}(100)$$

$$C = \frac{L_o - L_w}{L_w}(100)$$

$$L_o = \frac{L_o - L_w}{T}(100)$$

$$L_o = L_w \frac{(C + 100)}{100}$$

$$\frac{L_o - L_w}{T}(100) = L_w \frac{(C + 100)}{100}$$

$$T = \frac{(L_o - L_w)(100)}{L_w \dfrac{(C + 100)}{100}}$$

Since

$$\frac{L_o - L_w}{L_w}(100) = C$$

then

$$T = \frac{C(100)}{C + 100}$$

To express crimp in terms of take-up, solve the above for C.

$$T = \frac{C(100)}{C + 100}$$

$$TC + T(100) = C(100)$$

$$C(100) - TC = T(100)$$

$$C = \frac{T(100)}{100 - T}$$

EXAMPLE. For a take-up of 7 per cent, what is the crimp?

$$C = \frac{100T}{100 - T} = \frac{100(7)}{100 - 7} = 7.5\,\%$$

The calculation to determine the yards to slash (the length between cut marks) for a given cut length uses the basic formula,

$$\text{Crimp, } \% = \frac{\text{straightened length} - \text{woven length}}{\text{woven length}}\,(100)$$

For example, if a fabric has a predetermined crimp of 8 per cent, what should the slashed length be to give 120 yards of fabric for each double cut?

$$8 = \frac{\text{straightened length} - 120}{120}\,(100)$$

$$\text{Straightened length} = 129.6 \text{ yards}$$

Crimp, Take-up, Moisture Regain, and Moisture Content Tables. In Appendix I are given tables which can be used for converting crimp to take-up and *vice versa*. These tables can also be used for regain and moisture content.

References

1. Schiefer, H. F., Textile Section, Natl. Bur. Standards, Washington, D. C.

CHAPTER XXXII

Fabric Strength

The strength of a fabric may be determined from three approaches: its resistance to a tensile load, its resistance to a tearing force, or its resistance to a bursting force. Each of these three measurements has its own usefulness. Whether one of these determinations, two, or all three are made on any particular sample depends on the type of fabric, its end use, and the specifications set up by the manufacturer or user.

The breaking strength is a measure of the resistance of the fabric to a tensile load or stress in either the warp or filling direction. The tearing strength is a measure of the resistance to tearing of either the warp or filling series of yarns. The third measure of fabric strength is the bursting strength, which is used primarily for knit fabrics, but which is also used for light-weight woven fabric. The bursting tests measure a composite strength of both the warp and filling yarns simultaneously and indicate the extent to which a fabric can withstand a bursting type of force with the pressure being applied perpendicular to the surface of the fabric.

To measure the breaking strength, there are three tests that may be used, as follows: the grab, raveled-strip, and cut-strip methods. The cut-strip method (in which specimens are cut to an exact 1-inch width) is generally used only for coated fabrics, and the raveled-strip is used only on uncoated fabrics. The grab test, which is in widest use, is applicable for coated and uncoated fabrics; however, it is primarily used for uncoated material. Fig. 32.1 illustrates the sample preparation details for grab and raveled-strip tests.

Grab Test

Comparison Analysis. The grab method is preferred over the raveled-strip for several reasons. One of the main advantages of the grab method is the ease and speed with which the samples can be prepared for testing. The 4 by 6 inch sample for the grab test is cut from the fabric sample and placed with no further delay in the tester. For the raveled-strip-method, the test specimen must be cut either $1\frac{1}{4}$ or $1\frac{1}{2}$ inches wide, depending upon the ends per inch in the fabric, and raveled to a 1-inch width. The raveling of the sample is tedious and time-consuming, which means that an operator will make fewer raveled-strip tests than grab tests in the same length of time. This obviously increases

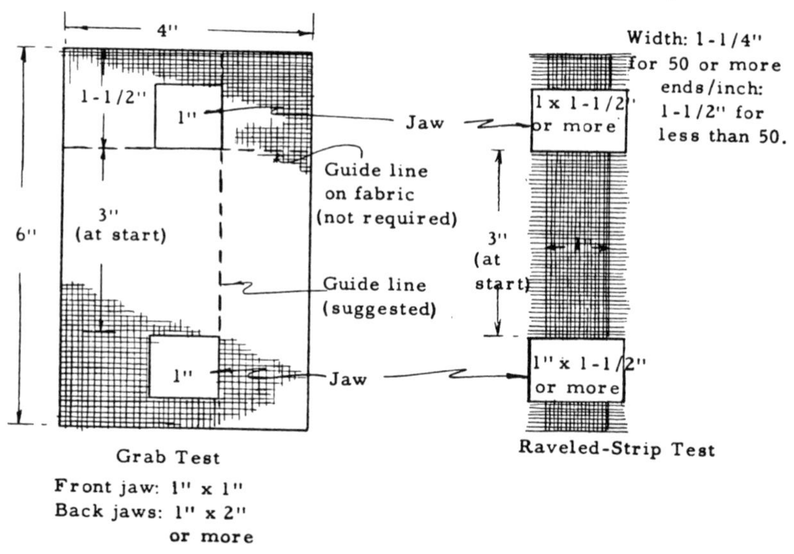

Fig. 32.1. Sample details: grab and raveled-strip tests.

the cost of testing considerably. Another advantage of the grab test is that it more nearly simulates wearing or field conditions for more fabrics than does the raveled-strip test.

When using both the grab and raveled-strip tests on the same fabric, the grab method will always give a higher average strength value for the fabric than the raveled-strip. This is due to the size and design of the samples and the size of the clamps used on the tester. For example, as the load is applied to the grab sample, as shown in Fig. 32.1, the yarns held by both the bottom and top jaws resist the stress. As the stress is continued, the yarns immediately adjacent to the jaws also absorb some of the load. Thus, when the sample breaks, the total load required for rupture of the one-inch width held by the jaw is increased due to the support given by the yarns immediately adjacent to the jaws. For this reason, the grab test will indicate a strength which is greater than that actually required to rupture 1 inch of fabric. However, in actual wear of field conditions, this support is generally present, which means that the accurate measurement of the strength of a 1-inch strip is not absolutely necessary.

On the raveled-strip sample, as the load is applied, the yarns in the center of the strip absorb the load, and the yarns on each side of the strip will support very little if any of the stress. Since the sample was raveled, there is no weave interlacing to hold the outer yarns, so that as the load increases, these outer yarns lose crimp rather than support the load. As a result, the inner yarns supporting most of the load will break first. After the center yarns break, the outer yarns will then tear or slip. This type of reaction to load does not give the actual breaking strength of the 1-inch strip. The raveled-strip method will always indicate a strength less than the actual strength for a 1-inch strip. With the grab test receiving support from the outer yarn and the strip not receiving support from the yarns on the edges, there always will be a substantial difference between the two strength averages. The magnitude of the difference is influenced by such factors as fabric weight and fabric construction. When reporting fabric strengths, it is necessary that the type of test—whether grab or strip—be recorded with the test results to explain any differences which may exist from using one of the two methods.

When making either the grab or strip test, a stress-strain curve may be made, or the strength may be read directly from the dial of the instrument. On most instruments there is no provision for reading the strain directly, and if this property is of interest, a stress-strain curve must be made.

Machine and Specimen Details. The specimen and jaw details used for the grab test are shown in Fig. 32.1. It should be noted that the front jaw on both the top and bottom clamps is 1 by 1 inch. In setting up the testing machine, the clamps must be set 3 inches apart. The lower jaw moves at a rate of 12 inches per minute.

Preparation of Test Samples. Test samples 4 by 6 inches are cut from the master laboratory sample. The 6 inch length is parallel to the series of yarns to be tested. For example, if the warp yarns are to be tested, the 6-inch length would be in the warp direction, and the 4-inch in the filling direction.

If more than one test is to be made, the samples should be cut so that no two samples contain the same yarns. If the warp yarns are to be tested, the samples should be taken across the fabric. If the filling yarns are to be tested, the samples should be taken lengthwise of the fabric, and if possible, only one sample should be taken from each filling change.

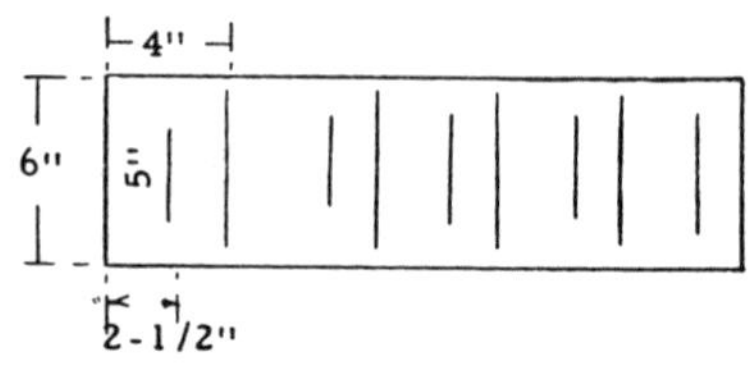

Fig. 32.2. Template for grab test samples.

To facilitate the proper marking of the test specimens on the master sample, a template similar to the one shown in Fig. 32.2 should be used. With such a template, it is possible to mark five samples within a short length of time. The longer lines spaced 4 inches apart on the template are for marking the edge of the sample. The shorter line $2\frac{1}{2}$ inches from the edge is for marking the sample so as to facilitate placing it properly in the jaws of the tester. When preparing the sample either with or without the use of a template, it is important that the sample be cut perfectly parallel with the yarns in the fabric, especially in the longer dimension.

Procedure for Grab Strength Test

1. Inspect the tester for the proper size clamps, distance between clamps, and any other parts or settings deemed necessary. If a pendulum-type tester is to be used, select the proper pendulum weight so that the pendulum will be between

approximately 9 and 45 degrees with the vertical when the sample breaks.

2. If a stress-strain curve is to be made, place the chart and pen in position and align the chart properly.
3. Place the sample in the clamps. It is important that the sample be held in the clamps in such a way that both the upper and lower clamps are holding the same yarns. When using the template shown in Fig. 32.2, if the mark which is $1\frac{1}{2}$ inches from the right edge of the sample is placed on the right edge of the front clamps, the sample will be spaced properly. If elongation of the sample is to be measured, a tensioning load of 6 ounces should be placed on the sample before the lower clamp is closed.
4. Apply the load to the sample. When the sample breaks, reverse the movement of the lower clamp and raise the pen from the chart if a stress-strain chart is being made.
5. If a stress-strain chart is not being made, record the breaking strength and return the pendulum to the zero position.

Clamp Breaks. If the fabric sample breaks in the clamps or at the edge of the clamps, it is classed as a "clamp break." In most instances, if the results of such breaks are below the standard or specification for the fabric, these breaks are discarded and additional tests made. If clamp breaks occur frequently, the condition of the clamps and the technique of the machine operator should be investigated. Some fabrics, especially brittle fabrics, have a marked tendency for clamp breaks.

Calculations. If a stress-strain curve is not made, the only calculations involved are usually the average, range, standard deviation, and coefficient of variation of the test data.

When a stress-strain curve is made, the elongation is of interest in addition to the strength. The stress-strain curve for fabric is analyzed in the same manner as for yarn. This is covered in detail in the chapter on yarn testing.

The curve shown in Fig. 32.3 made by the grab method shows a strength of 250 pounds and a stretch of approximately 0.59 inches. If more than one test is made, the average of the tests would be calculated. The average of the strength tests would be reported as the strength of the fabric. The average stretch in inches would be used to calculate the average per cent elongation of the fabric. In recording the stretch of the fabric, there is some disagreement as to whether the distance from the origin of the

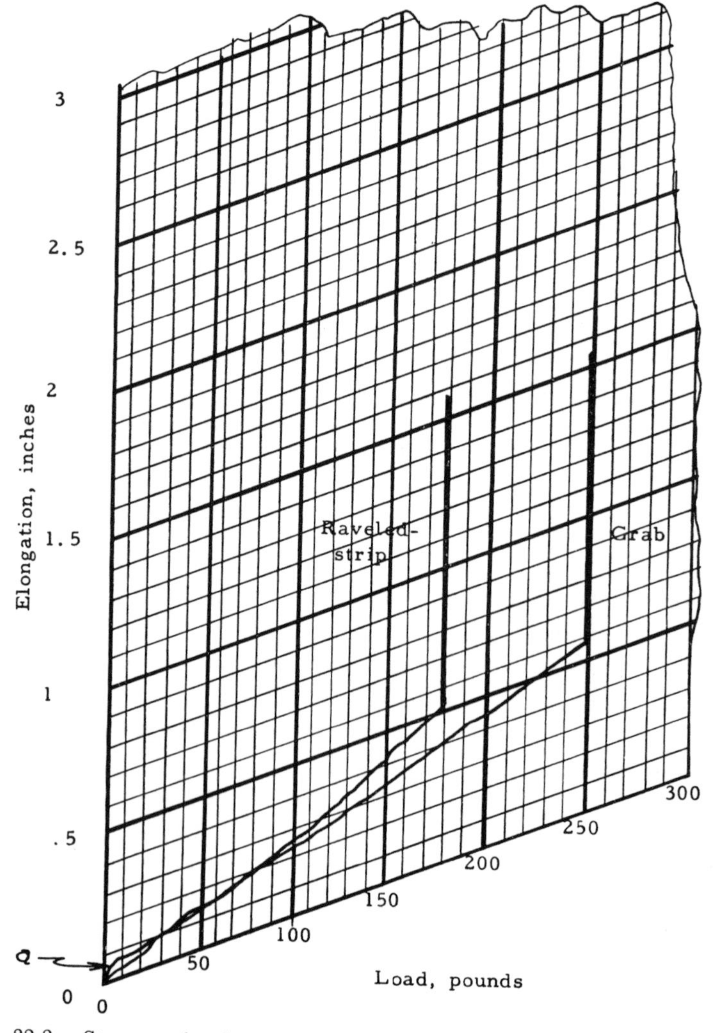

Fig. 32.3. Stress-strain chart made on a pendulum tester. The two curves show the grab and raveled-strip test results from the same sample of fabric.

curve should be used or whether the distance from the point where the curve deviates from the vertical axis should be used (point a). The argument in favor of using the point of origin is that if the sample has been pretensioned properly, this portion of the curve represents actual fabric performance such as tightening of ends and removing crimp. The major difficulty with this is determining what constitutes proper pretensioning. The other method is to measure stretch from the point where the curve deviates from the vertical line because that portion of the curve represents a tensioning action in the sample and is not a true measure of the stress-strain characteristics of the specimen. Another point is that as this portion of a stress-strain curve gives some indication of the stiffness and draping qualities of a material, it should not be completely disregarded. Actually, whether to use or not to use this section of the curve can become a rather complex problem, which depends to a great extent on the end use of the fabric. For example, if the fabric is a heavy duck that is to be coated with rubber prior to fabrication into a belt, the fabric will be subjected to a certain amount of tension during the calendering process. This would alter the initial portion of the stress-strain curve, and this is the condition in which the fabric would be used. On the other hand, a broadcloth shirting material would not be subjected to any excessive tensions after finishing, and the stress-strain properties of the fabric as it is tested would be the actual properties of the material in the end use.

There can be a considerable difference between the two methods of calculating the elongation. From the stress-strain curve previously discussed, the total stretch is approximately 0.59 inch. Based on a distance between the jaws of three inches, the per cent elongation would be:

$$\text{Per cent elongation} = \frac{0.59}{3}(100)$$

$$= 19.7$$

Using the other method, the approximate stretch would be $0.59-0.04$, or 0.55 inch. The per cent elongation would be:

$$\text{Per cent elongation} = \frac{0.55}{3}(100)$$
$$= 18.3$$

In addition to simply expressing the average strength of fabric, it is occasionally necessary to correct the strength for moisture content. This is especially true in industrial fabrics where strength is one of the primary quality characteristics. There are several tables available for correcting strength, one of which is in the *Handbook of Industrial Fabrics* by George B. Haven. Most technicians find it more satisfactory to develop a correction method based on the material manufactured by their company. This can be done by analyzing strength results and moisture content data. By using correlation techniques, it is possible to derive an equation for such corrections.

Number and Frequency of Tests. Fabric breaking strength tests are made daily, and it is recommended that five warp breaks and five filling breaks be made from each sample. Using five as the standard number of breaks also facilitates the use of $\bar{X}$ and R control charts. The use of such control charts has proven very satisfactory from both the producer and consumer points of view.

Raveled-Strip Test

Machine and Specimen Details. Specimen and jaw details for the raveled-strip test are shown in Fig. 32.1. It should be noted that the jaws, 1 by $1\frac{1}{2}$ inches (or more) for both top and bottom clamps, differ from the front jaws on the grab test. The rate of jaw separation is 12 inches per minute; initial spacing is 3 inches between clamps.

Test samples with a minimum length of 6 inches and a width of either $1\frac{1}{4}$ or $1\frac{1}{2}$ inches, depending upon the ends per inch, are cut from the master laboratory sample. If the fabric has less than 50 ends per inch, the sample is cut $1\frac{1}{2}$ inches wide. If there are 50 or more ends per inch, the strip is cut $1\frac{1}{4}$ inches wide. The 6 inch dimension is cut parallel to the series of ends to be tested. If more than one test is to be made, the samples should be cut so that no two samples contain the same yarns.

To facilitate the proper marking of test specimens, it is recommended that templates be used. It is necessary that two sizes of templates be available; one template should be $1\frac{1}{4}$ inches wide and the other $1\frac{1}{2}$ inches wide. In marking the samples, the longer dimension of the template should be placed exactly parallel to the yarns to be tested.

After the samples have been cut, each sample is raveled to 1 inch in width regardless of whether the original width was $1\frac{1}{4}$ or $1\frac{1}{2}$ inches. Approximately the same number of ends should be raveled from each side of the specimen so that the remaining one inch of woven fabric is centered in the sample.

Procedure for Raveled-Strip Test

1. Inspect the tester for the proper size clamps, distance between clamps, and any other parts or setting deemed necessary. If a pendulum-type tester is to be used, select the proper pendulum weight so that the pendulum will be between approximately 9 and 45 degrees with the vertical when the sample breaks.
2. If a stress-strain curve is to be made, place the chart and pen in position and align the chart properly.
3. Place the sample in the clamps. The $1\frac{1}{2}$ inch dimension of the clamps is perpendicular to the direction of the load. The $1\frac{1}{2}$ inch length clamp is used so that the full width of the test specimen will be clamped firmly across its entire width. The specimen should be placed in the clamps in such a manner that the yarns to be broken are exactly perpendicular to the load.
4. Apply load to the sample. When the sample breaks, reverse movement of the lower clamp and raise the pen from the chart if a stress-strain chart is being made.
5. If a stress-strain chart is not being made, record the breaking strength and return the pendulum to the zero position.

Calculations. The calculations for this test are the same as for the grab test.

Number and Frequency of Tests. The same schedules apply for the strip tests as for the grab tests.

Cut Strip Method

The cut-strip method of testing fabric strength is identical to the raveled-strip technique with one exception: the test specimens are cut one inch in width; no raveling of the sample is necessary. (Since this technique is used only for coated or heavily sized fabrics, it would be impractical to attempt any raveling of the fabric.) In cutting the sample, special care must be taken to cut parallel to the series of yarns in the longer dimension. Other procedures follow those above.

Tear Tests, Trapezoid and Tongue

The amount of resistance of a material to tearing is quite frequently of importance, particularly with mechanical fabrics. It is of prime importance in many military fabrics, such as those

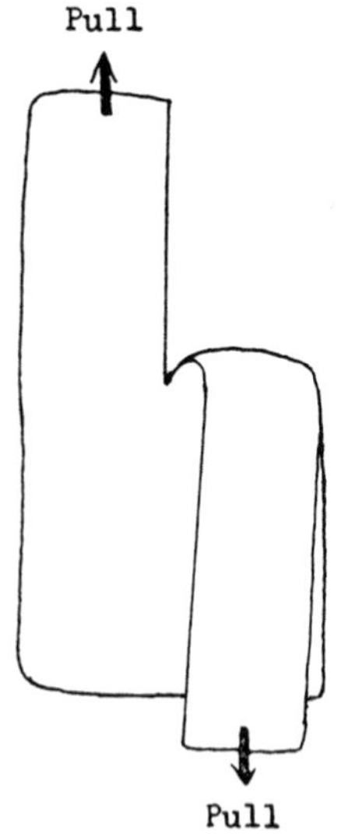

Fig. 32.4. Tongue test sample cut nearly halfway through.

used for aircraft and parachutes. The measurement of this resistance to tear is made on the same laboratory equipment used for tensile tests, with only slight modification of the instrument, and, of course, with samples prepared specifically for the test. The pendulum-type testers, set up with unrestrained pendulums (pawls disengaged) plus the stress-strain attachment, are most frequently used.

There are two accepted methods of making tear tests, called the trapezoid test and the tongue test from the method of preparing the samples and applying the load. Possibly the differences between the two tests can be understood by explaining the principle by a simulated hand method. If a small piece of fabric is nicked slightly on one edge with a pair of scissors, and then the sample is held in both hands and torn starting at the place where the small cut was made, we have the basic idea of making the trapezoid tear test. On the other hand, if we take another small piece of fabric, then cut it about halfway in two so that we can tear the fabric by holding onto each "tongue," then we have the basic idea of the tongue test (see Fig. 32.4). These hand methods are actually adopted for application to machine use in practice.

As with other types of tests where different methods of testing are employed to get an index of a given property, differences exist in the results obtained from the two tests. In the case of the tongue test, a disadvantage is found when tearing unbalanced fabrics (i.e., fabrics with predominating warp or filling). For fabrics with a high number of warp ends per inch and a low number of picks per inch to be torn in the warp direction, the direction of the tear frequently will change during the test and go in the direction of the least resistance. If the direction of the tear does not change through a ninety degree angle in unbalanced fabrics, there is a likelihood that the strength recorded will be a composite strength. For this reason the trapezoid test is generally preferred for most woven fabrics.

Trapezoid Test

Machine and Specimen Details. For the trapezoid test, test samples with a width of 3 inches and a minimum length of 6 inches are cut from the master sample. The long dimension is parallel to the series of yarns to be tested. Samples taken from the master sample should be selected so that no two samples contain the same yarns.

After the samples have been cut, each sample is marked with the aid of a template which has the form of an isosceles trapezoid. The trapezoid has an altitude of 3 inches and bases of 1 inch and 4 inches respectively. The trapezoid is placed on the fabric in such a way that this 1 inch dimension is parallel with the longer dimension of the sample and on the edge of the sample. Then

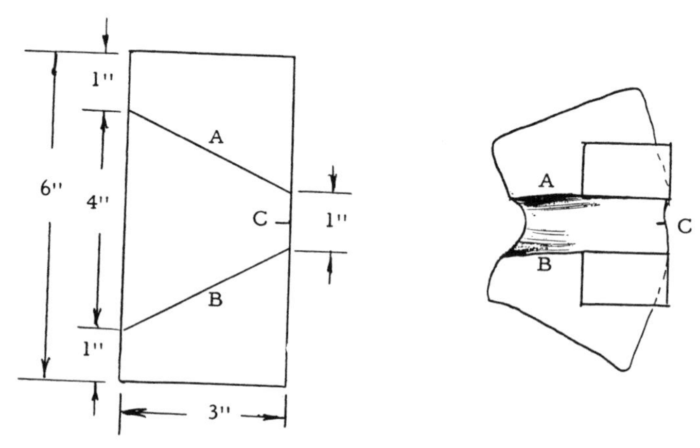

Fig. 32.5. Specimen for trapezoid test.

lines *A* and *B* are marked on the sample, as shown in Fig. 32.5. The lines *A* and *B* are to facilitate placing the sample in the jaws of the testing machine, and the small cut *C* is made so that when the load is applied, the fabric begins to tear at the cut.

For the trapezoid test, the clamps are set 1 inch apart, using 1 by 3 inch front and back jaws for both the top and bottom clamps. The rate of jaw separation should be 12 inches per minute. A pawl on the pendulum quadrant must be disengaged in order to give the continuous record of tear resistance.

Procedure for the Trapezoid Tear Test

1. Inspect the tester for the proper size clamps, distance between clamps, and any other parts deemed necessary. If a pendulum-type tester is to be used, select the proper pendulum weight so that the fabric will tear between one-fourth and three-fourths the capacity of the machine.

2. Place the chart and pen in position and align the chart.
3. Disengage the pawls from the rachet.
4. Place the sample in the clamps. This is done by aligning one side of the trapezoid along the lower edge of the upper clamp and the other side of the trapezoid along the upper edge of the lower clamp. The short base should be taut and the long base should lie in folds.
5. Apply the load to the sample. When the tearing is complete, the pendulum will return to the zero position and the pen should rest on the vertical zero line of the chart. If it does not, the chart is not aligned properly.
6. Return the platen to the starting position and repeat the procedure for any additional tests.

Tongue Test

Machine and Specimen Details. A sample 3 inches wide and 8 inches long is used for the tongue method. If more than one test is to be made, the samples should be cut so that no two samples contain the same yarns. The 3 inch dimension is parallel to the yarns to be broken. A longitudinal cut 3 inches long is made in the sample starting at the center of one of the short edges and cutting parallel to the longer dimension. See Fig. 32.4. One of the resulting tongues is placed in each clamp of the tester.

The clamps are set 3 inches apart. The minimum dimensions of front and back jaws in both the top and bottom clamps are 1 by 2 inches. The rate of jaw separation is 12 inches per minute.

Procedure for the Tongue Tear Test. The same procedure is used for this test as for the trapezoid test except for placing the specimen in the clamps. The tongue tear specimen is centered in the clamps with one tongue of the sample in each of the clamps. The remainder of the sample is allowed to hang freely.

Calculations: Trapezoid and Tongue Tests

The calculating of the tearing strength from the chart is a point of disagreement among the technologists. By examining Fig. 32.6, the type of curve resulting from a tear test may be seen.

From this type of curve, the average, the minimum, and the maximum tearing strength can be calculated. To calculate the

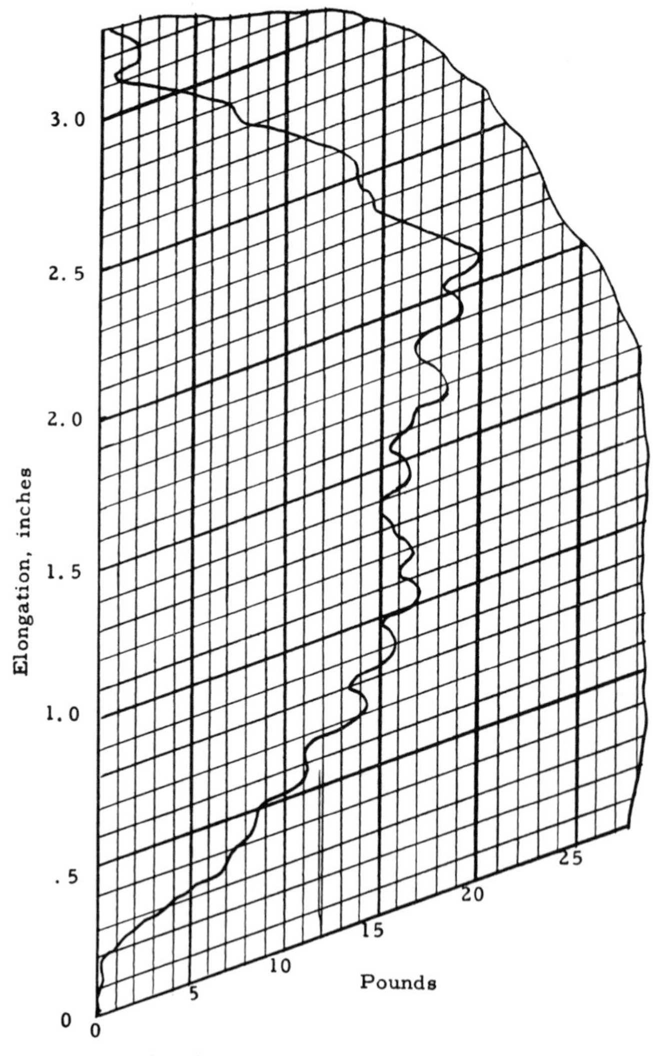

Fig. 32.6. Trapezoid tear test chart.

average minimum tearing strength, the mean of the minimum strength points is calculated. The average maximum strength is calculated by using the maximum points. The average tearing strength is calculated by using both the maximum and minimum points and calculating the mean of these values.

It has also been suggested that the area under the curve be measured. This is a measure of the work necessary to tear the material; however, it would be a more time-consuming technique and would be justifiable only on a research basis.

In reporting the tearing strength of fabrics, the type of test — whether trapezoid or tongue — and the method of calculating strength — whether minimum, average, or maximum — should be included with the results.

Bursting Strength

Bursting strength has its widest application for knit fabrics, where the fabric construction does not lend itself to the usual

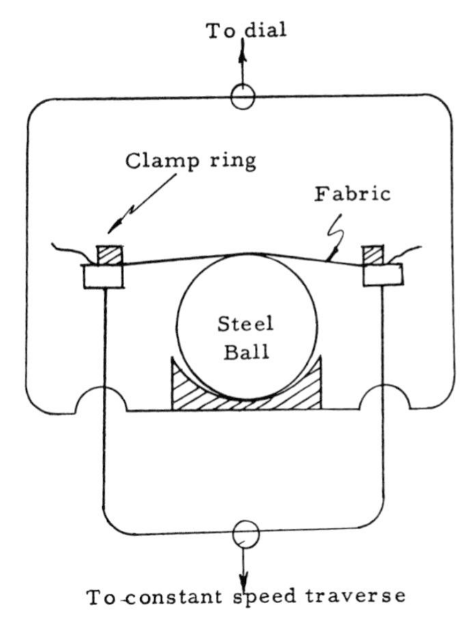

Fig. 32.7. Ball burst test principle.

tensile type of test used for woven materials, and to some other industrial and military fabrics, particularly those for parachute use. In addition, bursting strength is used for non-woven fabrics.

In the paper industry, bursting strength is one of the prime tests made, and in fact, the diaphragm type of tester used for fabric was designed primarily for the paper trade.

As is the case with many other laboratory methods of evaluating some particular physical property, there is more than one approach to the specific way of getting the desired results. In the case of bursting strength, two methods accepted by the industry have come into being. These are known as the constant rate of traverse test with a ball burst attachment and the diaphragm test.

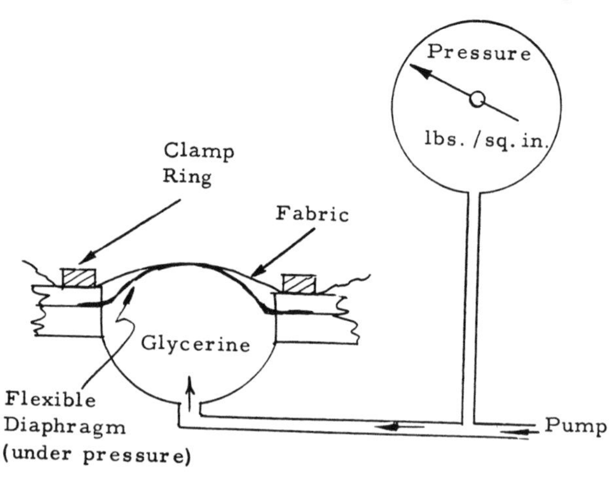

Fig. 32.8. Diaphragm burst test principle.

The first of these two tests, the constant rate of traverse with the Scott ball burst attachment, is performed on a regular type of pendulum tester. The attachment for making the bursting test consists of a device for holding a steel ball and a ring-clamp mechanism for holding the sample. The attachment is fitted to the pendulum tester in such a way that the downward movement of the traverse forces the fabric against a 1 inch ball. The load necessary to burst the ball through the fabric gives the bursting strength. The principle is illustrated diagrammatically in Fig. 32.7. It will be noted in referring to this sketch that the steel ball is supported from the pendulum and dial mechanism and that the fabric is held in a clamp-ring mechanism that is attached to the constant speed traverse drive. Thus, the movement of the clamp ring forces the fabric against the steel ball, with the resultant pressure registered on the dial.

The second bursting test is performed on a Mullen diaphragm burst tester, for which the principle is illustrated in Fig. 32.8. As with the ball test, the fabric is securely clamped in a clamp-ring device. In place of a steel ball, however, a rubber diaphragm is forced against the fabric by liquid pressure in back of the diaphragm developed by a pump. The pressure can be increased until rupture of the fabric occurs. The total pressure exerted by the diaphragm is measured on a Bourdon tube type pressure gauge.

Machine and Specimen Details: Scott Ball Burst Test. The circular specimens for the ball burst test should be cut with the minimum dimension at least half an inch greater than the outside diameter of the clamp ring. Specimens should be selected from the fabric to avoid inclusion of the same ends in any group of samples. Generally speaking, at least ten test samples should be made for any given fabric sample.

The ring-clamp mechanism should be 1.750 inches in internal diameter. The polished steel ball should be 1.0 inch in diameter. The traverse rate of the machine is the same as that for fabric strength tests, 12 inches per minute.

Machine and Specimen Details: Diaphragm Test. The specimens for this test should be cut so that the smallest dimension is $\frac{1}{2}$ inch greater than the outside diameter of the clamp ring. Ten specimens are selected from the master sample, avoiding inclusion of the same ends in the different test specimens.

The instrument used should have a constant rate of speed capable of giving uniform displacement of 6 ± 0.25 cubic inches per minute. The clamp ring should have an internal diameter of 1.20 inches. For proper operation, the machine must be stopped at the instant of rupture in order to avoid additional application of pressure and load. The instrument must be constructed so as to hold the contents of the pressure chamber when the sample has been burst. Thus, it is possible to read the maximum pressure developed to rupture the sample.

There has been some disagreement on the method of measuring the net bursting pressure. The questions revolve about the concept of the tare loading of the instrument. When a sample of fabric bursts, the diaphragm projects through the broken fabric. At this time, a reading of the gauge will most always indicate that some pressure is being exerted by the diaphragm. Therefore, it is contended that the true bursting pressure is the gross pressure at the moment of rupture (indicated by a floating reset type pointer on the gauge) less the tare pressure necessary to distend the diaphragm to the size attained just after rupture.

The other concept is that the diaphragm is restrained by the fabric to a minimum distortion during the test and up to rupture. (This can be verified by experimental tests). Therefore, any distortion of the diaphragm immediately after rupture has no influence on the gross bursting pressure developed. Thus, the true reading should be the gross pressure realized. Certainly this is the simplest approach and the one that should give the most consistent results. For this reason, it is recommended.

Procedure. The procedure for the use of the bursting instruments is extremely simple. The samples are placed in the clamps, and the pressure applied by the bursting element, that is, the ball or diaphragm, depending upon the instrument used. When using the diaphragm type of tester, one important point is that the pump must be stopped immediately after the sample ruptures.

Wet Strength Test

When measuring the breaking or tearing strength of a wet specimen, it is very important that the samples be wet throughout the entire area of the sample. If the fabric has been coated with size, finishes, repellents, or any other materials which resist wetting, a wet strength test will be of little value unless these materials are removed. In removing any of the sizings or coatings, techniques and chemicals should be used which will not affect in any way the physical properties of the fabric. Shrinkage may result from the desizing process. This is usually unavoidable; however, if the fabric is tendered by the treatment, then any strength tests are useless.

In preparing samples for wet tests only, they are cut and prepared the same as for the dry tests. If both the wet and dry tests are to be made, the samples are cut so that the same group of yarns will be broken in both the wet and dry tests. For example, the grab sample would be cut 12 inches long, and then cut into two 6-inch specimens. The specimens would be marked or coded in such a way that the wet and dry strengths of the same original twelve-inch length could be kept together.

After the specimens are prepared, the ones to be tested wet are immersed in water at room temperature containing a neutral wetting agent. The samples are left in the solution until they are thoroughly wet. The samples are then removed and tested immediately according to the type of test to be made. This can be either the grab or raveled-strip strength test for strength, or the trapezoid or tongue tear test for tear.

The average of five tests is generally considered sufficient for the wet strength test for any one master sample.

Fabric Strength Testing Machines

The same types of machines are used for testing fabric strength as are used for yarn strength. The pendulum testers, such as the Scott and Suter machines, are the most common types used in quality control laboratories. In addition to the pendulum machines, the strain gauge type, such as the Instron, is also used. This type, however, is used mostly in the research laboratories for work of a more fundamental nature than for straight quality control.

Calibration of Testers

The basic calibration of the testers is the same for fabric as it is for yarn. For the general calibration procedure, refer to page 357.

When calibrating a tester for fabric testing, in addition to the basic verification of the weighting system, the jaws and recorder should also be checked. In checking the clamps for a test, both the distance between the clamps and the surface contact should be verified. The distance between the bite of the clamps is established for each test and should be set as accurately as is possible. The distance is controlled by the upper collar on the return lever.

To verify the clamping action of the clamps, place a sheet of white paper on a very soft pencil carbon between the clamps. Close the clamps with the same pressure as is normally used on the fabric sample. Remove the paper and examine the carbon imprint. If the carbon deposit is uniform, the clamping surface is even and uniform. If the deposit is not uniform, the clamps should be removed, resurfaced, and realigned.

When a measure of the elongation of a material in the form of a stress-strain curve is desirable, the recorder unit should be checked. The following suggestions are for the type of recorder used on the Scott testers.

Pen slides must be kept free of dirt, rust, or foreign matter that would add undue friction. All moving parts between the weighing head and the pen or marking point must be clean. On pendulum-type testers, the slide bar should be wiped with an oily cloth at least once daily. Avoid using more than a drop of light oil at any one point of contact in the mechanism.

To verify the elongation record obtained from the recorder, return the clamps to normal starting position. Place a chart in the holder so that the pen indicates zero load and zero elongation. Measure the distance between the butts of the clamps (or some other bench mark) to the nearest 0.01 inch, calling it *Cn* (normal clamp distance).

Test the location of the chart in the holder by operating the machine without any sample. The pen line should coincide with the chart axis showing zero load and variable elongation. If necessary, reposition the chart to obtain this coincidence. Then, without any sample in the clamps, operate the tester to 1.0 inch elongation on the chart. Measure the distance between the bench marks, which should equal *Cn* plus the elongation on the chart. Repeat this at 2.0 and 3.0 inches on the chart. Any discrepancy between the recorded elongation and the actual increase in distance between the clamps is a recording error due to faulty chart spacing.

To test for further errors, locate a chart correctly in the platen as previously described. Return the clamps to starting position and fix both firmly to a member, non-extensible under the load applied, such as a strip of steel. As the load is applied, the pen line should parallel the chart axis, indicating variable load and zero elongation. Non-parallelism indicates a variable error in elongation due to improper geometry of the tester or incorrect angle in printing the charts. It may also be due to a bent pen slide bar or use of the wrong chart, as elaborated below.

Possible causes of failure to record elongation properly are:

1. The ruling of the chart is not the proper angle for the recorder being used.

2. Where chart is held in guides, errors may be traced to faulty cutting of the chart, which should be a perfect rectangle with edges cut exactly at the line as a guide.

3. Errors can be due to failure of platen, pen, and clamp to start moving at the same time, usually caused by excessive wear in the sliding mechanism.

If errors still persist after the above procedure, it may be advisable to return the tester to the factory for correction.

CHAPTER XXXIII

Fabric Grading and Quality Control

The grading of fabric, whether in the grey or finished state, has two primary functions: first, to classify the fabrics according to the different qualities such as first and second, based on the demands of the market and customer, and second, to supply information to the proper levels of management as to the qualities being produced.

The classifying or grading of fabric can be a most difficult and controversial task. A length of fabric that may appear to one person as being a perfectly acceptable first-quality piece may appear to some one else as being acceptable only as second quality. This situation is more complicated from the mill point of view because of the fact that any given type or style of fabric may be destined for three or four different customers, each of whom has his own particular quality requirements. One of the main reasons for this situation is that no official industry-wide standards have been established or accepted whereby fabric may be graded quantitatively. Some segments of the industry have moved in this direction by installing what is commonly referred to as the "point system" of grading. In using this system, each defect is assigned a certain value or number of points so that when a certain total number of points is exceeded the piece is classed as second quality. This system has proven very successful and different versions of the basic idea are being adopted by a large number of organizations. This system will be discussed in detail under the heading, Point System of Grading.

The second reason for grading fabric, that is, to supply management with information pertaining to quality levels, is to gather information for one of the most powerful tools of any quality control system as well as one of the most important reports available to all levels of management. Generally speaking, the profits of a mill are determined by the per cent of first-quality fabric produced by the weave room because there is very little if any profit realized in the sale of the fabric classified as seconds. Also, any fabric that is sold as first quality but which is not accepted as firsts by the customer could result in a claim for adjustment that might cost the company more in settlement than the profit realized from the entire shipment. In addition to cost aspects, the data collected from fabric inspection supplies the quality control department and weave room management with a complete and unbiased tabulation of the magnitude and frequency of defects in the woven fabric. If the data is properly collected and analyzed, it will not only provide information concerning the magnitude and frequency of the defects, but it will also indicate their causes and sources. For example, the defects may be separated by style, by weaving defects in the warp or in the filling, by yarn defects in the warp or in the filling, by type of loom, by the weaver's set, by the fixer's set, by the slasher set, or by the shift run.

Defects

Fabric defects are generally classified as either "major" or "minor." The definition of exactly what constitutes a major defect and minor defect depends upon the type of fabric and the end use, as well as whether the fabric is being graded in the grey or finished state. For example, a defect that would be considered as serious (a major defect) in a high-quality combed poplin would most likely not be classified in the same way in a low-quality carded print cloth. Also, some defects that are major in grey fabric become minor or completely disappear when the fabric is finished. It is common practice in some organizations to classify grey fabric as first quality for dark shades, first quality for light shades, second quality for dark shades, etc. In this way, some fabrics when dyed a dark shade become first quality due to the obliterating of defects by the finishing, whereas if it were finished in a light shade, it would be classed as a second. By the same token some defects such as warp streaks become more pronounced after finishing and the defects are not detected in the grey fabric inspection.

From the foregoing discussion, it becomes apparent that one grading system cannot be used for all fabrics, which means that the description of a major and minor defect will also vary with the types of fabrics to which they apply, the end use of the fabric, as well as market conditions.

In grey fabrics the following general descriptions have been used (1).

Major defect. A defect that cannot be repaired in the grey so that it would not be obvious in the finished fabric.

Minor defect. A defect that can be corrected in the grey or will be covered in finishing so that it will not be detected in the finished fabric.

For finished fabrics the following descriptions have been in use by one of the larger textile organizations with a high degree of success (2).

Sub-minor. A defect which is not obvious and may not be noticeable at first glance. It would not likely cause a garment to be defective to such an extent that the garment would have to be sold as a second. No grading points would be assigned to these defects, but if they occur with a high frequency, this fact should be called to the attention of the responsible personnel. If an excessive number of this type defect occurs in a single piece of fabric, consideration should be given to grading the entire cut as seconds.

Minor. A fairly obvious defect which is noticeable more or less at first glance and might easily cause a defective garment. From 1 to 3 grading points would be assigned to such defects, depending upon length.

Major. An obvious or very obvious defect which can easily be seen from a considerable distance and would most likely cause a defective garment. From 2 to 4 points would be assigned this type defect, depending upon the length.

Critical defect. This is a classification used for defects of such severity that would cause a garment not to be saleable even as a second. For this type defect, 6 to 12 points would be assigned, depending upon the length. A warpwise major defect over

TYPICAL FABRIC INSPECTION FORM

Style No.__________ Shift__________ Mill No.________ Date________

Total Yards________ Loom No.________ Grader______________

Classification________

	Defects	Major	Minor	Total Points	Comments
Filling Defects	Yarn: Singlings Doublings Dirty Broken Filling Bar Chaffed Kinky Mixed Slubby Wild Sloughed-Off Oily Weaving: Bad Start-Up Jerked-In Filling Incorrect Stripe Mispick Mixed Filling Thick Place Thin Place Float				
Warp Defects	Yarn: Singlings Doublings Uneven Dirty Broken Chaffed Kinky Mixed Slubby Oily Weaving: Draw Back Doubled End Harness Skips Reed Marks Bad Selvage Temple Marks Thread Out Wrong Draw Float Broken End Slack Thread				
Misc.	Hard Size Holes Oil-Grease Waste				

Fig. 33.1. Typical fabric inspection form.

12 inches long is automatically classified as a critical defect.

A relatively complete tabulation of fabric defects will be found in Fig. 33.1. Some of the more common and serious defects will be discussed at this point. Filling slubs and warp slubs are probably the most common of all defects, especially in the lighter weight garment fabrics, and in some cases amount to fifty per cent of the defects in a piece. Following closely behind slubs in frequency of occurrence and importance are holes, broken picks, jerked-in filling, coarse picks, thick and thin places, and broken selvages.

In finished fabric, typical finishing defects which appear with the highest frequency are: overbleaching, stains, streaks, dye specks, overdrying, over or under shrinkage, creases, selvage to selvage shading, and end to end shading.

Common defects for fabrics woven from filament yarns would include, in addition to some of those listed above, faults such as: mixed yarn (either mixed deniers or filament counts, or both), shiners, twist variation, broken filaments, and reed marks.

Point System of Grading

In order to rank each defect in the proper order of importance and also to reduce variation from inspector to inspector, the defects of different magnitudes are assigned specific point values. Once the proper weighting or ranking of the defects has been established, then the allowable number of defect points acceptable for a first-quality piece of fabric is established. A second-quality piece of fabric would be one that has more than the allowable number of major or minor defects or points in the piece. The allowable number of defects per piece varies with the quality requirements; as the quality requirements increase, the number of allowable defects or points is reduced. Also, the totals vary with the width of the fabric; obviously the same number of total defects would not be the same for a 36-inch fabric as for a 72-inch fabric.

The two major considerations in grading fabric are the frequency and seriousness of the defects. Before any point values can be assigned to the defects, these two factors must be clearly described. The degree of seriousness of a defects is determined by the intensity (obviousness or visibility) and the size or length (2). The intensity determines whether the cutter will see the defect or not, and if so, whether he will cut it out or allow it to remain and cause a defective or second-quality garment. The length of a defect also affects the obviousness of a defect, but, more important, it determines how many panels or garments may be affected. Therefore, length is of greater importance than intensity.

The simplest possible system that will include both factors (intensity and length) is to use two basic categories, i.e.,

A. Intensity
   1. Minor: obvious defects
   2. Major: very obvious defects

B. Length
   1. Short: up to 6 inches long
   2. Long: 6 to 18 inches long

In order to place the defects in their proper perspective and at the same time to keep the numerical values as low as possible, the values in Table 33.1 are assigned.

TABLE 33.1
Point System of Grading Defects

Length of defect	Minor	Major	Critical
Short (0–6 ins.)	1	2	6
Long (6–18 ins.)	3	4 [a]	12

[a] Warpwise major defects over 12 inches in length are automatically classed as critical defects and assigned 12 points.

TABLE 33.2
Another Point System of Grading

Length of defect	Points
0–3 ins.	1
3–6 ins.	2
6–9 ins.	3
9 ins. and over	4

Another point system that has been suggested (3) is shown in Table 33.2. In using this system, no one-linear-yard of 40-inch width would be penalized more than 4 points. The reason for this is that a 4-point defect by itself would render the yard practically worthless, and to assign further points would uselessly penalize the average for the full piece.

One disadvantage of the point system of grading is that it merely places an upper limit upon the defects and does not specify any definite distributions of defects. For example, first-quality fabric from a mill producing 10 per cent seconds will have more defects per piece than the same fabric from a mill producing only 5 per cent seconds.

Inspecting, Burling, and Grading

The three operations of inspecting, burling (repairing defects), and grading are generally performed in one operation by the inspectors.

As the grey fabric is removed from the loom, it is either delivered directly to the inspection department or it is sheared, seamed, and rolled into rolls of 1,000 to 1,500-yard lengths. The procedure used depends upon the flow of the material after inspection. If the material is to be finished on a continuous range, it will most likely be put into rolls. If the fabric is to be shipped after grading, it will most likely be handled in the loom lengths or even reduced to cuts of 100 or 120 yards.

Most mills have power-driven inspection machines. These machines are so designed that the rolls of fabric can be mounted behind the inspection table and the fabric threaded through calender rolls, across the inspection table under adequate light, and re-rolled as it leaves the table. The machines are also equipped to accurately measure the length of each piece of fabric. In addition to the power-driven inspection machines, some mills have similar units except that the inspectors pull the fabric over the table by hand. This arrangement is most common for fancy patterns.

The inspectors, usually women, inspect each yard of fabric as it is drawn across the table. The defects are located and recorded on the inspection form.

If the burling, such as removing jerked-in filling and slubs, is to be performed as a part of inspection, it is done as the defects appear on the inspection table. The amount of burling or repairing that is done depends entirely upon the economics of the situation. Obviously a very cheap low-quality print cloth would receive little if any burling; however, extremely high-priced fabrics would receive all possible attention.

After the inspection is completed, the piece of roll is classed by the inspector according to the number, frequency, and magnitude of the defects. At this point in the operation, the flow of the fabric varies with the different mills. In some mills all of the fabric is conveyed immediately to the cloth storage area where it is placed in bins according to style, width, quality designation, ultimate destination, and so forth. In other mills, all second-quality fabric is reinspected by the head grader to make sure that the piece is actually second quality. One very effective common practice is to have the superintendent, overseers of weaving and spinning, and the quality control manager to inspect all seconds to observe the causes of the off-quality so that the necessary corrective action can be taken to prevent or reduce reoccurrences.

Reinspection of Fabric

The reinspection section in an inspection department is one of the most important groups in the quality control department of any organization. The reinspectors, who should be under the quality control department, actually see the mill's quality exactly as it is seen by the customer. Their primary duty is to inspect lengths of fabric that have been inspected and classified by the regular inspectors. The reinspectors select samples at random from the cloth room and reinspect the pieces grading against the fabric standards. From these results they can predict the per cent of the production that is being misgraded. They not only inspect from first-quality production, but also from second quality. It is just as important not to have first-quality material classed as second as it is to have seconds classed as first. The method of selecting samples varies, but the one thing that should always be remembered is the samples must be selected at random. In some plants the reinspectors select at random a given number of cuts per day from each inspector and rate each inspector as to excellent, good, fair, and poor. In this way all inspectors are sampled with the same frequency. Another system is to select at random cases in the shipping department that have been prepared for shipment and reinspect the fabric in the case.

In addition to the above duties, the reinspection department can be used to examine fabric that has been returned by customers. Regardless of the exact scope of the reinspection, the importance of this group cannot be overemphasized.

Reports

As stated previously, fabric inspection has two primary functions: (1) to classify fabric according to different qualities and (2) to supply information to the proper levels of management as to the qualities being produced. The second function is accomplished by means of frequent and well-designed reports. There are three reports that will supply most of the necessary information: Repeating defect ticket, daily second quality, and the weekly quality report. The repeating defect ticket is filled in by the inspector when she locates a running or repeating defect that occurs throughout a length of fabric. As soon as such defects are found in a piece, a ticket which lists the loom number and type of defect is delivered immediately to the weave room overseer. The overseer notifies the second hand who with the fixer examines the loom and takes the necessary steps to prevent further trouble.

The daily second-quality report is designed to inform the weaving overseer of the number of second-quality pieces, causes of the downgrading, and the looms involved. This report is actually a tabulation of the inspection forms for the second-quality pieces. In some cases carbon copies which have been made of the inspection forms at the time of inspection are delivered to the weave room overseer. The original copies of the inspection forms are forwarded to the quality control department for examination and preparation of the weekly report.

The weekly report is designed to give a complete description of the inspection results. While it can be prepared in several different ways, this report should include over-all per cent seconds and the percentage of the total defects contributed by each defect. For example, it may be found that 40 per cent of the defects are slubs, 20 per cent are holes, 20 per cent jerked-in filling, etc. In this way everyone concerned will know where to spend the most effort to improve quality. The report can also include a tabulation of defects by styles, shift, loom, weaver, fixer, or other headings that will isolate causes of off-standard fabric.

In addition to the above, weekly reports should be issued by the reinspection group. These reports should list per cent misgraded by styles and inspectors. This type of information is most valuable in the locating of ineffective inspectors and in the training or retraining of the inspectors.

On-the-Loom Inspection

In addition to the above inspection, it is frequently advisable to have an on-the-loom inspection. This consists of patrolling the weave room and inspecting the fabric as it is being woven. The object of this inspection is not to grade the fabric but to observe any defects that are being made, especially, the running and repeating defects. The inspector immediately notifies the fixer or weaver of defects being made. In addition, he should be given authority to stop a loom if it is producing substandard fabric. For these reasons, the inspector assigned to this job should have had some weaving experience.

This type of inspection is used where the quality and price level of the fabric demands such added attention and where the weavers and other weave room personnel have not become quality-conscious to the point of sacrificing quantity for quality.

Sample Size

It is standard practice in the mills throughout the textile industry to inspect all fabric produced. To the statistician this may appear to be an unnecessary expense; however, when comparing the relatively low cost for inspection with the price per yard of the fabric and the cost of customer rejections and claims, the 100 per cent inspection is easily justified. Many of the larger customers of the mills do use sampling plans and as a result do not inspect 100 per cent. One of the originators of sampling plans in the textile industry was the U.S. Army, and Military Standard 105A is now in standard use in textiles as well as other products. Since the reinspection group in a mill to some extent represents the customer, it could use the standard sampling plans. Common practice, however, is to reinspect from 5 to 10 per cent of the production on a routine basis.

Source, Cause, and Responsibility for Fabric Defects

Table 33.3 is a list of the defects most frequently found in grey fabric. Included in the list with the defects is a description of the defect, some suggestions as to the source or cause, and the person or persons responsible for the occurrence of the imperfection. (The information in this list was tabulated by W. L. Clement, Jr.) See also Appendix II for descriptions of defects.

References

1. Clement, Jr., W. L., Textile Div. Conf., Am. Soc. Quality Control, Feb. (1956).
2. Hailes, G., Textile Div. Conf., Am. Soc. Quality Control, Feb. (1955).
3. Jenks and Lowe, Cluett Peabody & Company, Inc.

TABLE 33.3

Defects, Source, Cause, and Responsibility

Defect	Description	Source and cause	Responsibility
Balks	Too many or too few picks in a pattern	Loom: Box motion not working properly; improper matching of picks; chain built incorrectly.	Weaver Loom fixer
Bad reed	Sprung dent in reed.	Loom: Large object has been pushed through reed, either accidentally or on purpose, such as a reed hook; loom has been started with an object protruding past fell of cloth and braced against breast beam.	Weaver Loom fixer
Bad selvage	Wrong draw in selvage; selvage damaged by temples.	Loom: Too much tension on filling; a defective temple; bad harness setting; pick motion not timed right.	Weaver Loom fixer
Breakout	Four or more warp ends that have broken and then been redrawn.	Loom: Splintered shuttle; shuttle eye or spring bolt has come loose; frog packing worn down; poor or no protection on loom; objects have fallen between harness and reed; change motion not working right.	Weaver Loom fixer
Bristles	Shuttle bristles in cloth.	Loom: Bristle peg becomes loose, so that bristles fall out of shuttle.	Loom fixer Weaver
Broken picks	Separation of picks of cloth.	Loom: Burrs on shuttle and boxes; bobbin out of line; wrong filling tension. Spinning: High reserve bunches; bad piecings; slubs in filling; careless cleaning or blowing of frames.	Weaver Loom fixer Spinning room
Broken warp ends or loose ends on face of fabric	Long ends loose on face of fabric.	Loom: Long or short ends of warp not removed by weaver after piecing up; careless weaving.	Weaver

Coarse thread	Coarse warp thread in single yarn or in one of plies of plied yarns.	Carding, Drawing, Roving, Spinning: The defect could be due to mix in carding or spinning; a careless piecing at drawing; a weight off of rolls; roving doubling up instead of one end; spinning too many ends in the trumpet. Dress room, Spooling, Warping: Coarse end from spinning should be caught here.	Weaver Card room Spinning room Slasher room
Cotton slubs	Loose cotton caught in spun yarn.	Spinning: Wild blowing or cleaning of frame; spinner piecing up carelessly; cleaning lap sticks carelessly.	Spinning room
Double ends	Two ends weaving as one.	Loom: Wrong draw; loose thread coming up and into harness. Slashing: Two ends stuck together.	Weaver Slasher room Tie-in operator
Doubling	Filling twice the count $\frac{1}{4}''$ to $1\frac{1}{2}''$ in cloth.	Spinning: Two ends running where one should be; an extra roving feeding to trumpet.	Spinning room
Drawbacks	Tight ends	Dress room: Slasher tender crossed ends in slasher. Loom: Crossed ends on loom; mat up. Drawing-in: Tie in hands fail to brush warp out straight.	Slasher room Drawing-in Weaver Warper room Spooler room
Filling hanging	Loose ends hanging from left hand side of cloth.	Loom: Filling fork improperly set causes filling to hang; burrs on leather on front boxes.	Weaver Loom fixer
Filling skips	Filling skipping over warp ends.	Loom: Harness not set correctly or improperly timed; high or low picker; warped shuttle.	Weaver Loom fixer
Floats	Misweaves of warp or filling or both, with ends floating usually $\frac{1}{2}''$ to $2''$ square.	Loom: End or ends break, mat up between harness and reed; cotton or waste blown in to warp; gouts on yarn cause breakage behind reed.	Weaver Slashing room Warping room Spooler room Spinning room

Table continued

TABLE 33.3 (*Continued*)

Defect	Description	Source and cause	Responsibility
Floating warp end	Warp end not weaving in spots.	Loom: Mat up between drop wires and harness and reed; warp thread not drawn in harness eye.	Weaver
Hard size	Hard gummed spots in warp.	Dress room: Size too cold; cld size used; improper cleaning of size box.	Slasher room
Holes	Cuts or tears through fabric.	Loom: Careless handling of cloth rod at doffing; rough sand roller.	Cloth doffers Loom fixer
Jerkin filling	Loose ends of filling drawn into cloth, making part of extra pick.	Loom: Stafford thread cutter motion out of adjustment; catch cords on Harris motion broken or not on loom; leather or steel burrs on boxes; shuttle not boxing.	Weaver Loom fixer
Kinked filling	Kinks showing on face of cloth.	Spinning: Too much twist in filling; Loom Rough box front; shuttle bouncing; filling not properly conditioned.	Loom fixer Weaver Yarn conditioning department
Lint balls	Accumulations of loose yarn or fibers around one end between harness and reed.	Dress room: Size not penetrating. Loom: Harness eyes chafing; harness not set properly; wrong humidity.	Slasher room Loom fixer
Mispicks	Two picks in same shed for the entire width of fabric.	Spinning: Improper reserve bunch. Loom: Burrs on shuttle; shuttle in poor condition; bobbin out of line; too much tension on filling. Improper setting of feeler.	Weaver Loom fixer
Misreed	End in wrong dent of reed.	Loom: Weaver's carelessness in drawing-in pieced-up end.	Weaver
Mixed filling	Wrong count of filling in cloth.	Loom: Filling lots mixed. Spinning: Filling marked wrong; filling put on wrong quill.	Battery hand Spinning room

Oil and grease	Oil and grease spots: black or yellow.	Loom: Oil thrown out by bad loom bearing; too much oil used; wild yarn and foreign material blown into the loom; roll of cloth dropped on floor.	Oiler Loom fixer Cloth doffer
Oily filling	Filling with oil spots.	Spinning: Careless oiling of frames; excessive oil on top rolls; excessive oil on bottom roll bearings.	Battery hand Spinning room
Rust	Brown rust, colored spots, or stains.	Weave room: Too much humidity; whip roll not lubricated: pipes sweating.	Weaver Overseer
Selvage skips	Loose filling not weaving in selvage.	Loom: Selvage motion not timed right; loom throwing bad shuttle.	Weaver Loom fixer
Singlings	Filling half the count it should be.	Spinning: One end of roving broken back in creel.	Spinning room
Slack thread	Warp end running in loosely.	Dress room, Slasher: Wrong number of ends per dent; too many per dent on slasher. Loom: Warp bobbin not put up properly.	Warper room Slasher room Weaver
Sluffed filling	Extra filling snarled in same shed of cloth.	Spinning: Improper traveler; wrong bobbin build; loose tape. Loom: Too much power on pick.	Spinning room Loom fixer Weaver
Temple marks	Warp threads pulled out of line.	Loom: Bad temple burrs; temples not aligned.	Weaver Loom fixer
Thick place	More picks than desired in cloth.	Loom: Let-off or take-down motion not working right.	Weaver Loom fixer
Thin place	No filling interlacing with warp.	Loom: Let-back motion not right; filling fork out of line with grate; filling grate clogged; filling fork knock-off motion not right.	Weaver Loom fixer
Torn selvage	Damaged at selvage.	Loom: Careless cloth handling when doffing; after breakout is fixed, selvage sticks to sand rolls causing tear later.	Weaver Loom fixer

Table continued

TABLE 33.3 (*Continued*)

Defect	Description	Source and cause	Responsibility
Turned selvage	Selvage turned or folded under.	Loom: Temple too close so as to crowd edge of cloth.	Weaver Loom fixer
Uneven filling	Light and heavy places in filling yarn.	Card room: Uneven roving. Spinning room: Bad drafting; bent steel rolls; dry top rolls.	Weaver Loom fixer
Warp end out	Thread not weaving in cloth 3″ or more.	Loom: Warp stop motion not working; clogged-up drop wires.	Weaver Loom fixer
Warp skips	Warp ends not weaving in.	Loom: Harness not timed properly; harness too loose; harness not set properly; trash or object between harness and reed; loom throwing bad shuttle.	Weaver Loom fixer
Warp slubs	Wild fibers or yarn twisted into warp yarn.	Spinning: Careless cleaning; careless piecing up.	Spinning room Spooler room
Waste, blow-off	Thread or fibers caught between shed and woven into cloth.	Loom: Lay not on front center when blowing off looms.	Weaver Cleaner
Wavy cloth	Alternately thick and thin places in cloth.	Loom: Let-off or take-up motion not working properly.	Weaver Loom fixer
Wild filling	Loose yarn caught between shed in loom.	Loom: Wild blowing off of looms; ends of quills not blown out at quill machine.	Quill machine operators
Wrong draw	Warp end drawn in at wrong place.	Loom: Careless weaving.	Weaver

APPENDIX I

TABLE I.1

Areas Under the Normal Probability Curve [a]

X/σ	.00	.02	.04	.06	.08	.09
0.0	.0000	.0080	.0159	.0239	.0319	.0359
0.1	.0398	.0478	.0557	.0636	.0714	.0753
0.2	.0793	.0871	.0948	.1026	.1103	.1141
0.3	.1179	.1255	.1331	.1406	.1480	.1517
0.4	.1554	.1628	.1700	.1772	.1844	.1879
0.5	.1915	.1985	.2054	.2123	.2190	.2224
0.6	.2257	.2324	.2389	.2454	.2518	.2549
0.7	.2580	.2642	.2704	.2764	.2823	.2852
0.8	.2881	.2939	.2995	.3051	.3106	.3233
0.9	.3159	.3212	.3264	.3315	.3365	.3389
1.0	.3413	.3461	.3508	.3554	.3599	.3621
1.1	.3643	.3686	.3729	.3770	.3810	.3830
1.2	.3849	.3888	.3925	.3962	.3997	.4015
1.3	.4032	.4066	.4099	.4131	.4162	.4177
1.4	.4192	.4222	.4251	.4279	.4306	.4319
1.5	.4332	.4357	.4382	.4406	.4430	.4441
1.6	.4452	.4474	.4495	.4515	.4535	.4545
1.7	.4554	.4573	.4591	.4608	.4625	.4633
1.8	.4641	.4656	.4671	.4686	.4699	.4706
1.9	.4713	.4726	.4738	.4750	.4762	.4767
2.0	.4773	.4783	.4793	.4803	.4812	.4817
2.1	.4821	.4830	.4838	.4846	.4854	.4857
2.2	.4861	.4868	.4875	.4881	.4887	.4890
2.3	.4893	.4898	.4904	.4909	.4913	.4916
2.4	.4918	.4922	.4927	.4931	.4934	.4936
2.5	.4938	.4941	.4945	.4948	.4951	.4952
2.6	.4953	.4956	.4959	.4961	.4963	.4964
2.7	.4965	.4967	.4969	.4971	.4973	.4974
2.8	.4974	.4976	.4977	.4979	.4980	.4981
2.9	.4981	.4983	.4984	.4985	.4986	.4986
3.0	.49865	.4987	.4988	.4989	.4989	.4990
3.1	.49903	.4991	.4992	.4992	.4993	.4993
3.2	.4993129					
3.3	.4995166					
3.4	.4996631					
3.5	.4997674					
3.6	.4998409					
3.7	.4998922					
3.8	.4999277					
3.9	.4999519					
4.0	.4999683					
4.5	.4999966					
5.0	.4999997					

[a] This table has been adapted by permission from Rugg's *Statistical Methods Applied to Education* published by Houghton Mifflin Company.

TABLE I.2

Table of Values of t for Given Degrees of Freedom (n) and at Specified Levels of Significance (P)[a]

n	Level of Significance (P)							
	.8	.4	.3	.1	.05	.02	.01	.001
1	.325	1.376	1.963	6.314	12.706	31.821	63.657	636.619
2	.289	1.061	1.386	2.920	4.303	6.965	9.925	31.598
3	.277	.978	1.250	2.353	3.182	4.541	5.841	12.941
4	.271	.941	1.190	2.132	2.776	3.747	4.604	8.610
5	.267	.920	1.156	2.015	2.571	3.365	4.032	6.859
6	.265	.906	1.134	1.943	2.447	3.143	3.707	5.959
7	.263	.896	1.119	1.895	2.365	2.998	3.499	5.405
8	.262	.889	1.108	1.860	2.306	2.896	3.355	5.041
9	.261	.883	1.100	1.833	2.262	2.821	3.250	4.781
10	.260	.879	1.093	1.812	2.228	2.764	3.169	4.587
11	.260	.876	1.088	1.796	2.201	2.718	3.106	4.437
12	.259	.873	1.083	1.782	2.179	2.681	3.055	4.318
13	.259	.870	1.079	1.771	2.160	2.650	3.012	4.221
14	.258	.868	1.076	1.761	2.145	2.624	2.977	4.140
15	.258	.866	1.074	1.753	2.131	2.602	2.947	4.073
16	.258	.865	1.071	1.746	2.120	2.583	2.921	4.015
17	.257	.863	1.069	1.740	2.110	2.567	2.898	3.965
18	.257	.862	1.067	1.734	2.101	2.552	2.878	3.922
19	.257	.861	1.066	1.729	2.093	2.539	2.861	3.883
20	.257	.860	1.064	1.725	2.086	2.528	2.845	3.850
21	.257	.859	1.063	1.721	2.080	2.518	2.831	3.819
22	.256	.858	1.061	1.717	2.074	2.508	2.819	3.792
23	.256	.858	1.060	1.714	2.069	2.500	2.807	3.767
24	.256	.857	1.059	1.711	2.064	2.492	2.797	3.745
25	.256	.856	1.058	1.708	2.060	2.485	2.787	3.725
26	.256	.856	1.058	1.706	2.056	2.479	2.779	3.707
27	.256	.855	1.057	1.703	2.052	2.473	2.771	3.690
28	.256	.855	1.056	1.701	2.048	2.467	2.763	3.674
29	.256	.854	1.055	1.699	2.045	2.462	2.756	3.659
30	.256	.854	1.055	1.697	2.042	2.457	2.750	3.646
40	.255	.851	1.050	1.684	2.021	2.423	2.704	3.551
60	.254	.848	1.046	1.671	2.000	2.390	2.660	3.460
120	.254	.845	1.041	1.658	1.980	2.358	2.617	3.373
∞	.253	.842	1.036	1.645	1.960	2.326	2.576	3.291

[a] Abridged from Table III of Fisher and Yates: *Statistical Tables for Biological, Agricultural, and Medical Research,* published by Oliver and Boyd, Ltd., Edinburgh, by permission of the authors and publisher.

TABLE I.3

Areas Under the Normal Probability Curve Included by the Average Plus and Minus the Indicated Standard Deviations[a]

Standard deviation	Area	Standard deviation	Area
0.0	.0000	2.1	.9642
0.1	.0796	2.2	.9722
0.2	.1586	2.3	.9786
0.3	.2358	2.4	.9836
0.4	.3108	2.5	.9876
0.5	.3830	2.6	.9906
0.6	.4514	2.7	.9930
0.7	.5160	2.8	.9948
0.8	.5762	2.9	.9962
0.9	.6318	3.0	.99730
1.0	.6826	3.1	.99806
1.1	.7286	3.2	.9986258
1.2	.7698	3.3	.9990332
1.3	.8064	3.4	.9993262
1.4	.8384	3.5	.9995348
1.5	.8664	3.6	.9996818
1.6	.8904	3.7	.9997844
1.7	.9108	3.8	.9998554
1.8	.9282	3.9	.9999038
1.9	.9426	4.0	.9999386
2.0	.9546	4.5	.9999932

[a] This table was calculated from Table I.1.

TABLE I.4
Variance Ratio, 5 Per Cent Significance Limits of F

n_2 [c]	n_1 [b]					
	2	4	6	12	24	∞
2	19.2	19.3	19.3	19.4	19.5	19.5
4	6.9	6.4	6.2	5.9	5.8	5.6
6	5.1	4.5	4.3	4.0	3.8	3.7
8	4.5	3.8	3.6	3.3	3.1	2.9
10	4.1	3.5	3.2	2.9	2.7	2.5
12	3.9	3.3	3.0	2.7	2.5	2.3
14	3.7	3.1	2.9	2.5	2.3	2.1
16	3.6	3.0	2.7	2.4	2.2	2.0
20	3.5	2.9	2.6	2.3	2.1	1.8
30	3.3	2.7	2.4	2.1	1.9	1.6
40	3.2	2.6	2.3	2.0	1.8	1.5
60	3.2	2.5	2.3	1.9	1.7	1.4
120	3.1	2.5	2.2	1.8	1.6	1.3
∞	3.0	2.4	2.1	1.8	1.5	1.0

TABLE I.5
Variance Ratio, 1 Per Cent Significance Limits of F

n_2 [c]	n_1 [b]					
	2	4	6	12	24	∞
2	99.0	99.3	99.4	99.4	99.5	99.5
4	18.0	16.0	15.2	14.4	13.9	13.5
6	10.9	9.2	8.5	7.7	7.3	6.9
8	8.7	7.0	6.4	5.7	5.3	4.9
10	7.6	6.0	5.4	4.7	4.3	3.9
12	6.9	5.4	4.8	4.2	3.8	3.4
14	6.5	5.0	4.5	3.8	3.4	3.0
16	6.2	4.8	4.2	3.6	3.2	2.8
20	5.9	4.4	3.9	3.2	2.9	2.4
30	5.4	4.0	3.5	2.8	2.5	2.0
40	5.2	3.8	3.3	2.7	2.3	1.8
60	5.0	3.7	3.1	2.5	2.1	1.6
120	4.8	3.5	3.0	2.3	2.0	1.4
∞	4.6	3.3	2.8	2.2	1.8	1.0

[a] Tables I.4 and I.5 are abridged from Table V of Fisher and Yates; *Statistical Tables for Biological, Agricultural, and Medical Research,* published by Oliver and Boyd, Ltd., Edinburgh, by permission of the authors and publisher.

[b] Degrees of freedom-numerator.

[c] Degrees of freedom-denominator.

TABLE I.6

Table of t Values for Significance Limits

n	Level of significance	
	5 %	1 %
1	12.706	63.657
2	4.303	9.925
3	3.182	5.841
4	2.776	4.604
5	2.571	4.032
6	2.447	3.707
7	2.365	3.499
8	2.306	3.355
9	2.262	3.250
10	2.228	3.169
11	2.201	3.106
12	2.179	3.055
13	2.160	3.012
14	2.145	2.977
15	2.131	2.947
16	2.120	2.921
17	2.110	2.898
18	2.101	2.878
19	2.093	2.861
20	2.086	2.845
21	2.080	2.831
22	2.074	2.819
23	2.069	2.807
24	2.064	2.797
25	2.060	2.787
26	2.056	2.779
27	2.052	2.771
28	2.048	2.763
29	2.045	2.756
30	2.042	2.750
40	2.021	2.704
60	2.000	2.660
120	1.980	2.617
∞	1.960	2.576

TABLE I.7

Relative Humidity Table (for Use with the Sling Psychrometer)

Dry bulb reading	Difference between dry bulb and wet bulb reading												
°F	1	2	3	4	5	6	7	8	9	10	11	12	13
60	94	89	83	78	73	68	63	58	53	48	43	39	34
61	94	89	84	78	73	68	63	58	54	49	44	40	35
62	94	89	84	79	74	69	64	59	54	50	45	41	36
63	94	89	84	79	74	69	64	60	55	50	46	42	37
64	95	90	84	79	74	70	65	60	56	51	47	43	38
65	95	90	85	80	75	70	66	61	56	52	48	44	39
66	95	90	85	80	75	71	66	61	57	53	48	44	40
67	95	90	85	80	75	71	66	62	58	53	49	45	41
68	95	90	85	80	76	71	67	62	58	54	50	46	42
69	95	90	85	81	76	72	67	63	59	55	51	47	43
70	95	90	86	81	77	72	68	64	59	55	51	48	44
71	95	90	86	81	77	72	68	64	60	56	52	48	45
72	95	91	86	82	77	73	69	65	61	57	53	49	45
73	95	91	86	82	78	73	69	65	61	57	53	50	46
74	95	91	86	82	78	74	69	65	61	58	54	50	47
75	96	91	86	82	78	74	70	66	62	58	54	51	47
76	96	91	87	82	78	74	70	66	62	59	55	51	48
77	96	91	87	83	79	74	71	67	63	59	56	52	48
78	96	91	87	83	79	75	71	67	63	60	56	53	49
79	96	91	87	83	79	75	71	68	64	60	57	53	50
80	96	91	87	83	79	75	72	68	64	61	57	54	50
81	96	92	88	84	80	76	72	69	65	61	58	55	51
82	96	92	88	84	80	76	72	69	65	61	58	55	51
83	96	92	88	84	80	76	73	69	66	62	59	56	52
84	96	92	88	84	80	76	73	69	66	62	59	56	52
85	96	92	88	84	80	77	73	69	66	63	60	57	53
86	96	92	88	84	81	77	73	70	66	63	60	57	53
87	96	92	88	85	81	77	74	70	67	64	61	57	54
88	96	92	88	85	81	77	74	70	67	64	61	57	54
89	96	92	88	85	81	77	74	70	67	64	61	57	54
90	96	92	89	85	81	78	74	71	68	65	61	58	55
91	96	92	89	85	82	78	75	72	68	65	62	59	56
92	96	92	89	85	82	78	75	72	68	65	62	59	56
93	96	93	89	85	82	79	75	72	69	66	63	60	57
94	96	93	89	85	82	79	75	72	69	66	63	60	57
95	96	93	89	85	82	79	75	72	69	66	63	60	57
96	96	93	89	86	82	79	76	73	69	66	63	61	58
97	96	93	89	86	82	79	76	73	69	66	63	61	58
98	96	93	89	86	83	79	76	73	70	67	64	61	58
99	96	93	89	86	83	80	77	73	70	68	65	62	59

TABLE I.8

Cloth Take-Up to Crimp or Moisture Content to Regain [a]

	Take-up or moisture content %				
	0.0	0.2	0.4	0.6	0.8
	Crimp or regain %				
0	0.000	0.200	0.401	0.603	0.806
1	1.010	1.214	1.419	1.626	1.832
2	2.040	2.250	2.459	2.669	2.880
3	3.092	3.305	3.519	3.734	3.950
4	4.166	4.384	4.602	4.821	5.042
5	5.263	5.485	5.708	5.932	6.157
6	6.382	6.609	6.838	7.066	7.296
7	7.525	7.758	7.991	8.225	8.459
8	8.695	8.932	9.170	9.409	9.649
9	9.890	10.132	10.375	10.619	10.864
10	11.111	11.358	11.607	11.856	12.107
11	12.359	12.612	12.866	13.122	13.378
12	13.636	13.895	14.155	14.416	14.678
13	14.942	15.207	15.473	15.740	16.009
14	16.279	16.551	16.822	17.096	17.370
15	17.647	17.924	18.203	18.483	18.764
16	19.047	19.331	19.617	19.904	20.192
17	20.481	20.772	21.065	21.359	21.654
18	21.951	22.249	22.547	22.850	23.152
19	23.456	23.762	24.069	24.378	24.688
20	25.000	25.313	25.628	25.945	26.262

[a] Example: For a take-up (or moisture content) value of 9.4 per cent, the corresponding value for crimp (or per cent moisture regain) is 10.38 per cent.

TABLE I.9

Cloth Crimp to Take-Up or Moisture Regain to Content[a]

	Crimp or regain %				
	0.0	0.2	0.4	0.6	0.8
	Take-up or moisture content %				
0	0.000	0.200	0.398	0.596	0.794
1	0.990	1.186	1.381	1.575	1.768
2	1.961	2.153	2.344	2.534	2.724
3	2.913	3.101	3.288	3.475	3.661
4	3.846	4.031	4.215	4.398	4.580
5	4.762	4.943	5.123	5.303	5.482
6	5.660	5.838	6.015	6.191	6.367
7	6.542	6.716	6.890	7.063	7.236
8	7.407	7.579	7.749	7.919	8.088
9	8.257	8.425	8.592	8.759	8.925
10	9.090	9.256	9.420	9.585	9.747
11	9.909	10.072	10.233	10.394	10.555
12	10.714	10.873	11.032	11.190	11.348
13	11.504	11.661	11.817	11.972	12.127
14	12.281	12.434	12.587	12.740	12.892
15	13.043	13.194	13.345	13.495	13.644
16	13.793	13.941	14.089	14.237	14.384
17	14.530	14.676	14.821	14.966	15.110
18	15.254	15.398	15.541	15.683	15.825
19	15.966	16.107	16.248	16.388	16.528
20	16.667	16.805	16.944	17.081	17.219

[a] Example: For a crimp (or moisture regain) value of 8.6 per cent, the corresponding value for take-up (or moisture content) is 7.92 per cent.

APPENDIX II

TABLE II.1

Defects Classification Chart for Woven Cotton Fabrics which require a Severe Evaluation of Defects (issued by the Clothing Supply Office, U. S. Naval Supply)

Defects	No.	Major	Minor
FILLINGWISE DEFECTS			
Bad set mark	2	Clearly noticeable.	Slightly noticeable.
Barre mark	18	Clearly noticeable.	Slightly noticeable.
Broken pick / Missing pick	11	One or more broken or missing picks more than $\frac{1}{3}$ across the width of the fabric.	One broken pick less than $\frac{1}{3}$ the width of the fabric.
Bunch	5	Clearly noticeable affecting serviceability.	
Coarse or heavy pick / Coarse or heavy filling	1	Single pick more than twice the thickness of the yarn normally used in the fabric or several adjacent picks of larger diameter.	Single heavy pick which is not more than twice the thickness of the yarn normally used in the fabric.
Crack	12	Extending $\frac{1}{3}$ or more the width of the fabric.	Extending less than $\frac{1}{3}$ the width of the fabric.
Curl	6		Clearly noticeable.
Dirty filling	24	One or more picks discolored or dirty.	
Double pick	4	Two or more picks in the same shed across the full width of the fabric or any portion of it.	
Fine or light filling / Fine or light pick	8	One pick less than $\frac{1}{2}$ the thickness of yarn normally used in fabric or several picks	One pick smaller in diameter than normally being used in the fabric.

Table continued

TABLE II.1 (*Continued*)

Defects	No.	Major	Minor
FILLINGWISE DEFECTS (continued)			
Light place Fine filling bar Light filling bar	9	of yarn smaller in diameter than normally used in the fabric.	
Float Skip Overshot	13	Multiple floats, skips, or overshots concentrated in a small area.	Two or three floats, skips, or overshots spread over a large area.
Finger mark Pressure mark	22	Clearly noticeable affecting serviceability and appearance.	
Hang pick Hang shot	17		Noticeable and resulting in a hole.
Heavy filling bar Heavy place	2	Several adjacent picks containing yarn of larger diameter than normally used in the fabric or place containing more than the normal number of picks.	
Jerked-in filling Lashed-in or pull-in filling	3	Extending more than 2 inches from selvage.	Extending 2 inches or less from selvage.
Kink Knot	6 7	Large and bulky, seriously affecting serviceability and/or appearance.	Small and flat, not seriously affecting serviceability and/or appearance.
Lump	5	Large and bulky.	Small and flat.
Loop	6		
Mispick	10	Clearly noticeable.	Slightly noticeable.
Mixed filling Change in filling Shade bar	14	Clearly noticeable filling yarn of wrong twist, wrong number of plies, wrong color, mixed yarn lots.	
Pick-out mark Cleaning place	23	Clearly noticeable.	
Shuttle mark Box mark	19 20	Clearly noticeable due to injury of adjoining ends.	Slightly noticeable.
Slack pick or filling Stretched filling	6 5		Causing a few filling loops.

TABLE II.1 (*Continued*)

Defects	No.	Major	Minor
FILLINGWISE DEFECTS (continued)			
Slub or slug / Slough-off	5	Large, bulky, and loose.	Small and tight, not seriously affecting serviceability and/or appearance.
Stop mark	2	Clearly noticeable.	
Tear drop or teariness	21	Clearly noticeable.	
Tight pick or filling	15		Noticeable.
Uneven filling	25	Clearly noticeable filling threads of varying thickness.	
WARPWISE DEFECTS			
Balling up / Fuzz balls	45	Large and bulky.	Small and flat.
Broken or missing end / End out	33	Two or more adjacent ends broken or missing.	One broken end two or more inches in length or one end out.
Coarse end	26	Single end more than twice the thickness and/or containing more plies than normally being used in fabric.	
Draw-back / Hitch-back / Tie-back / Warp holding place	35		Clearly noticeable.
Fine end / Drop ply	31 / 32	Yarn considerably smaller in diameter or containing fewer plies than specified.	
Floats	34	Multiple warp floats concentrated in a small area.	Two or three floats scattered and not seriously affecting appearance or serviceability.
Hard size or sticker	43		Clearly noticeable.
Kink / Loop	28		Clearly noticeable.
Mixed end	36	Yarn definitely of wrong twist, wrong number of plies, wrong color.	

TABLE II.1 (*Continued*)

Defects	No.	Major	Minor
WARPWISE DEFECTS (continued)			
Reed mark	41	Clearly noticeable, crowded, or open warp threads.	
Slub, Bunch	27	Large and bulky.	Small and flat.
Knot	29	Large and bulky.	Small and flat.
Hang or loose threads	30	Large and bulky.	Small and flat.
Slack end	38		Clearly noticeable.
Slack warp	39	Definitely having a crimped or cockled appearance.	
Soiled or dirty ends	46	One or more warp ends soiled or discolored.	
Temple marks	42	Clearly noticeable holes, impressions, or marks adjacent to the edge of the fabric.	
Tight end	37	Two or more tight ends causing a definite puckering of the fabric.	One tight end clearly noticeable.
Uneven ends	47	Clearly noticeable warp threads of varying thickness.	
Warp streak	44	Clearly noticeable narrow streak caused by differences in color from adjacent ends.	
Wrong draw	40	Clearly noticeable.	
BAD SELVAGES:			
Curled, Rolled, Folded, Doubled	48	Width inside defective selvages below specified tolerances.	Width inside defective selvages below specified width but within specified tolerances.
Scalloped	51		
Wavy	52		
Slack	49	Slack selvages causing the fabric to bulge halfway across the width or more.	Causing the fabric to bulge slightly.

TABLE II.1 (*Continued*)

Defects	No.	Major	Minor
BAD SELVAGES (continued)			
Tight	50	Considerably tighter than the body of the fabric.	Slightly tighter than the body of the fabric.
Carded, Beaded, Rough, Loopy	54	Selvage considerably heavier than the body of the fabric and/or containing a great number of irregular filling loops extending beyond the outside of selvage.	
Cut, Broken	55	Width of fabric inside defect less than specified tolerance.	Width of fabric inside defect within a specified tolerance.
Torn	56		
FINISHING DEFECTS			
Clip marks, Tentering marks, Pin marks	73	Width of fabric inside defects less than specified tolerances.	Width of fabric inside defects less than specified but within tolerances allowed.
Crease, Mill wrinkle	59	Creases or mill wrinkles sharp and permanent.	Soft crease or wrinkles which will not be permanent.
Crowsfeet	67	Clearly noticeable.	Faintly noticeable,
Dyestains	69	Large dyestain clearly noticeable.	Small dyestain hardly noticeable.
Dye streak	70	Clearly noticeable.	Faintly noticeable.
Finishing bar	57	Clearly noticeable.	
Rope marks	68	Clearly noticeable.	Faintly noticeable.
Sanforizing mark, Blanket mark	58	Clearly noticeable crimped, rippled, or cockled place in the cloth causing distortion of the weave.	
Seams	61 A	One or more seamed pieces in the roll or bolt.	
Seam mark, Wrinkle mark	61	Clearly noticeable and permanent.	Soft and not permanent.

Table Continued

TABLE II.1 (*Continued*)

Defects	No.	Major	Minor
FINISHING DEFECTS (continued)			
Selvage mark	60		Clearly noticeable mark along the selvage caused by folded or doubled selvage.
Uneven dyeing	62	Definite variations in shade.	Faint variations in shade.
Uneven napping	64	Nap definitely uneven.	
Uneven shrinkage	72	Definite wavy warpwise condition in the fabric, preventing it from lying flat on a table.	
GENERAL DEFECTS:			
Abrasion mark / Chafe mark / Bruise	79	Clearly noticeable.	
Bad odor	80	Definite objectionable odor.	
Bias filling / Bowed filling	88	Filling yarn definitely not at right angle to the warp.	
Burrs	82		Several burrs concentrated in a small area.
Cut	91	Any cut.	
Flyers, loom fly	81		Numerous waste fibers in fabric or yarn concentrated in a small area.
Gout	77		Large and bulky lint or waste woven into fabric.
Hole	90	Any hole as a result of broken yarns.	A small hole as a result of separated yarns.
Naps	85	Numerous fiber tangles, large or small, concentrated in a small area.	Numerous fiber tangles spread over a large area.
Neps	84		
Neppiness	86		
Smash	78	Texture definitely ruptured.	
Shift mark	87	Clearly noticeable light and heavy spots.	

TABLE II.1 (*Continued*)

Defects	No.	Major	Minor
GENERAL DEFECTS (continued)			
Spots	74	Numerous conspicuous rust, oil, or grease spots.	Small not conspicuous rust, oil, or grease spots.
Stains	75	Large and/or conspicuous rust, oil, or grease stains.	
Streak	76	Clearly noticeable rust, oil, or grease streak.	
Tear	92	Any tear.	
Trash	83	Motes, seed coat fragments and leaf particles appearing as specks in a concentrated area.	
Weak spots / Tender spots	89	Clearly noticeable weak and/or tender spots definitely affecting serviceability.	
Width	—	Less than specified tolerances at any one place.	

TABLE II.2

Defects Classification Chart for Woven Cotton Fabrics which require a Less Severe Evaluation of Defects (issued by the Clothing Supply Office, U. S. Naval Supply)

Defects	No.	Major	Minor
FILLINGWISE DEFECTS:			
Bad set mark	2		Clearly noticeable.
Barre mark	18		Clearly noticeable.
Broken pick Missing pick	11	Two or more broken or missing picks more than $\frac{1}{3}$ across the width of the fabric.	One broken or missing pick more than $\frac{1}{3}$ the width of the fabric.
Bunch	5		Clearly noticeable, affecting serviceability.
Coarse or heavy pick Coarse or heavy filling	1		Single or several adjacent picks of larger diameter than normally used in the fabric.
Crack	12		Extending $\frac{1}{3}$ or more the width of the fabric.
Dirty filling	24		Several picks in a concentrated area, discolored or dirty.
Fine or light filling	8		
Fine filling bar Light filling bar Light place	9	Several picks of yarn smaller in diameter than normally used in fabric.	
Float Skip Overshot	13		Multiple floats, skips, or overshots concentrated in a small area.
Hang pick Hang shot	17		Clearly noticeable and resulting in a hole.
Heavy place Heavy filling bar	2		Several adjacent picks containing yarn of a larger diameter than normally used in fabric or place containing more than normal number of picks.

TABLE II.2 (*Continued*)

Defects	No.	Major	Minor
FILLINGWISE DEFECTS (continued)			
Kink Knot	6 7		Large and bulky, seriously affecting serviceability.
Lump	5		Large and bulky.
Loop	6		
Mispick	10		Clearly noticeable.
Mixed filling Change in filling Shade bar	14		Clearly noticeable filling yarn of wrong twist, wrong number of plies, wrong color, or mixed yarn lots.
Pick-out mark Cleaning place	23		Clearly noticeable.
Shuttle mark Box mark	19 20		Clearly noticeable due to injury of adjoining ends.
Slug or slub Slough off Snarl	 5 6		Large, loose, and bulky.
Stop mark	2		Clearly noticeable.
Uneven filling	25		Clearly noticeable filling threads of varying thickness.
WARPWISE DEFECTS:			
Broken or missing end End out	33	Two or more adjacent ends broken or missing.	One broken end two or more inches in length or one end out.
Coarse end	26		Single end more than twice the thickness and/or containing more plies than normally being used in fabric.
Fine end Drop ply	31 32		Yarn considerably smaller in diameter or containing fewer plies than specified.

Table Continued

TABLE II.2 (*Continued*)

Defects	No.	Major	Minor
WARPWISE DEFECTS (continued)			
Floats	34	Multiple warp floats concentrated in a small area.	Two or three floats scattered.
Kink Loop	28		Clearly noticeable and large.
Reed mark	41	Clearly noticeable open or crowded warp threads.	
Slub Bunch	27		
Knot	29		Large and bulky.
Hang or loose threads	30		
Slack warp	39		Definitely having a crimped or cockled appearance.
Soiled or dirty ends	46		One or more warp ends soiled or discolored.
Temple marks	42	Clearly noticeable holes, impressions, or marks adjacent to edge of fabric.	
Tight ends	37	Two or more tight ends causing a definite puckering of the fabric.	
BAD SELVAGES			
Curled Folded Rolled Doubled	48	Width inside defective selvages below specified tolerances.	Width inside defective selvages below specified width but within specified tolerances.
Scalloped	51		
Wavy	52		
Slack	49	Slack selvages causing the fabric to bulge halfway across the fabric.	Causing the fabric to bulge or balloon slightly.
Tight	50	Considerably tighter than the body of the fabric.	

TABLE II.2 (*Continued*)

Defects	No.	Major	Minor
BAD SELVAGES (continued)			
Corded Beaded Rough Loopy	54	Selvage considerably heavier than the body of the fabric and/or containing a great number of irregular filling loops extending beyond the outside of selvage.	
Cut Broken Torn	55 56	Width of fabric inside defect less than specified tolerance.	Width of fabric inside defect within specified tolerance.
FINISHING DEFECTS:			
Clip marks Tentering marks Pin marks	73	Width of fabric inside defects less than specified tolerances.	Width of fabric inside defects less than specified but within specified tolerances.
Dye strain	69		Clearly noticeable.
Dye streak	70		
Seams	61 A	One or more seamed part pieces in roll or bolt.	
Uneven dyeing	62		Definite variations in shade.
GENERAL DEFECTS			
Abrasion mark Chafe mark Bruise	79	Clearly noticeable.	
Bad odor	80	Definite objectionable odor.	
Bias filling Bowed filling	88	Filling yarn running definitely not at right angle to the warp	
Cut	91 92	Any cut.	
Gout	77		Large and bulky lint or waste woven into fabric.
Hole	90		Any hole as a result of broken yarns.

Table Continued

TABLE II.2 (*Continued*)

Defects	No.	Major	Minor
GENERAL DEFECTS (continued)			
Neppiness	86		Numerous fiber tangles large or small, concentrated in a small area.
Smash	78	Texture definitely ruptured.	
Stains Streaks	74 76	Large and/or conspicuous rust, oil, or grease stains or streaks.	
Tear	92	Any tear.	
Trash	83		Motes, seed coat fragments and leaf particles appearing as specks in a concentrated area.
Weak spots Tender spots	89		Clearly noticeable weak and/or tender spots definitely affecting serviceability.
Width	—	Less than specified tolerances at any one place.	

TABLE II.3

Defects Classification Chart for Woven Cotton Fabrics which require a Considerably Less Severe Evaluation of Defects (issued by the Clothing Supply Office, U. S. Naval Supply)

Defects	No.	Description
FILLINGWISE DEFECTS		
Broken picks / Missing picks	11	Two or more broken or missing picks more than ⅓ the width of the fabric.
Floats	13	Multiple floats, skips, or overshots concentrated in a small area.
Knots	7	A number of large and bulky knots concentrated in a small area.
WARPWISE DEFECTS		
Broken or missing ends / Ends out	33	Two or more broken or missing ends.
Floats	34	Multiple warp floats concentrated in a small area.
Knots	29	A number of large and bulky knots concentrated in a small area.
FINISHING DEFECTS		
Seams	61 A	One or more seamed part pieces in a roll or bolt.
GENERAL DEFECTS		
Bad odor	80	Definite objectionable odor.
Cut	91	Any cuts and/or tears.
Tear	92	
Hole	90	Large holes as a result of broken yarns.
Smash	78	Texture definitely ruptured.
Weak spots / Tender spots	89	Clearly noticeable weak and/or tender spots definitely affecting serviceability,
Width	—	Less than specified tolerance at any one place.

APPENDIX III

GLOSSARY OF SYMBOLS AND TERMS

A_2	A factor used in calculating control limits for the $\bar{X}$ chart.
E	Allowable sampling error. Expressed in per cent plus and minus.
c	Defects per unit.
$\bar{c}$	Average number of defects per unit.
CL	Center line.
d_2	A factor used in estimating the standard deviation using ranges.
D_3	A factor used in calculating the lower control limit for the R chart.
D_4	A factor used in calculating the upper control limit for the R chart.
k	Skewness or lopsidedness of the frequency distribution.
LCL	Lower control limit.
N	Lot size.
n	Sample size, number of units inspected or number of measurements.
P	The probability associated with any given set of conditions.
p	Fraction defective (also per cent defective).
$\bar{p}$	Average fraction defective.
pn	Number of defective units in a sample of size n.
$\bar{p}n$	Average number of defective units in a sample of size n.
R	Range of a subgroup (difference between highest and lowest value).
$\bar{R}$	Average range.
r	Coefficient of correlation.
σ	Standard deviation for individuals.
σ'	Estimated standard deviation calculated by use of ranges.
$\sigma_{\bar{X}}$	Standard deviation for averages.
$\sigma_{\bar{X}n=4}$	Standard deviation for averages of subgroups of 4.
$\sigma_{\bar{X}n=5}$	Standard deviation for averages of subgroups of 5.
σ_{xs}	Standard deviation of the x values about the regression line.
σ_{ys}	Standard deviation of the y values about the regression line.
σ_D	Standard deviation for the distribution of differences.
UCL	Upper control limit.
V	Coefficient of variation.
X	An individual measurement.
$\bar{X}$	Average of a subgroup.
$\bar{\bar{X}}$	Average of two or more averages. Grand average.
Σ	Sum of two or more measurements.
ΣX	Sum of individual measurements. $X_1 + X_2 + X_3 + \ldots X_n = \Sigma X$.

The following definitions are reprinted from *Industrial Quality Control*, the official journal of the American Society for Quality Control, by permission of the president of the Society.

Attributes, method of	Measurement of quality by the method of attributes consists of noting the presence or absence of some characteristic (attribute) in each of the units in the group under consideration and counting how many do or do not possess it. Example: Go and No-Go Gauging of a dimension.
Central line	A line on a control chart representing the average or expected value of the statistical measure being plotted.
Control chart	A graphical chart with upper and lower control limits and plotted values of some statistical measure for a series of samples or subgroups. A central line is shown frequently.
Control chart – no standard given	A control chart whose limits are based on the data of the samples or subgroups for which values are plotted on the chart.
Control chart – standard given	A control chart whose control limits are based on adopted standard values of the statistical measure (*s*) for which values are plotted on the chart.
Control limits	Limits on a control chart which are used as criteria for action or for judging the significance of variations between samples or subgroups.
Sample	A group of units, or portion of material, taken from a larger collection of units, or quantity of material, which serves to provide information that can be used as a basis for judging the quality of the larger quantity, or as a basis for action on the larger quantity or on the production process. Also used in the sense of a "sample of observations."
Sample size	The number of units in a sample. Also used in the sense of the number of observations in a sample.

Statistical measure	A mathematical function of a set of numbers or observations. Common statistical measures are the arithmetic mean or average, the standard deviation, the range (for variables), and the relative frequency (for attributes).
Subgroup	One of a series of groups of observations obtained by subdividing a larger group of observations; alternatively, the data obtained from one of a series of samples taken from one or more universes.
Rational subgroups	Subgroups *within* which variations may for engineering reasons be considered to be due to non-assignable chance causes only, but *between* which there may be variations due to assignable causes whose presence is considered possible. (One of the essential features of the control chart method is to break up inspection data into rational subgroups.)
Unit	One of a number of similar items, objects, individuals, etc.
Universe or population	The total collection of units from a common source; the conceptual total collection of units from a process, such as a production process. Also used in the sense of a "universe (or population) of observation." Note: "Universe," "population," and "parent distribution" are synonymous terms. Statistical methods are based on the concept of a distribution of an exceedingly large number of observations, termed an infinite universe or population. An individual observation, the $\overline{X}$ of a sample, etc. may be thought of as one coming from a parent distribution or infinite population of like items.
Variables, method of	Measurement of quality by the method of variables consists of measuring and recording the numerical magnitude of a quality characteristic for each of the units in the group under consideration. This involves *reading a scale* of some kind.

APPENDIX IV

Proposed ASQC STD. A2 on Definitions and Symbols for Acceptance Sampling

The following definitions relate particularly to one of the most common uses of acceptance sampling, namely, acceptance inspection of individual lots of product, comprising individual units. The general intent of the definitions, however, should be considered applicable also to the inspection of bulk material, to the acceptance or rejection of a process on the basis of sampling results, etc. The alternative to acceptance is termed "rejection" for the purposes of definition, although, in practice, the alternative may take some form other than out-right rejection.

	Term	Symbol or abbreviation [a]	Definition
1.	GENERAL TERMS RELATING TO ACCEPTANCE SAMPLING.		
1.1	Acceptance Sampling		The art or science that deals with procedures in which decisions to accept or reject lots or processes are based on the examination of samples.
1.2	Sampling Plan		A specific plan which states (a) the sample sizes, and (b) the criteria for accepting, rejecting, or taking another sample, to be used in inspecting a lot.
1.3	Inspection Lot		A specific quantity of similar material, or a collection of similar units, offered for inspection and acceptance at one time.
1.4	Lot Size	N	The number of units in the lot.
1.5	Sample		A portion of material or a group of units, taken from a lot, which serves to provide information for reaching a decision regarding acceptance.
1.6	Sample Size	n	The number of units in the sample.

Continued

	Term	Symbol or abbreviation [a]	Definition
1.7	Single Sampling		Sampling inspection in which a decision to accept or to reject is reached after the inspection of a single sample.
1.8	Double Sampling		Sampling inspection in which the inspection of the first sample leads to a decision to accept, to reject, or to take a second sample and the examination of a second sample, when required, always leads to a decision to accept or to reject.
1.9	Multiple Sampling		Sampling inspection in which, after each sample, the decision may be to accept, to reject, or to take another sample, but in which there is usually a prescribed maximum number of samples, after which a decision to accept or to reject is reached. *Note:* Multiple sampling as defined here is sometimes called "sequential sampling" or "group sequential sampling". The term "multiple sampling" is preferred.
1.10	Sequential Sampling		Sampling inspection in which, after each unit is inspected, the decision is made to accept, to reject, or to inspect another unit. *Note:* Sequential sampling as defined here is sometimes called "unit sequential sampling".
1.11	Curtailed Inspection		Sampling inspection in which, as soon as a decision is certain, the inspection of the sample is stopped. Thus, as soon as the rejection number for defectives is reached, the decision is certain and no further inspection is necessary. *Note:* Commonly a first sample is always completed for the purpose of estimating the process average.

	Term	Symbol or abbreviation [a]	Definition
1.12	Normal Inspection		Inspection in accordance with a sampling plan that is used under ordinary circumstances.
1.13	Reduced Inspection		Inspection in accordance with a sampling plan requiring smaller sample sizes than those used in normal inspection.
1.14	Tightened Inspection		Inspection in accordance with a sampling plan that has more strict acceptance criteria than normal inspection.
1.15	Process Average Quality		Expected quality of product from a given process, usually estimated from first sample inspection results of past lots.
1.16	Acceptable Quality Level	AQL	The maximum per cent defective (or the maximum number of defects per 100 units) which can be considered satisfactory as a process average.
1.17	Probability of Acceptance	*Pa*	Probability that a lot or process will be accepted.
1.18	Operating Characteristic Curve for Acceptance Sampling Plan	OC Curve	A curve showing the relation between the probability of acceptance and either lot quality or process quality, whichever is applicable.
1.19	Consumer's Risk		The probability or risk of accepting a lot, for a given lot quality or process quality, whichever is applicable. Usually applied only to quality values that are relatively poor.
1.20	Producer's Risk		The probability or risk of rejecting a lot, for a given lot quality or process quality, whichever is applicable. Usually applied only to quality values that are relatively good.

Continued

	Term	Symbol or abbreviation[a]	Definition
1.21	Average Sample Number	ASN	The average number of sample units inspected per lot in reaching a decision to accept or to reject.
1.22	Average Total Inspection	ATI	The average number of units inspected per lot including all units in rejected lots (applicable when the procedure calls for 100 % inspection of rejected lots.)
2.	TERMS RELATING TO ATTRIBUTES SAMPLING PLANS		
2.1	Acceptance Number	Ac	The largest number of defectives (or defects) in the sample or samples under consideration that will permit the acceptance of the inspection lot.
2.2	Rejection Number	Re	The smallest number of defectives (or defects) in the sample or samples under consideration that will require the rejection of the inspection lot.
2.3	Average Outgoing Quality	AOQ	The average quality of outgoing product after 100 % inspection of rejected lots, with replacement by good units of all defective units found in inspection.
2.4	Average Outgoing Quality Limit	AOQL	The maximum average outgoing quality (AOQ) for a sampling plan.
2.5		P_{95}, P_{50}, P_{10}, P_{05}, etc.	Lot quality or process quality for which the probability of acceptance is 0.95, 0.50, 0.10, 0.05, etc. respectively for a given acceptance sampling plan.

[a] Of the entries in the second column, only N, n, Pa, P_{95}, P_{50}, P_{10}, and P_{05} are intended as mathematical symbols. All other entries are abbreviations for practical use.

APPENDIX V

AMENDED OFFICIAL COTTON STANDARDS OF THE UNITED STATES FOR THE GRADE OF AMERICAN UPLAND COTTON (UNIVERSAL STANDARDS), EFFECTIVE AUGUST 1, 1960[1]

On May 22, 1959, a notice of proposed rule making was published in the FEDERAL REGISTER (24 F.R. 4147) regarding proposed revisions in the Official Cotton Standards of the United States for the Grade of American Upland Cotton (7 CFR 28.401-28.427), also termed "Universal Standards for American Cotton," pursuant to authority contained in section 10 of the United States Cotton Standards Act, as amended (42 Stat. 1519; 7 U.S.C. 61), and in section 4854 of the Internal Revenue Code of 1954 (68A Stat. 580; 26 U.S.C. 4854).

The amended Official Cotton Standards of the United States for Grade of American Upland Cotton, as hereinafter set forth, incorporate the following changes to be effective on and after August 1, 1960:

1. New descriptive standards for Plus cotton in the White grades Middling through Good Ordinary.

2. New descriptive standards for Light Spotted cotton in the grades Good Middling Light Spotted through Low Middling Light Spotted.

3. New physical standards for Spotted cotton in the grades Strict Middling Spotted through Low Middling Spotted to replace the present descriptive standards for these grades. Good Middling Spotted will continue to be a descriptive standard.

4. New descriptive standards for Light Gray cotton in the grades Good Middling Light Gray through Strict Low Middling Light Gray.

5. Redefined descriptive standards for Gray cotton in the grades Good Middling Gray through Strict Low Middling Gray.

6. A procedure for describing below grade cotton in cotton classification.

For the purposes of the aforesaid acts and pursuant to the authority contained therein, and after consideration of all relevant written and oral data, views, and arguments presented pursuant to the notice, including those presented by delegates and cotton industry groups at the 1959 Universal Cotton Standards Conference, said standards are hereby amended to read as follows, effective on and after August 1, 1960:

[1] Reprinted from the Federal Register of June 25, 1959, Title 7, Chapter I, Part 28, Subpart C.

Sec. WHITE COTTON

28.401 Strict Good Middling.
28.402 Good Middling.
28.403 Strict Middling.
28.404 Middling Plus.
28.405 Middling.
28.406 Strict Low Middling Plus.
28.407 Strict Low Middling.
28.408 Low Middling Plus.
28.409 Low Middling.
28.410 Strict Good Ordinary Plus.
28.411 Strict Good Ordinary.
28.412 Good Ordinary Plus.
28.413 Good Ordinary.

LIGHT SPOTTED COTTON

28.420 Good Middling Light Spotted.
28.421 Strict Middling Light Spotted.
28.422 Middling Light Spotted.
28.423 Strict Low Middling Light Spotted.
28.424 Low Middling Light Spotted.

SPOTTED COTTON

28.430 Good Middling Spotted.
28.431 Strict Middling Spotted.
28.432 Middling Spotted.
28.433 Strict Low Middling Spotted.
28.434 Low Middling Spotted.

TINGED COTTON

28.440 Good Middling Tinged.
28.441 Strict Middling Tinged.
28.442 Middling Tinged.
28.443 Strict Low Middling Tinged.
28.444 Low Middling Tinged.

YELLOW STAINED COTTON

28.450 Good Middling Yellow Stained.
28.451 Strict Middling Yellow Stained.
28.452 Middling Yellow Stained.

LIGHT GRAY COTTON

28.460 Good Middling Light Gray.
28.461 Strict Middling Light Gray.
28.462 Middling Light Gray.
28.463 Strict Low Middling Light Gray.

GRAY COTTON

BELOW GRADE COTTON

GENERAL

AUTHORITY: §§28.401 to 28.481 issued under sec. 10, 42 Stat. 1519; 7 U.S.C. 61. Interpret or apply sec. 6, 42 Stat. 1518, as amended, sec. 4854, 68A Stat. 580; 7 U.S.C. 56, 26 U.S.C. 4854.

WHITE COTTON

§ 28.401 Strict Good Middling. Strict Good Middling is American upland cotton which in color, leaf and preparation is better than Good Middling.

§ 28.402 Good Middling. Good Middling is American upland cotton which in color, leaf, and preparation is within the range represented by a set of samples in the custody of the United States Department of Agriculture in the District of Columbia in a container marked "Original Official Cotton Standards of the United States, American Upland, Good Middling, effective August 1, 1954."

§ 28.403 Strict Middling. Strict Middling is American upland cotton which in color, leaf, and preparation is within the range represented by a set of samples in the custody of the United States Department of Agriculture in the District of Columbia in a container marked "Original Official Cotton Standards of the United States, American Upland, Strict Middling, effective August 15, 1953."

§ 28.404 Middling Plus. Middling Plus is American upland cotton which is Middling in leaf and preparation with Strict Middling color.

§ 28.405 Middling. Middling is American upland cotton which in color, leaf, and preparation is within the range represented by a set of samples in the custody of the United States Department of Agriculture in the District of Columbia in a container marked "Original Official Cotton Standards of the United States, American Upland, Middling, effective August 15, 1953."

§ 28.406 Strict Low Middling Plus. Strict Low Middling Plus is American upland cotton which is Strict Low Middling in leaf and preparation with Middling color.

§ **28.407 Strict Low Middling.** Strict Low Middling is American upland cotton which in color, leaf, and preparation is within the range represented by a set of samples in the custody of the United States Department of Agriculture in the District of Columbia in a container marked "Original Official Cotton Standards of the United States, American Upland, Strict Low Middling, effective August 15, 1953."

§ **28.408 Low Middling Plus.** Low Middling Plus is American upland cotton which is Low Middling in leaf and preparation with Strict Low Middling color.

§ **28.409 Low Middling.** Low Middling is American upland cotton which in color, leaf, and preparation is within the range represented by a set of samples in the custody of the United States Department of Agriculture in the District of Columbia in a container marked "Original Official Cotton Standards of the United States, American Upland, Low Middling, effective August 15, 1953."

§ **28.410 Strict Good Ordinary Plus.** Strict Good Ordinary Plus is American upland cotton which is Strict Good Ordinary in leaf and preparation with Low Middling color.

§ **28.411 Strict Good Ordinary**. Strict Good Ordinary is American upland cotton which in color, leaf, and preparation is within the range represented by a set of samples in the custody of the United States Department of Agriculture in the District of Columbia in a container marked "Original Official Cotton Standards of the United States, American Upland, Strict Good Ordinary, effective August 15, 1953."

§ **28.412 Good Ordinary Plus.** Good Ordinary Plus is American upland cotton which is Good Ordinary in leaf and preparation with Strict Good Ordinary color.

§ **28.413 Good Ordinary.** Good Ordinary is American upland cotton which in color, leaf, and preparation is within the range represented by a set of samples in the custody of the United States Department of Agriculture in the District of Columbia in a container marked "Original Official Cotton Standards of the United States, American Upland, Good Ordinary, effective August 15, 1953."

Light Spotted Cotton

§ **28.420 Good Middling Light Spotted.** Good Middling Light Spotted is American upland cotton which in leaf and preparation is Good Middling, but which in spot or color, or both, is between Good Middling and Good Middling Spotted.

§ **28.421 Strict Middling Light Spotted.** Strict Middling Light Spotted is American upland cotton which in leaf and preparation is Strict Middling, but which in spot or color, or both, is between Strict Middling and Strict Middling Spotted.

§ **28.422 Middling Light Spotted.** Middling Light Spotted is American upland cotton which in leaf and preparation is Middling, but which in spot or color, or both, is between Middling and Middling Spotted.

§ **28.423 Strict Low Middling Light Spotted.** Strict Low Middling Light Spotted is American upland cotton which in leaf and preparation is Strict Low Middling, but which in spot or color, or both, is between Strict Low Middling and Strict Low Middling Spotted.

§ **28.424 Low Middling Light Spotted.** Low Middling Light Spotted is American upland cotton which in leaf and preparation is Low Middling, but which in spot or color, or both, is between Low Middling and Low Middling Spotted.

SPOTTED COTTON

§ **28.430 Good Middling Spotted.** Good Middling Spotted is American upland cotton which in color, leaf, and preparation is better than Strict Middling Spotted.

§ **28.431 Strict Middling Spotted.** Strict Middling Spotted is American upland cotton which in color, leaf, and preparation is within the range represented by a set of samples in the custody of the United States Department of Agriculture in the District of Columbia in a container marked "Original Official Cotton Standards of the United States, American Upland, Strict Middling Spotted, effective August 1, 1960."

§ **28.432 Middling Spotted.** Middling Spotted is American upland cotton which in color, leaf, and preparation is within the range represented by a set of samples in the custody of the United States Department of Agriculture in the District of Columbia in a container marked "Original Official Cotton Standards of the United States, American Upland, Middling Spotted, effective August 1, 1960."

§ **28.433 Strict Low Middling Spotted.** Strict Low Middling Spotted is American upland cotton which in color, leaf, and preparation is within the range represented by a set of samples in the custody of the United States Department of Agriculture in the District of Columbia in a container marked "Original Official Cotton Standards of the United States, American Upland, Strict Low Middling Spotted, effective August 1, 1960."

§ **28.434 Low Middling Spotted.** Low Middling Spotted is American upland cotton which in color, leaf, and preparation is within the range represented by a set of samples in the custody of the United States Department of Agriculture in the District of Columbia in a container marked "Original Official Cotton Standards of the United States, American Upland, Low Middling Spotted, effective August 1, 1960."

Tinged Cotton

§ **28.440 Good Middling Tinged.** Good Middling Tinged is American upland cotton which in color, leaf, and preparation is better than Strict Middling Tinged.

§ **28.441 Strict Middling Tinged.** Strict Middling Tinged is American upland cotton which in color, leaf, and preparation is within the range represented by a set of samples in the custody of the United States Department of Agriculture in the District of Columbia in a container marked "Original Official Cotton Standards of the United States, American Upland, Strict Middling Tinged, effective August 15, 1953."

§ **28.442 Middling Tinged.** Middling Tinged is American upland cotton which in color, leaf, and preparation is within the range represented by a set of samples in the custody of the United States Department of Agriculture in the District of Columbia in a container marked "Original Official Cotton Standards of the United States, American Upland, Middling Tinged, effective August 15, 1953."

§ **28.443 Strict Low Middling Tinged.** Strict Low Middling Tinged is American upland cotton which in color, leaf, and preparation is within the range represented by a set of samples in the custody of the United States Department of Agriculture in the District of Columbia in a container marked "Original Official Cotton Standards of the United States, American Upland, Strict Low Middling Tinged, effective August 15, 1953."

§ **28.444 Low Middling Tinged.** Low Middling Tinged is American upland cotton which in color, leaf, and preparation is within the range represented by a set of samples in the custody of the United States Department of Agriculture in the District of Columbia in a container marked "Original Official Cotton Standards of the United States, American Upland, Low Middling Tinged, effective August 15, 1953."

Yellow Stained Cotton

§ **28.450 Good Middling Yellow Stained.** Good middling Yellow Stained is American upland cotton which in leaf and preparation is Good Middling Tinged, but which in color is deeper than Good Middling Tinged.

§ **28.451 Strict Middling Yellow Stained.** Strict Middling Yellow Stained is American upland cotton which in leaf and preparation is Strict Middling Tinged, but which in color is deeper than Strict Middling Tinged.

Light Gray Cotton

§ 28.460 **Good Middling Light Gray.** Good Middling Light Gray is American upland cotton which in color is Middling and which in leaf and preparation is Good Middling.

§ 28.452 **Middling Yellow Stained.** Middling Yellow Stained is American upland cotton which in leaf and preparation is Middling Tinged, but which in color is deeper than Middling Tinged.

§ 28.461 **Strict Middling Light Gray.** Strict Middling Light Gray is American upland cotton which in color is Strict Low Middling and which in leaf and preparation is strict Middling.

§ 28.462 **Middling Light Gray.** Middling Light Gray is American upland cotton which in color is Low Middling and which in leaf and preparation is Middling.

§ 28.463 **Strict Low Middling Light Gray.** Strict Low Middling Light Gray is American upland cotton which in color is Strict Good Ordinary and which in leaf and preparation is Strict Low Middling.

Gray Cotton

§ 28.470 **Good Middling Gray.** Good Middling Gray is American upland cotton which in color is Strict Low Middling and which in leaf and preparation is Good Middling or better.

§ 28.471 **Strict Middling Gray.** Strict Middling Gray is American upland cotton which in color is Low Middling and which in leaf and preparation is Strict Middling or better.

§ 28.472 **Middling Gray.** Middling Gray is American upland cotton which in color is Strict Good Ordinary and which in leaf and preparation is Middling or better.

§ 28.473 **Strict Low Middling Gray.** Strict Low Middling Gray is American upland cotton which in color is Good Ordinary and which in leaf and preparation is Strict Low Middling or better.

Below Grade Cotton

§ 28.475 **Below Grade Cotton.** Below Grade cotton is American upland cotton which is lower in grade than Good Ordinary, or Low Middling Light Spotted, or Low Middling Spotted, or Low Middling Tinged, or Middling Yellow Stained, or Strict Low Middling Gray. In cotton classification, the official designation for such cotton is Below Grade. The term Below Good Ordinary, or Below Low Middling Light Spotted, or Below Low Middling Spotted, or Below Low Middling Tinged, or Below Middling Yellow Stained, or Below Strict Low Middling Gray may be entered in the remarks space on classification memorandums or certificates as an additional explanatory description.

General

§ **28.481 General.** American upland cotton which in color, leaf, and preparation is within the range of the standards established in this part, but which contains a combination of color, leaf, and preparation not within any one of the standards set out in this part, shall be designated according to the standard which is equivalent to, or if there be no exact equivalent is next below, the average of all the factors that determine the grade of the cotton: *Provided*, That in no event shall the grade assigned to any cotton or sample be more than one grade higher than the grade classification of the color or leaf contained therein.

§ **28.481 Alternate title for standards.** Since these standards have been agreed upon and accepted by the leading European cotton associations and exchanges, they may also be termed and referred to as the "Universal Standards for American Cotton."

Done at Washington, D.C., this 22d day of June 1959.

Roy W. Lennartson,

Deputy Administrator,
Agricultural Marketing Service,
United States Department of Agriculture.

INDEX